112016

LIBRARY
LEWIS-CLARK STATE COLLEGE
LEWISTON, IDAHO

D1505960

WITHDRAWN

LIBRARY

Hard-to-Teach BIOLOGY CONCEPTS

Revised 2nd Edition

Designing Instruction Aligned to the *NGSS*

Hard-to-Teach BIOLOGY CONCEPTS

Revised 2nd Edition

By Susan Koba and Anne Tweed

Designing Instruction Aligned to the *NGSS*

NSTApress

National Science Teachers Association

Arlington, Virginia

National Science Teachers Association

Claire Reinburg, Director
Wendy Rubin, Managing Editor
Andrew Cooke, Senior Editor
Amanda O'Brien, Associate Editor
Amy America, Book Acquisitions Coordinator

ART AND DESIGN
Will Thomas Jr., Director, cover and interior design

PRINTING AND PRODUCTION
Catherine Lorrain, Director

NATIONAL SCIENCE TEACHERS ASSOCIATION
David Evans, Executive Director
David Beacom, Publisher
1840 Wilson Blvd., Arlington, VA 22201
www.nsta.org/store

Copyright © 2014 by the National Science Teachers Association.
All rights reserved. Printed in the United States of America.
17 16 15 14   4 3 2 1

NSTA is committed to publishing material that promotes the best in inquiry-based science education. However, conditions of actual use may vary, and the safety procedures and practices described in this book are intended to serve only as a guide. Additional precautionary measures may be required. NSTA and the authors do not warrant or represent that the procedures and practices in this book meet any safety code or standard of federal, state, or local regulations. NSTA and the authors disclaim any liability for personal injury or damage to property arising out of or relating to the use of this book, including any of the recommendations, instructions, or materials contained therein.

PERMISSIONS
You may photocopy, print, or email up to five copies of an NSTA book chapter for personal use only; this does not include display or promotional use. Elementary, middle, and high school teachers *only* may reproduce a single NSTA book chapter for classroom or noncommercial, professional-development use only. For permission to photocopy or use material electronically from this NSTA Press book, please contact the Copyright Clearance Center (CCC) (*www.copyright.com*; 978-750-8400). Please access *www.nsta.org/ permissions* for further information about NSTA's rights and permissions policies.

LIBRARY OF CONGRESS CATALOGING-IN-PUBLICATION DATA
Koba, Susan, author.
  Hard-to-teach biology concepts : designing instruction aligned to the NGSS / by Susan Koba and Anne Tweed. -- Revised second edition.
       pages cm
Includes bibliographical references and index.
  ISBN 978-1-938946-48-6
1. Biology--Study and teaching (Secondary) I. Tweed, Anne, author. II. Title.
QH315.K58 2014
570.76--dc23
                                    2014022577

Cataloging-in-Publication Data for the e-book are also available from the Library of Congress.

Contents

About the Authors vii

Chapter Contributors viii

Introduction ix

Part I: The Toolbox: A Framework, Strategies, and Connections

Chapter 1: *The Instructional Planning Framework: Addressing Conceptual Change* 3

Chapter 2: *Understanding the Next Generation Science Standards and How They Link to the Instructional Planning Framework* 19

Chapter 3: *Implementation of the Hard-to-Teach Instructional Planning Framework* 43

Chapter 4: *Connections to the Instructional Planning Framework and Resulting Lessons* 141

Part II: Toolbox Implementation: The Framework and Strategies in Practice

Chapter 5: *Matter and Energy in Organisms and Ecosystems (Cynthia Long)* 171

Chapter 6: *Ecosystems: Interactions, Energy, and Dynamics: An Issue-Based Approach (Nancy Kellogg)* 189

Chapter 7: *Variations of Traits (Ravit Golan Duncan and Brian J. Reiser)* 229

Chapter 8: *The Role of Adaptation in Biological Evolution (Sue Whitsett)* 255

Reflection and Next Steps 287

References 291

Appendix 1: *Glossary of Acronyms* 309

Appendix 2: *Hints and Resources for the Design Process* 311

Appendix 3: *Planning Template for Proteins and Genes Unit* 317

Index 325

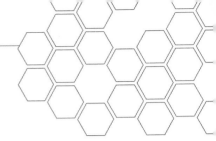

About the Authors

Susan Koba, a science education consultant, works primarily with the National Science Teachers Association (NSTA) on its professional development website, The NSTA Learning Center. She retired from the Omaha Public Schools (OPS) after 30 years, having taught on the middle and high school levels for more than 20 years and then having served as a curriculum specialist and district mentor. Koba ended her service to OPS as project director and professional development coordinator for the OPS Urban Systemic Program serving 60 schools.

Koba has been named an Alice Buffett Outstanding Teacher, Outstanding Biology Teacher for Nebraska, Tandy Technology Scholar, and Access Excellence Fellow. She is also a recipient of a Christa McAuliffe Fellowship and a Presidential Award for Excellence in Mathematics and Science Teaching. She received her BS degree in biology from Doane College, an MA in biology from the University of Nebraska–Omaha, and a PhD in science education from the University of Nebraska–Lincoln.

Koba has published and presented on many topics, including school and teacher change, effective science instruction, equity in science, inquiry, and action research. She has developed curriculum at the local, state, and national levels and served as curriculum specialist for a U.S. Department of Energy Technology Innovation Challenge Grant. A past director of coordination and supervision on the NSTA Board and a past president of her state NSTA chapter, she currently serves NSTA on the Budget and Finance Committee. Other past NSTA work includes serving as the chairperson of the Professional Development Task Force, scope author for the *NGSS* SciPack currently in development, and the conference chairperson for the 2006 Area Conference in Omaha. She is also a past president of the National Science Education Leadership Association (NSELA) and served as NSELA's Interim Executive Director.

Anne Tweed is the Director of STEM Learning with McREL International in Denver. Her work at McREL is research-based and includes ongoing professional development workshops in the areas of science and standards, assessment systems, effective science and math instruction, and formative assessment. Additionally, she is a co-principal investigator on an IES Formative Assessment project that supports implementation in middle level classrooms and was the co-principal investigator for the recently completed NanoTeach NSF-funded project that supported teachers as they learned about nanoscience and technology content and effective science instructional strategies to help them design and implement lessons focused on this emerging area of content.

Tweed earned an MS in botany from the University of Minnesota and a BA in biology from Colorado College. A 28-year veteran classroom biology and environmental science teacher and department coordinator, Tweed also taught AP Biology and AP Environmental Science in addition to Marine Science and off-campus programs. Tweed is a past president of the National Science Teachers Association (2004–2005). She also served at a District Director and a High School Division Director for NSTA and chaired the 1993 and 1997 NSTA Regional Conferences in Denver. Additionally, Tweed chaired the life science program planning team revising the 2009 NAEP (National Assessment of Educational Progress) Framework for Science. Tweed has been recognized for her work in education and has received the Distinguished Service Award and the Distinguished High School Science Teaching Award from NSTA, and the Outstanding Biology Teacher Award for Colorado; she is also a state Presidential Award honoree.

Anne Tweed has published many articles, authored and co-authored several books (*Designing Effective Science Instruction*, 2009, NSTA Press), and given more than 250 presentations and workshops at state, national, and international conferences. Tweed has provided numerous webinars and conference presentations on the instructional shifts and changes in lesson design resulting from the *Next Generation Science Standards*.

Chapter Contributors

Ravit Golan Duncan
Associate Professor of
    Science Education
Rutgers University
New Brunswick, NJ

Nancy Kellogg
Science Education Consultant
Boulder, CO

Cynthia Long
COO & Director of Education
Ocean Classrooms
Boulder, CO

Brian J. Reiser
Professor, Learning Sciences
Northwestern University
Evanston, IL

Sue Whitsett
eCYBERMISSION Outreach Manager
with NSTA
Biology Teacher (1979–2012)
Oshkosh, WI

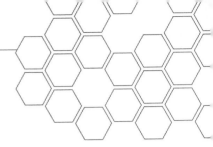

Introduction

"Biology has become the most active, the most relevant, and the most personal science, one characterized by extraordinary rigor and predictive power."
—John A. Moore, 1993

"A pessimist sees the difficulty in every opportunity; an optimist sees the opportunity in every difficulty."
—Winston Churchill (1874–1965)

Biology is a science in which the curriculum continuously changes. New knowledge and emerging content have an enormous impact on our lives. With each new discovery, biologists develop new questions, which lead to more new knowledge. As biology teachers, we constantly learn new content and develop not only our own understanding of biological concepts but also ways to best teach that content to our students.

In addition, we now have new standards in the *Next Generation Science Standards* (*NGSS*; NGSS Lead States 2013) and *A Framework for K–12 Science Education: Practices, Crosscutting Concepts, and Core Ideas* (*Framework*; NRC 2012) to inform the teaching and learning in classrooms. The *NGSS*, based on the *Framework*, bring significant conceptual shifts that must be reflected in our curriculum, instruction, and assessment. These changes, along with continually increasing content, bring new challenges to biology teachers—but these challenges bring opportunity to improve teaching and learning in our classrooms.

This book does not contain a recipe to follow as you plan and deliver lessons. Nor is it a set of predesigned lessons for use in biology classrooms. Instead, it features both an instructional framework you can use as you plan—our Instructional Planning Framework (for a visual representation of the framework, see Figure 1.1, p. 7)—and sets of strategies and resources you can select from to help your students learn. We believe that both new and veteran teachers can use the framework to develop students' conceptual understanding of hard-to-teach biology topics.

The *Next Generation Science Standards* were written to emphasize a teaching approach that blends science and engineering practices with disciplinary core ideas and crosscutting concepts. This represents a significant change with implications for teacher knowledge and practices. We recognize that you, as a biology teacher, will need to reflect on your beliefs, attitudes, and instructional approaches to address the shifts represented in the new documents. This second edition will expand on the previous edition to include the revised thinking and reflection that will be needed for you to make adjustments to your planning and teaching. We recognize that all states are in different places with their review of the *Next Generation Science Standards* but we hope this second edition will provide a perspective that supports your teaching of important biology ideas. Chapters 5–8 are chapters where the teaching and learning shifts are described from the perspective of contributing authors (who are educators like you). We will take this journey with you as you review your current practices and make adjustments needed to realize the vision for teaching the hard-to-teach biology concepts. One concept from the first edition is repeated to model how an existing unit can be modified to align with NGSS. The four concepts addressed by contributing authors are new to this edition.

We will begin this book by looking at what is known biologically and what is expected in the *NGSS*. From there we must determine what and how we should teach to develop our students' biological literacy (essential biology concepts) and appreciation of the living world. Obviously we all want students to understand ideas such as genetic engineering, stem cell research, and evolutionary biology. But for students to learn about genetic engineering, they also must understand how molecules in the cell work and how they provide the genetic information in all living things. To understand stem cells, students have to understand the process of cell division and differentiation. To understand evolutionary biology, they have to understand the processes that produced the diversity of life on Earth. And to understand the new standards, we need to appreciate and implement the intimate relationship among the disciplinary core ideas, the crosscutting concepts, and the science and engineering practices in the *NGSS*. Making sure that students understand the fundamentals of biology is not a simple process, and therein lies the dilemma we all face.

Learning biology is clearly a struggle for many of our students, as evidenced by biology achievement scores across the country. In other words, if you have trouble teaching your students the basic principles of biology, you're not alone! What might be the reasons for these difficulties? With the advent of state standards, adoption of *NGSS*, and high-stakes assessments, biology teachers are finding it difficult to teach in ways that worked for them in the past. A common complaint of both students and teachers is that there is so much content to cover that there is not enough time to do the investigations and activities that engage students with the ideas. Biology teachers know that laboratory experiences help students learn complex concepts (Singer, Hilton, and Schweingruber 2007), yet we get caught up in the attempt to cover so many topics and lists of vocabulary that, on average, students are only provided one laboratory investigation each week. "Science educators have decried the common practice of reading textbooks instead of doing investigations: the former is still alive and well" (Stage et al. 2013). In the classroom, we often focus on the names and labels for living organisms or steps in processes, and our students get lost in details without learning the important, essential biological principles and the scientific practices used to make sense of them.

With this book, we seek to help all biology teachers teach the hard-to-teach biology concepts that are found in the broader high school level, life science, and disciplinary core ideas. Although this book is not about providing teachers with scripted lessons, it does include much that we have learned from our own experiences and from recent research findings, as well as outlined in the *NGSS*. Science research that focuses on how students learn recommends certain strategies that teachers can use to help develop and implement effective instructional methods. In this book, we do not tackle all the issues in high school biology. Rather we focus on selected research that informs our Instructional Planning Framework.

We realize that teachers' implementation of selected instructional strategies impacts the effectiveness of a strategy in the classroom. Even with research-based strategies and tools, we need to figure out ways to use them in the best way possible. For example, we know that classroom discourse helps students think about their ideas and supports sense making. But if we just ask students to discuss a question or problem without setting a time limit, establishing the groups they will work with, and determining how they will report-out to the class, then classroom discourse won't help students make sense of the hard-to-teach biology concepts. And when planning with a focus on *NGSS*, teaching biological argumentation procedures that incorporate effective discourse strategies makes this a critical practice that also connects to literacy skills.

We love teaching biology, and we want to provide opportunities for you to meet the challenges posed when teaching hard-to-teach biology concepts. We were prompted to write the first edition of this book because guidance for teachers is located in so many different places; our hope was to put all of the findings together into a model that made sense to us and would support your work. Our hope in this second edition is that you are provided one way in which to interpret and implement the *NGSS*, while still using the best thinking from the first edition. This book presents a framework for planning, shares appropriate approaches to develop student understanding, and provides opportunities to reflect on and apply those approaches to specific concepts and topics. It is more about helping you learn how to improve your practice than it is about providing sample lessons that recommend a "best" way to provide instruction. Clearly, you must decide what works best for you and your students.

Science Education Reform and Conceptual Understanding

At that same time that our students struggle to master biology concepts, many states require students to pass high-stakes tests in order to graduate. Science reform efforts stress science understanding by all citizens; unfortunately, little impact is made on persistent achievement gaps (Chubb and Loveless 2002). However, the current cycle of science education reform that resulted in the *Next Generation Science Standards* (NGSS Lead States 2013) expects, among other things, meaningful science learning for all students at all grade levels—that is, students are able to build connections among ideas, moving past recall and into more sophisticated understandings of science. To meet the standards, it is critical that all of us work to implement strategies shown as effective to build these types of student understandings.

We know that serious change takes time, often 7–10 years to move from establishing goals to changing teacher practice and curriculum materials that meet the needs of our students (Bybee 1997). One major obstacle to change is the lack of support for teachers to fully understand ways to teach hard-to-teach concepts (Flick 1997). School structures in the United States do not adequately provide professional support for us to engage in new learning to improve our teaching. We are rarely provided the time to work individually or collaboratively to inquire into our own teaching and our students' learning (Fisher, Wandersee, and Moody 2000). So what makes current reform efforts any different from those in the past? Perhaps the standards, political influences, and the growing body of research provide an answer.

Hope for change begins with the *NGSS* because we now have standards that integrate a few core disciplinary ideas with crosscutting concepts and science and engineering practices. Integration of the practices, in particular, aligns with research about conceptual change since it calls for building understanding through models and explanations and requires discourse to argue, criticize, and analyze. With the review and revision process associated with the framework and *NGSS* documents, the teaching shifts needed to support conceptual change by students have been clearly identified. Brian Reiser identifies the following shifts as important and we will address them in the revised components of Instructional Planning Framework and the invited chapters.

- The goal of instruction needs to shift from facts to explaining phenomena.

- Inquiry is not a separate activity—all science learning should involve engaging in practices to build and use knowledge.

- Teaching involves building a coherent storyline across time.

- Students should see that they are working on answering explanatory questions and not just moving to the next topic.

- Extensive class focus needs to be devoted to argumentation and reaching consensus about science ideas.

- A positive classroom culture is necessary to support teaching and learning where students are intellectually motivated, where they actively share responsibility for learning and where they work cooperatively with their peers. (Reiser 2013)

The next ray of hope is that the political focus on science education has grown even more since the first edition of this book, as evidenced by the federal government's growing focus on the needs in mathematics and science, which has resulted in increased funding for science education efforts in support of science, technology, engineering and mathematics (STEM) education.

What should directly impact us, as educators, is a growing body of research on teaching and learning in general (Bransford, Brown, and Cocking 1999) and science teaching and learning in particular (NRC 2005; Banilower, Cohen, Pasley, and Weiss 2010; Banilower et al. 2013; Windschitl, Thompson, Braaten, and Stroupe 2012). Also, we now have access to a considerable body of research on the understandings and skills required for meaningful learning in biology (Fisher, Wandersee, and Moody 2000; Hershey 2004), inquiry (Anderson 2007; Windschitl, Thompson, and Braaten 2008), and the nature of science (Lederman 2007). Finally, there is an increasing understanding of conceptual change (Driver 1983; Hewson 1992; Lemke 1990; Minstrell 1989; Mortimer 1995; Scott, Asoko, and Driver 1992; Strike and Posner 1985; Darling-Hammond et al. 2008), as well as research on common misconceptions and strategies to address them (Coley and Tanner 2012; Committee on Undergraduate Science Education 1997; Driver, Squires, Rushworth, and Wood-Robinson 1994, Mortimer and Scott 2003; NAS 1998; Tanner and Allen 2005).

But hope, by itself, is not a method. Because biology is the most common entry course for science in secondary schools, it is essential that changes in science teaching and learning begin with us, the biology teachers. It is the goal of this book to support your walk down the path to more effective teaching and learning in biology as aligned with the *Next Generation Science Standards*. Even if your state has not adopted the *NGSS*, we believe that you will find the suggestions for instructional planning and the strategies recommended helpful.

Hard-to-Teach Biology Concepts—Why Are They Hard?

Traditionally students struggle to learn some of the basic ideas taught in high school biology classes. To understand why, we must analyze not only the content itself but also the classroom conditions and learning environment. One concern cited by biology teachers is the "overstuffed" biology curriculum. Because of the sheer amount of information that is taught related to each topic, even good students find it difficult to retain what they learn (NRC 2011b). Because of an emphasis on a fact-based biology curriculum, instruction often relies on direct instruction to cover all of the material. As a result, students have limited experiences with the ideas and rarely retain what they learned past the quiz or unit test.

Certain biology topics are hard for students to learn because students aren't given the time they need to think and process learning. We must give students multiple opportunities to engage with biology ideas. Research suggests that students need at least four to six experiences in different contexts with a concept before they can integrate the concept and make sense of what they are learning (Marzano, Pickering, and Pollock 2001; Dean, Hubbell, Pitler, and Stone 2012).

Another reason that there are hard-to-teach (and learn) topics relates to the prior knowledge of our students. High school students are far from being blank slates; they come to us with their own ideas and explanations about biology principles. After all, everyone knows something about biology and our students have had a variety of experiences both as they have grown up outside the school setting and in previous science classrooms. Student preconceptions can be incomplete and students often hold onto them tenaciously. One classic research study was captured in the video *A Private Universe: Minds of Our Own* (Harvard-Smithsonian Center for Astrophysics 1995). In one segment, researchers asked Harvard graduates where the mass of a log came from. The response was water and nutrients from the soil. Students and even college graduates hadn't learned the fundamental concept that photosynthesis requires carbon dioxide from the air to manufacture carbohydrates, which are the basis for the vast majority of a tree's mass.

This example relates to two additional reasons why some biology topics are hard to teach: (1) many biology lessons are highly conceptual and students can't visualize what is taking place on a microscopic level. And (2) some biology teachers are not aware of strategies that engage students with a scientific way of knowing (Banilower, Cohen, Pasley, and Weiss 2010; Lederman 2007). Such strategies include asking questions, building and using models to explain and argue, inferring from data, challenging each other's ideas, communicating results, and synthesizing student explanations with scientific explanations.

When we consider these various impeding factors, it is no wonder that students struggle in our biology classes.

Why Aren't Students Learning?

Science research helps us answer this question.

- Students may not learn because of their learning environments. The meta-analyses of the research in *How People Learn: Brain, Mind, Experience, and School* (Bransford, Brown, and Cocking 1999) and *How Students Learn: Science in the Classroom* (NRC 2005) report that the instructional environment must be learner-, not teacher-, centered. Students come to school with conceptions of biological phenomena from their everyday experiences and teachers need to take into account such preconceptions. Furthermore, what we teach is often too hard for students because they lack the necessary backgrounds on which the hard-to-teach topics are based.

- Several studies have shown that high school students perceive science knowledge as either right or wrong (NRC 2005). Unfortunately, biology concepts are rarely this clear-cut and the body of knowledge in biology is ever-changing. Biological systems are dynamic, and long-term observations are often needed to understand and make sense of the evidence. The norm in many classrooms, however, is to come up with a correct answer, which is not reasonable or possible in biology classrooms, where we look at probabilities, changes over time, and trends. Quantitative and qualitative data can be ambiguous. This can be very uncomfortable for students who ask us, "Why don't you just tell me the answer?" While biologists, like other scientists, give priority to evidence to justify explanations, students think that we should have *the* answer to biology questions and problems. Students may believe that biology is really a collection of facts because we often use direct instruction to cover the biology facts and vocabulary that may be addressed in state assessments.

- Students learn best when they are able to work collaboratively with other students. With only one investigation per week in the average biology classroom, students may not receive sufficient opportunities to engage in interactive work, where, as explained in the *NGSS* documents, learning should be driven by questions about the phenomena and ideas.

Organization of the Book

Hard-to-Teach Biology Concepts: Designing Instruction Aligned to the NGSS is designed to support biology teachers as they plan and implement *NGSS*-aligned lessons that will intellectually engage students with the biology concepts that most students find challenging. To develop successful learners, teachers must identify prior student conceptions and research-identified misconceptions related to the concept being taught and then select instructional approaches to dispel those misconceptions and promote students' conceptual understanding.

The book is made up of two parts: Part I, The Toolbox: A Framework, Strategies and Connections (Chapters 1–4), and Part II, Toolbox Implementation: The Framework and Strategies in Practice (Chapters 5–8). In Part I, we share our instructional planning framework and tools and outline the connection between our framework and the *NGSS*. In addition, we share a process to implement our framework and describe other connections that enhance learning by all students. Chapter 1 introduces our research-based framework to address conceptual change—the Instructional Planning Framework—and gives an overview of (1) the identification of conceptual targets and preconceptions, (2) the importance of confronting preconceptions, (3) sense-making strategies to address preconceptions, and (4) best ways in which students can demonstrate understanding. Chapter 2 outlines some of the major instructional shifts in the

NGSS and the connections of the standards to the instructional framework. It also introduces a process to use during development of instruction. Chapter 3 uses the topic Proteins and Genes to model the process outlined in Chapter 2 and discusses specific instructional approaches that teachers might use to dispel preconceptions: metacognitive approaches, standards-based approaches, and specific strategies for sense making. Chapter 4 introduces research related to formative assessment, the *Common Core State Standards*, STEM, and Universal Design for Learning (UDL) and then builds connections for each to the unit of study developed in Chapter 3. Though our framework can be followed in a linear manner, it is not really intended as a stepwise process. Instead, it is important for you to reflect on the framework presented in Chapter 1, adapt it for your use, and select strategies from Chapter 3 most appropriate for your own classroom.

Part II is organized to model use of our framework through its application in the analysis of four additional hard-to-teach topics not covered in the first edition of this book. The topics were carefully chosen to include those related to each of the *NGSS* disciplinary core ideas. Each chapter is developed based on Part I, but through the interpretation of a contributing author. Recommended resources, including technology applications and websites, will be found at the end of each chapter in Part II. The Part II chapters focus respectively on the following disciplinary core ideas:

- Chapter 5: From Molecules to Organisms: Structures and Processes
- Chapter 6: Ecosystems: Interactions, Energy, and Dynamics
- Chapter 7: Heredity: Inheritance and Variation of Traits
- Chapter 8: Biological Evolution: Unity and Diversity

The appendixes found in the *NGSS* enhance our understanding of our framework and its application. We will discuss several of this book's appendixes in Chapter 4 when we address connections to *NGSS*.

PART I
The Toolbox:
A Framework, Strategies, and Connections

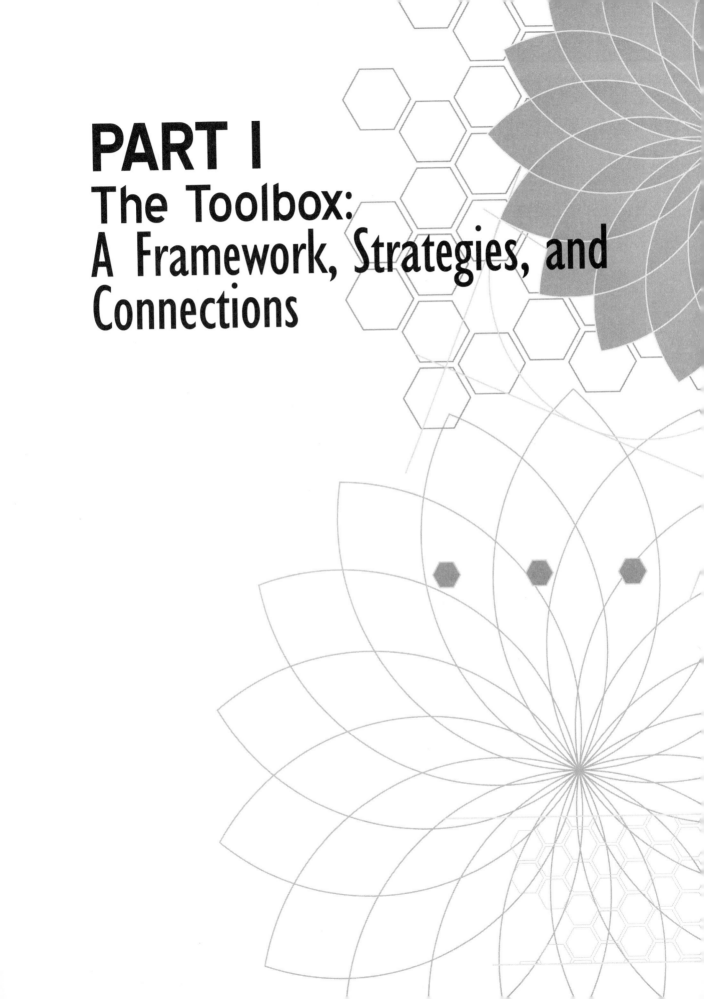

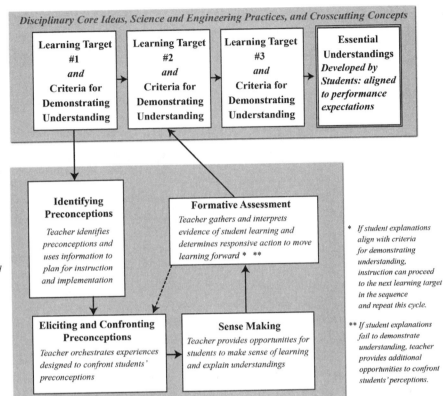

Phase 1: Identifying Essential Content

The teacher selects performance expectations that inform a unit's essential understandings and develops success criteria to determine student progress.

A sequence of learning targets is identified based on disciplinary core ideas, science and engineering practices and crosscutting concepts.

Phase 2: Planning for Responsive Action

Building on the foundation provided in the identifying essential content phase, the teacher plans for and implements instruction during the responsive action phase, one target at a time.

Disciplinary Core Ideas, Science and Engineering Practices, and Crosscutting Concepts

Learning Target #1 *and* **Criteria for Demonstrating Understanding**

Learning Target #2 *and* **Criteria for Demonstrating Understanding**

Learning Target #3 *and* **Criteria for Demonstrating Understanding**

Essential Understandings *Developed by Students: aligned to performance expectations*

Identifying Preconceptions

Teacher identifies preconceptions and uses information to plan for instruction and implementation

Formative Assessment

Teacher gathers and interprets evidence of student learning and determines responsive action to move learning forward * **

Eliciting and Confronting Preconceptions

Teacher orchestrates experiences designed to confront students' preconceptions

Sense Making

Teacher provides opportunities for students to make sense of learning and explain understandings

* *If student explanations align with criteria for demonstrating understanding, instruction can proceed to the next learning target in the sequence and repeat this cycle.*

** *If student explanations fail to demonstrate understanding, teacher provides additional opportunities to confront students' perceptions.*

Chapter 1

The Instructional Planning Framework: Addressing Conceptual Change

"Students come to the classroom with preconceptions about how the world works. If their initial understanding is not engaged, they may fail to grasp the new concepts and information that are taught, or they may learn them for purposes of a test but revert to their preconceptions outside the classroom."
— *Donovan, Bransford, and Pellegrino 1999, p. 10*

As biology teachers, we know that certain foundational biology concepts are challenging for our students. And these same students depend on us to teach, assess, and provide feedback to them so that they can understand these concepts. Teachers also know that worldwide shifts resulting in societal, economic, and technological changes shape the skills and understanding required in the future (Bransford, Brown, and Cocking 1999; Bureau of Labor Statistics 2000). With the release of *A Framework for K–12 Science Education* (NRC 2012) and the *Next Generation Science Standards* (NGSS Lead States 2013), science teachers are focused on standards and on identifying what students should know and be able to do in core subjects. The *Next Generation Science Standards* are based on the *Framework* document. The 2012 document provides an explanation of the conceptual shifts included in the visionary approach to science and engineering education designed to engage students with science and engineering practices and apply crosscutting concepts in ways that deepen conceptual understanding of the disciplinary core ideas. The *Framework* emphasizes the integrated nature of the knowledge and practices that should be incorperated in the K–12 learning experiences. Many states reviewed their standards documents and are aligning the performance expectations with *NGSS* and the *Framework*. Other states are including an emphasis on the science and engineering practices with their existing state science standards. Whether your state has adopted *NGSS* or is reviewing the document, we must consider the additional emphasis on advancing STEM learning and the new emphasis on science practices that is central to the *NGSS*. These new directions in science mean that teachers must maintain the classroom focus on learning important concepts and science ideas rather than on rote fact-based learning. This change is particularly significant in biology because of the difficulty that students often have with abstract biological ideas. The Instructional Planning Framework (Figure 1.1, p. 7) initially introduced in the first edition of this book, incorporates research findings and implications for biology teachers in regard to the four hard-to-teach biology concepts (abstract biological ideas). The selected topics arise from the four major disciplinary core ideas in life science defined in the *Framework* and include: proteins and genes, cellular respiration, ecosystems and human impacts, and inheritance and natural selection.

Why Are There Hard-to-Teach Biology Concepts?

As we discussed in the introduction, learning biology is challenging for many students. Students often report that learning biology is like learning a foreign language—that mastering the vocabulary alone is a struggle. This is not surprising since much of the new vocabulary terms represent key biology ideas. Some students are quick to give up and will say that they just don't understand science and they were never any good at it anyway. What can we do to help students with unfamiliar terminology and motivate students who have a mindset that they are not capable of learning biology? How can we engage our students to meet their needs as well as our own? Research-based strategies such as those described in this book offer answers to these important questions.

According to *Teaching With the Brain in Mind*, "Since what is challenging for one student may not be challenging for another … [teachers must provide] more variety in the strategies used to engage learners better" (Jensen 1998, p. 39). As you think about the differences among your students, we're sure that it is obvious that all students do not face the same challenges. What makes certain biology concepts more challenging for some students? In some instances, it is the content itself that cannot be studied directly. So students say that the ideas are just too conceptual and abstract and they can't visualize what is happening.

To further complicate matters, biological concepts frequently require understanding chemistry ideas. As a result, students' preconceptions (those ideas that arise prior to instruction) may include incomplete foundational knowledge that causes them to struggle to understand complex biological concepts. Our approach here is not only to engage students intellectually with the ideas in ways that get them to think about their thinking but also to provide strategies that will increase student motivation to learn and bring about conceptual change. Table 1.1 (p. 6) organizes these ideas and reminds us that our students may have difficulty for a variety of reasons, so reflecting on student needs is where we should start.

Introducing the Instructional Planning Framework

We draw from the research base that supports a conceptual change framework. For our purposes, we understand *conceptual change* to be a process where students access their own thinking, and as needed, alter their existing understanding to generate scientific explanations for the science phenomena being studied. Change in students' preconceptions can occur if teachers use a conceptual change process that addresses the following conditions:

- Students must be aware of their personal conceptual understandings.
- Students must become dissatisfied with their existing views through the introduction of new evidence.
- New conceptions (scientific viewpoints) must appear somewhat plausible.

Table 1.1

Reasons Why Biology Concepts Are Hard to Teach and Learn

Concepts are abstract.	Students can't visualize or directly study some biological concepts. In their attempts to learn from models and representations, misconceptions occur.
Biological systems are complex.	When studying biological systems, they may be complex and composed of integrated components that each need to be understood separately before putting them together. These often include multiple levels of organizations (e.g., protein synthesis and protein functions, flow of matter and energy in ecosystems, and so on).
Students come with limited prior knowledge.	Foundational knowledge may not have been taught so it wasn't learned and/or appropriate learning experiences were not provided in elementary and middle school classrooms.
Students struggle to understand content representations.	Students may struggle to decode/read models and representations of the learning (e.g., diagrams; physical, graphical and mathematical models; and so on).
Preconceptions and misconceptions are persistent among students.	Students may prefer their explanations and ideas to the science ideas. Also some concepts don't make sense to students so their misconceptions are difficult to change.

- New conceptions must be more attractive in order to replace previous conceptions. (Strike and Posner 1985)

Students develop their own ideas about and explanations for many of our hard-to-teach biology concepts. Their learning is an additive, not a replacement, process and the experiences that teachers provide must be incorporated into their existing conceptions as students restructure their knowledge (Chi 1992: diSessa 2006; diSessa, Gillespie and Esterly 2004; Slotta and Chi 2006; Smith, diSessa and Roschelle 1993). Students' early and sometimes naïve preconceptions may contain faulty reasoning that they developed during previous classes or from their own observations. Sometimes misunderstandings arise spontaneously during learning, and sometimes students hold more than one pre-conception about a topic. Ultimately, students must reconstruct their ideas and revise their mental models and conceptual frameworks. Revealing student thinking and adding to their ideas is an essential part of our instructional framework.

From the research, simply providing our students with the current concept explanation does not work (Chinn and Brewer 2001). The research around conceptual change shows that there are a variety of processes students can use to restructure their knowledge. Some studies have identified the importance of linking to students' prior knowledge as a foundation for adding new ideas, particularly when shifting from non-scientifically based ideas to understanding aligned with current science knowledge (Clark 2000, 2006). We recommend a conceptual change approach to learning and instruction, which shows strong evidence of effectiveness when teaching biology concepts that are difficult for students.

In this chapter, we elaborate on the steps of our research-based Instructional Planning Framework. Figure 1.1 provides a diagrammatic representation of our framework. Using this framework and the research cited throughout the book, you can select strategies to support development of student conceptual understanding.

Figure 1.1

Instructional Planning Framework

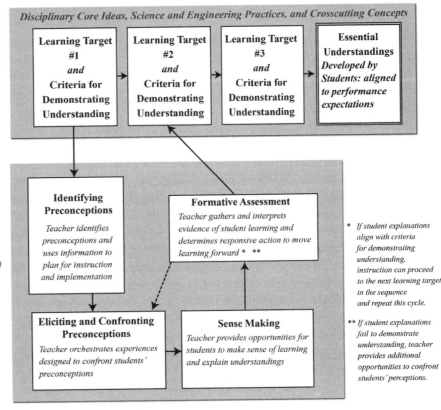

Phase 1: Identifying Essential Content

The teacher selects performance expectations that inform a unit's essential understandings and develops success criteria to determine student progress. A sequence of learning targets is identified based on disciplinary core ideas, science and engineering practices and crosscutting concepts.

Phase 2: Planning for Responsive Action

Building on the foundation provided in the identifying essential content phase, the teacher plans for and implements instruction during the responsive action phase, one target at a time.

Disciplinary Core Ideas, Science and Engineering Practices, and Crosscutting Concepts

Learning Target #1 *and* **Criteria for Demonstrating Understanding**

Learning Target #2 *and* **Criteria for Demonstrating Understanding**

Learning Target #3 *and* **Criteria for Demonstrating Understanding**

Essential Understandings Developed by Students: aligned to performance expectations

Identifying Preconceptions
Teacher identifies preconceptions and uses information to plan for instruction and implementation

Formative Assessment
Teacher gathers and interprets evidence of student learning and determines responsive action to move learning forward * **

Eliciting and Confronting Preconceptions
Teacher orchestrates experiences designed to confront students' preconceptions

Sense Making
Teacher provides opportunities for students to make sense of learning and explain understandings

* *If student explanations align with criteria for demonstrating understanding, instruction can proceed to the next learning target in the sequence and repeat this cycle.*

** *If student explanations fail to demonstrate understanding, teacher provides additional opportunities to confront students' perceptions.*

From the book *Taking Science to School,* we know that there are four areas of proficiency that link the content and practices of science. With the new emphasis on science and engineering practices found in the *NGSS* and linking them to the disciplinary core ideas and the crosscutting concepts, the second edition of our book will discuss approaches that you can take to plan your curriculum lessons in ways that actively incorporate the practices of science. As you start thinking about the planning process, review the following student competencies that influenced the *Framework* and subsequent *NGSS* documents. The student proficiencies for science include:

- Know, use, and interpret scientific explanations of the natural world;

- Generate and evaluate scientific evidence and explanations;

- Understand the nature and development of scientific knowledge; and

- Participate productively in scientific practices and discourse. (Duschl, Schweingruber and Shouse 2007, p. 2)

In Chapter 2 we will describe the process represented by the Instructional Planning Framework and describe how it aligns with and incorporates the components found in the *NGSS* and the *Framework*. Chapter 3 provides instructional tools that support specific instructional approaches that have been shown to address preconceptions.

Instructional Planning Framework

The Instructional Planning Framework has two phases: Phase 1: *Identifying Essential Content* and Phase 2: *Planning for Responsive Action*. We will start with Phase 1.

Phase 1. Identifying Essential Content

Figure 1.1 represents the stages found in Phase 1 and 2. The framework acts as a logic model diagram. Refer to the top portion of Figure 1.1. This is the *Identifying Essential Content* section of the framework. During this phase, we first clarify the disciplinary core ideas (DCIs) from the *NGSS* or from your state standards—that is, the essential understandings we want students to develop, as well as the knowledge and skills to be taught and learned in order to develop these understandings. We want to also identify the crosscutting concepts (CCs) and science and engineering practices (SEPs) that logically link to the core ideas. We are then able to identify the steps in the learning sequence (the learning targets) needed to build student understanding. Finally, it is critical to provide the criteria for demonstrating conceptual understanding by students so that they can know what successful performance looks like. Performance Expectations (PEs) can be found for each foundation box in the *NGSS* and this is where the developers suggest we focus our attention—on student learning outcomes. To measure progress to the desired outcome, we create checkpoints for the students and provide feedback so that they know how to improve. The final steps in Phase 1 remind us to determine appropriate connections to the Nature of Science and identify meta-cognition goals to include in the teaching and learning experiences.

Phase 2. Planning for Responsive Action

The stages included in Phase 2 are similar to those included in our framework in the first edition. The following descriptions provide an overview of the important aspects of the planning and implementation that occur when teachers engage in responsive action needed to improve student learning.

1. Research common student misconceptions.

This step is designed to help you study the research-based ideas on student misconceptions so you can outline instruction focused on the conceptual target. Operationally then, the first part of the *Planning for Responsive Action* phase is to identify misconceptions, including common, research-identified misconceptions that your students might bring with them. This is essential as a foundation for planning instruction as we move forward with the framework steps.

2. Identify student preconceptions.

We recognize that, in the ideal, teachers should elicit their students' existing ideas (both preconceptions and research-based misconceptions) and use these ideas to plan instruction. Here we want to emphasize the importance of planning experiences and selecting strategies that allow teachers to identify your students' initial ideas.

3. Elicit and confront preconceptions.

We begin to implement instruction by eliciting our students' preconceptions and discussing those preconceptions with our students. Although this step may require you to make some modifications to your initial plan, it should serve you reasonably well because now your plan uses research-identified misconceptions and preconceptions[1] as well as those revealed by your students. Once you know what conceptual understandings your students already have related to the concepts, you can use the tools in Chapter 3 to select strategies that address the conceptual change process.

4. Use sense-making strategies to address preconceptions.

This step in the process is at the core of our framework. Provide opportunities for your students to make sense of the ideas they learn because it is unlikely that they will draw the appropriate conclusions on their own. Select strategies that engage students through questioning, discussions, and other methods so that students can make connections between their ideas and what they are meant to learn. Later, you will ask students to think about their initial ideas, relate them to their current learning, and then determine how their thinking has changed.

5. Formatively assess student learning and determine responsive actions.

To complete the framework, student learning must match the criteria for developing student conceptual understanding. To determine student progress toward the learning goal, you can formatively assess students to gather evidence that informs your responsive action. You can provide feedback to students, modify instruction, or both. If students do not meet the criteria, you will need to revisit the final steps in the *responsive* phase and select additional tools to help students understand the concepts.

Many of you are familiar with the well-known BSCS 5E Instructional Model (Bybee 1997) and may even use a curriculum built upon it. If that is the case, the framework and strategies found in this book serve as a nice supplement to your curriculum. If you use a more traditional curriculum, then this book will help you modify your lessons to promote inquiry and your students' conceptual understanding. The Instructional Planning Framework incorporates and builds on both the 5E model and the Conceptual Change Model. Table 1.2 compares our framework to those models.

Notice that all three models build on students' current conceptions and provide experiences that let students express and challenge their thinking. Each model also clarifies student understandings and then builds on and extends their thinking with new experiences. The Instructional Planning Framework differs from the other two models in its *predictive* phase, during which teachers establish clear learning goals, learning sequences, and criteria to demonstrate understanding—all prior to instruction. The components of the Phase 2 of the Instructional Planning Framework occur in the classroom just as do the components of the 5E Instructional Cycle and the Conceptual Change Model.

Another difference, and unique to this book, is the rich set of Instructional Tools in Chapter 3 (pages 43–139) that are used in the process of putting the framework into practice. The Instructional Planning Framework builds strongly on conceptual change research, but also recognizes the importance of social interaction using language (discourse) and other multiple resources (e.g., writing, visual images, and physical actions) to mediate learning. Tools to facilitate social interaction are central to the sense-making approaches (e.g., see Instructional Tool 3.5, pp. 110–118). It is also important that students participate in experiences that resemble those of scientists, so each of the tools incorporates references to the practices. The tools and strategies discussed in Chapter 3 both support the Instructional Planning Framework and reflect aspects of each of these three perspectives (conceptual change, social interactions, and scientist-like experiences). Thus, you can select tools that best fit both the context of your instruction and your personal teaching philosophy.[2]

The Research Behind This Framework

Broad research supports both Phase 1: *Identifying Essential Content* and Phase 2: *Planning for Responsive Action* of the Instructional Planning Framework. The Instructional Tools (Chapter 3) build on this research and provide effective methods to support student learning. We don't make recommendations for specific strategies. Instead, we model the use of the tools in the content chapters (Chapters 3, 5, 6, 7, and 8), providing suggestions and inviting you to make your own selections based on the part of the framework addressed.

Table 1.2

A Comparison of the 5E Instructional Model, the Conceptual Change Model, and the Instructional Planning Framework

5E Instructional Model	Conceptual Change Model	Instructional Planning Framework
		Phase 1: Clarifies the conceptual target (essential understandings for the lesson), science and engineering practices, and crosscutting concepts; identifies and sequences learning targets; and provides criteria for demonstrating understanding aligned to performance expectations. Identify other connections to the science ideas and select metacognitive strategies.
		Phase 2—Identify Preconceptions: Provides input for the teacher about students' *pre-lesson* conceptions and aids in lesson development.
Engage Phase: Engages students in the topic and motivates them; identifies current conceptions; connects learning to past lessons	**Commit to an Outcome:** Students are expected to make predictions about an investigation, which makes them aware of their (pre)conceptions.	**Phase 2—Elicit and Confront Preconceptions:** Requires students to make evident their thinking about a concept *during instruction* and compares their ideas with those of others through various learning experiences that include classroom inquiries. Reasoning and argumentation are evident as components of classroom communication.
Explore Phase: Provides students with a common context of shared experiences connected to the content; lets them test their own ideas and compare them to those of their peers	**Expose Initial Conceptions:** Students share ideas in small groups and with the whole class.	
	Confront Conceptions: Students are given opportunities to test their ideas and discuss them in small and large groups.	

Table 1.2 (continued)

5E Instructional Model	Conceptual Change Model	Instructional Planning Framework	
Explain Phase: Allows students to demonstrate their conceptual understandings	**Accommodate the Concept:** Students try to resolve the differences between their own ideas and their observations or what they learned, develop reasoning based on evidence, and analyze and discuss multiple explanations to further clarify their understanding.	**Phase 2—Sense Making**	*Perceiving, interpreting, and organizing information* Asks students to use information from learning experiences to clarify their understandings and resolve discrepancies among explanations
Elaborate Phase: Challenges and extends students' conceptual understandings by providing new experiences	**Extend the Concept:** Students are given opportunities to connect their understandings to new experiences.		*Connecting information* Provides additional experiences for students and asks them to connect thinking to these experiences
	Go Beyond: Students pursue related questions and problems.		*Retrieving, extending, and applying information* using knowledge in relevant ways
Evaluate Phase: Provides opportunities for teachers to evaluate student understanding and for students to self-assess		**Phase 2—Demonstrating Understanding:** Engages students in peer and self-assessments. Engages teachers in formative assessment processes that inform further instruction to guide development of student understanding.	

At this point, we ask you to consider both cognitive and metacognitive processes (learning and thinking about your thinking) as we explore each part of the planning sequence. Cognitively, you need to determine your own conceptual understandings about how students learn biology concepts in order to consider how to teach students to make sense of their learning. We also invite you to use a self-reflection process in each of the content chapters to address the metacognitive aspect of the book. We ask you to start this process by being very clear about why you are teaching what you are teaching. In essence, we hope that using a self-reflection process helps you become more metacognitive. How have you taught in the past? How can you engage in a critical-thinking process as you also learn about the research findings that support the framework?

Phase 1: Identifying Essential Content

Let's look again at the *Identifying Essential Content* phase of the framework (Figure 1.1). Recall that this phase includees three important themes: the importance of (1) identifying essential learning(s) or learning goals in a topic aligned to the *NGSS* disciplinary core ideas, science and engineering practices, and crosscutting concepts; (2) focusing in depth on the progression of the related learning targets and determine Nature of Science connections; and (3) determining criteria for demonstrating understanding for each of the learning goals and metacognitive goals (Bransford, Brown, and Cocking 1999; Heritage 2008; NGSS Lead States 2013; Masilla and Gardner 2008; Michaels, Shouse, and Schweingruber 2007; NRC 2012; Vitale, Romance, and Dolan 2006).

Each of the content chapters (Chapters 3, 5–8) models and provides examples of planning that implements the Instructional Planning Framework. Chapter 3, for example, models the steps to plan for and implement a series of lessons focused on proteins and genes. Providing students with clear learning goals is at the heart of Phase 1 of the framework. It is a critical first step for teachers and students and, as an instructional approach, can lead to increased student achievement (Dean, Hubbell, Pitler, and Stone 2012). Many biology textbooks are beginning to identify big ideas and key concepts to help teachers with this process of identifying essential learning. Before students can grasp the essential understandings, we need to deconstruct the big ideas into a sequence of learning targets. The steps in this process are featured in Chapter 3 and reinforced in subsequent Chapters 5–8. Remember that the goal of the *NGSS* is for teachers to plan in ways that integrate the disciplinary core ideas with the practices and crosscutting themes. This will be a departure from the planning approach most of us have used in the past and will be challenging until you practice and implement several units this way.

Why is identifying essential understandings (including all three components of *NGSS*) and criteria for student success the important first step in our framework? The answer is that if teachers are not clear about the concepts, subconcepts, and crosscutting themes that lead to adult literacy in biology and the practices needed to promote learning, then students will be unsure about what they are supposed to learn and will cling to facts and vocabulary without developing a fundamental understanding of the concepts. According to Margaret Heritage's (2008) work on learning progressions, teachers need to identify subgoals and sequence them to help students progress to the ultimate learning goals. Unfortunately, most state science standards do not provide clear progressions to understanding (Heritage 2008). Sometimes standards don't even make it clear what students should learn. Learning progressions provide our students with a pathway to the ultimate learning goals or big ideas we want them to learn.

We want to be very clear about the distinction between learning *progressions* (learning trajectories across grade bands or progressions within a course or grade level) and what we are calling learning *sequences* (shorter-term goals within standards). Learning progressions are included in the *NGSS*, so looking at the broader understandings in biology is a necessary part of our understanding of how student mastery of the concepts develops. Within units of study, students need short-term goals (subgoals or learning targets) so the gaps between their current preconceptions and the desired learning goals are not too great. To engage students in learning, we must make sure that the gaps are neither so big that students give up nor so small that students quickly "get it" and are bored. But learning presented in a sequence, or as Kathy Roth (2011) describes it in her research, *content storylines*, helps students link their thinking to scientific ideas. Identifying the short-term goals should also be accompanied by determining the criteria for conceptual understanding, so students know the evidence of learning they need to show to move forward.

A Framework for K–12 Science Education: Practices, Crosscutting Concepts, and Core Ideas (NRC 2012) is built on the notion of learning as a developmental progression. It is designed for teachers to help students engage with fundamental questions about the world and with how scientists use investigative processes to determine the answers. Like scientists, students must continually build on and revise their knowledge and abilities, starting from their curiosity about what they see around them and their initial conceptions about how the world works (NRC 2012, p. 11). Integral to this effort is students who can metacognitively[3] determine where they are with answering questions and how they might move their learning forward. The last step in Phase 1 planning includes selecting goals and strategies that can be taught to students to increase their metacognitive abilities and promote their skills as autonomous learners.

We recommend that you think metacognitively about the logic model for teacher planning that we have proposed to provide teachers with a design process to identify the learning goals and learning targets, establish a learning sequence, and plan for assessments (both formative and summative) that provide evidence of student learning. To realize the vision for science education that is presented in the *NGSS*, we also use this book as a way to work with you as you plan for instruction that limits the number of topics taught so you have sufficient time to engage in investigations and argumentation to develop conceptual understanding of important science ideas. The stages described in Phase 1 are necessary components of our unit planning. Once the pathway is clear, you can use the rest of the framework to create instruction that engages students with important biology concepts and identifies and addresses their preconceptions.

Phase 2: Planning for Responsive Action

The conceptual targets, learning sequence, and criteria for determining understanding and the other integrated components of the unit plan are now all in place as a result of completing Phase 1 of the framework. Next we can flesh out a plan for instruction that aligns with the conceptual learning goal. "By taking the time to study a topic before planning a unit, teachers build a deeper understanding of the content, connections, and effective ways to help students achieve understanding of the most important concepts and procedures in that topic" (Keeley and Rose 2006, p. 5). We need to know and understand the content ourselves so we are clear about what students should learn.

During the planning phase, you began to clarify the content. But you also need to know *when* and *how* to best teach those concepts to your students. So we now look at effective instructional strategies in Phase 2: *Planning for Responsive Action*. The Instructional Strategy Sequencing Tool is the core tool we developed to assist you as you learn about and reflect on the selection and sequencing of strategies. The tool, further explained in Chapter 3, is based on our research review and organized by the phases of the Instructional Planning Framework.

We know that teachers are experts and students are novices in terms of understanding biology concepts. Based on our extensive teaching and learning experiences, we can anticipate many of the conceptual barriers our students face. We can also put our own understanding together in ways that our students cannot because they don't see the patterns and features that we do (Bransford, Brown, and Cocking 1999). But we are responsible for helping our students to develop these understandings. The keys to their successful learning are for us to plan experiences that let students grapple with important science concepts and for us to ensure that they make sense of the concepts (Weiss et al. 2003).

In Phase 2: *Planning for Responsive Action*, we first identify student preconceptions. Getting students to make their thinking visible is a critical next step. The research into how people in general learn—and in particular how students learn—recommends that teachers determine prior knowledge so they know where to begin instruction (Bransford, Brown, and Cocking 1999; Donovan and Bransford 2005). If you are not familiar with the research and want to learn more about how students learn science, the NRC book, *How Students Learn: Science in the Classroom,* is available online for free (*www.nap. edu/openbook.php?record_id=11102*). A key finding from the literature is that a teacher doesn't really know what his or her students are thinking unless that teacher uses strategies to find out. In Phase 2 of the Instructional Planning Framework, we examine possible student preconceptions, determine student prior knowledge, and plan and deliver instruction that engages students intellectually with the content in ways that both confront their existing ideas and help them make sense of their learning.

Once we identify the nature of any differences between students' thinking and the science viewpoint, it becomes easier to plan activities (using the Instructional Tools in Chapter 3). Identifying the gap in student conceptual understanding helps determine the specific instructional strategies we can use to scaffold our students' learning. Then using formative assessment processes we can gather feedback on student progress. "Probes" are among the many strategies described in Chapter 3 that serve as opportunities to formatively assess students and uncover their ideas. They, like other strategies, are useful at the beginning of a unit (to determine students' current conceptual understandings), in the middle of a unit (to determine where learning is clear and where students may be stuck or still clinging to misconceptions), and at the end of a unit (to determine if students are ready for a summative assessment). This type of assessment can be used to gather evidence of student learning and help determine a teachers' responsive action because it provides information to us about student thinking related to a concept in science. "Probes are concerned less with the correct answer or quality of the student response and focus more on what students are thinking about a concept or phenomenon and where their ideas may have originated" (Keeley and Eberle 2008, p. 207). Many of the Instructional Tools can be used formatively or as part of instruction to promote conceptual understanding.

It goes without saying that understanding is more than just knowing facts. Confronting student's preconceptions is very different from direct instruction, which calls on us to tell students the scientific viewpoints and expect that they will gain comprehension through this direct presentation of information. Instead, we know that students need to engage with the content, integrating the three dimensions of the *NGSS*, so that they can incorporate conceptual learning into their brains in ways that result in durable understanding. The Instructional Tools now includes an alignment with the science and engineering practices. Through scientific inquiry, and more broadly with the science and engineering practices, students make observations and gather evidence that can change their ideas; deepen their understanding of important scientific principles; and develop important abilities such as reasoning, careful observing, and logical analysis (Minstrell 1989). Incorporating the features of science and engineering practices engages students in the lesson and arouses their interest, promotes teamwork, makes sense out of what is otherwise mystifying, and prepares students to successfully defend findings before an audience of their peers (Layman, Ochoa, and Heikkinen 1996). The practices are featured in each of the invited content chapters. Students connect their thinking to the inquiry investigations and practices (through hands-on investigations or virtual simulations) and create mental models that with support and careful instruction by us, can lead to understanding. Students who use inquiry-based materials understand science concepts more deeply and thoroughly than students who learn through more traditional methods (Thier 2002).

Phase 2: *Planning for Responsive Action* is all about developing student understanding, and so it is during this phase that the teacher must get the students actively engaged in learning. Within the classroom, our major responsibility is to facilitate the work of the students. The *NGSS* focus on strategies where students gather, analyze, and engage in argumentation. This is an approach that we can learn to help students make sense of their learning experiences.

The Instructional Planning Framework represents an iterative process and many of the steps can either lead you to the next step or return you to a previous step, depending on the depth of your students' understandings. We cannot prescribe these steps for you. We instead encourage you to use the framework and tools to determine next steps in the learning for *your* students because the context in which you teach is unlikely to be the same as that of other biology teachers.

Connecting the Instructional Planning Framework and Instructional Planning Tools With *NGSS*

Designing effective science instruction that incorporates *NGSS* is going to be a different process for biology teachers and presents a challenge for all of us. In Chapter 2 we will provide some guidance about how teachers can begin to develop a greater understanding of the *NGSS* documents and appendixes, the instructional shifts needed, the learning progression information that teachers can access, how teachers can bundle or group the learning goals and a discussion of what a lesson would look like. The stages in the design process shared in the beginning of this chapter will be elaborated upon so that you can see a lesson planning model.

We realize that there are other hard-to-teach biology concepts that we could have included in this book, but we picked some of our favorites for which there was a clear research base about students' ideas. We hope that you will become sufficiently adept at the process that you will be able to apply our framework to other units that you teach. The planning framework is designed to help us, as teachers, but the greatest expected outcome is improved student achievement and understanding of five complicated biology concepts.

Endnotes

1. As revealed in a study of the research, various terms are used for the explanations that students create for themselves as they make sense of scientific phenomena. All of the terms relate to the understanding students have when they arrive in our classrooms. The most prevalent terms are the following:

Preconceptions. This term refers to student ideas that were formed through life experiences and earlier learning. We use the term in this book to also include the initial ideas students reveal at the start of a unit of study or a lesson.

Alternative conceptions. This term refers to the variety of ideas that students have that differ from scientific explanations.

Naïve conceptions. These are usually incompletely formulated or simplistic representations of student conceptual understanding.

Misconceptions. This term refers to students' incorrect explanations and errors in thinking. (Naive conceptions, preconceptions, and alternative conceptions may also include incorrect explanations, but students view those terms less negatively than they do *misconceptions.*)

For our purposes, we use the term *preconceptions* to refer to the thinking that we hope to get students to reveal as we implement the Instructional Planning Framework. This can include ideas that they generate from everyday experiences or from learning in previous science classrooms. We use the term *misconceptions* specifically when referring to misconceptions that have been identified by the considerable research into students' ideas at different grade levels. These ideas are ones that students often retain even after instruction.

2. Another key to successful learning is that students feel safe to share ideas and express themselves. For example, students need to be able to practice sharing the results of investigations, their individual or team visual representations, or the results of consensus discussions. So that they can do these things without fear of rebuke, teachers must teach the strategies, model them, practice them, and reinforce them so that safe dialogue occurs within the class. For that risk-free environment, we should also include processing time for student sense making.

3. Getting students to be metacognitive must be taught and practiced. How to develop student metacognitive strategies is discussed in Chapter 3; we include strategies such as providing time for journaling (thinking creatively), conducting self-assessments (thinking critically), and goal setting based on what students don't understand (self-regulated thinking) (Marzano and Pickering 1997). These metacognitive strategies are a key piece of the *Identifying Essential Content* phase because they guide students forward from what they understand for sure, to what they think they understand, to what they know is still confusing, and finally to knowing what they need to do next to develop conceptual understanding.

Performance Expectation

Students who demonstrate understanding can: **HS-LS2-6. Evaluate claims, evidence and reasoning that the complex interactions in ecosystems maintain relatively consistent numbers and types of organisms in stable conditions, but changing conditions may result in a new ecosystem.**

[*Clarification Statement:* Examples of changes in ecosystems conditions could include modest biological or physical changes, such as moderate hunting or a seasonal flood, and extreme changes, such as volcanic eruption or sea-level rise.]

Science and Engineering Practice (SEP) **Engaging in Argument from Evidence**	**Disciplinary Core Ideas (DCI)** **LS2C: Ecosystem Dynamics, Functioning, and Resilience**	**Crosscutting Concepts (CC)** **Stability and Change**
Engaging in argument from evidence in 9–12 progresses to using appropriate and sufficient evidence and scientific reasoning to defend and critique claims and explanations about the natural and designed world(s). Arguments may also come from current scientific or historical episodes in science. • Evaluate the claims, evidence, and reasoning behind currently accepted explanations or solutions to determine the merits of arguments. (HS-LS2-6)	A complex set of interactions within an ecosystem can keep its numbers and types of organisms relatively constant over long periods of time under stable conditions. If a modest biological or physical disturbance to an ecosystem occurs, it may return to its more or less original status (i.e., the ecosystem is resilient), as opposed to becoming a very different ecosystem. Extreme fluctuations in conditions or the size of any population however can challenge the functioning of ecosystems in terms of resources and habitat availability. (HS-LS2-6)	Much of science deals with constructing explanations of how things change and how they remain stable. (HS-LS2-6)

Chapter 2

Understanding the *Next Generation Science Standards* and How They Link to the Instructional Planning Framework

"The fundamental idea underlying the standards is to describe clear, consistent, and challenging goals for science education. Then, based on the standards, we need to reform the school science curricula and classroom instruction to enhance student learning. It seems to me that clear and consistent goals and greater coherence in curriculum, teaching, and assessments increase the possibilities for higher levels of achievement for all students."

—*Bybee 2013, p. 3*

Overview

Since their release, science teachers have had varying experiences with the *Next Generation Science Standards* (*NGSS*) and the document on which they are based, *A Framework for K–12 Science Education: Practices, Crosscutting Concepts, and Core Ideas* (*Framework*). The *Framework* provides a description of the full progression of the biological science ideas and has been a useful resource for unit planning. Many of you may be part of a group of teachers that are first adopters given that your states have already replaced their existing standards documents with the *NGSS* or are revising their state standards to incorporate the goals that *Next Generation Science Standards* have included. No matter where your state is in their standards revision process, it is up to science educators to align the state standards to the science curriculum and instruction. In this chapter you will have the chance to reflect on your current practice and consider possible revisions to your curriculum that incorporate *NGSS*. We will focus only on the high school life science disciplinary core ideas and integrated science and engineering practices and crosscutting concepts.

We think there will be readers who are not as familiar with *NGSS* as others so we are including this chapter to provide the background we believe is necessary prior to planning with an *NGSS* focus. For those of you that are familiar with both the *Framework* and the *NGSS*, we will let you decide if you are ready to move forward to Chapter 3. You can return to this chapter at any time to review the recommendations for dealing with the shifts incorporated in the documents.

With the advent of the *Common Core State Standards*, we have seen an era of standards-based reform. As Harold Pratt suggests in *The NSTA Readers Guide to the* Next Generation Science Standards (2013, p. 15), "In rather straightforward terms, the *NGSS* has only two specific purposes beyond its broad vision for science education, namely (1) to describe the essential learning goals, and (2) to describe how those goals will be assessed at each grade level or band. The rest—instruction, instructional materials, assessments, curriculum, professional development, and the university preparation of teachers—is up to the science education community." The *Next Generation Science Standards* learning goals were selected to represent the priorities for science literate students. Teachers have long complained that there are too many standards resulting in teaching and learning that focuses more on coverage and less on spending significant class time to develop conceptual understanding of the important science ideas. With *Next Generation Science Standards* we should be able to maintain a focus on rigorous learning experiences for students around a few core ideas that can be presented to students using practices and strategies that support learner-centered environments. This chapter will tie the *NGSS* to our Instructional Planning Framework.

Our intent is not to deepen your understanding of the *NGSS*, but rather to help you think about them as a resource necessary to inform your unit and lesson plans when you use the Instructional Planning Framework. If you are interested in learning more about *NGSS*, there are many books and articles available. The *NSTA Readers Guide* (Pratt 2013) previously cited is one way to get started and the NGSS@NSTA site (*www.nsta.org/ngss*) offers a variety of links to resources. When it comes to developing curriculum, we recommend that you work with other teachers or district teams to generate guides and curriculum maps that will provide consistent expectations across school environments. It is not an easy process and with persistent concerns about time, we can't expect to be able to do this alone.

The NSTA Reader's Guide and the Instructional Planning Framework

In this chapter, we build on the Instructional Planning Framework introduced in Chapter 1, we include a design process for implementing the framework, and we provide an overview of the stages in the design process modeled in Chapter 3 to put the framework into practice. We start here with a return to *The NSTA Reader's Guide to the* Next Generation Science Standards.

Figure 2.1 (p. 22) shows a representation from the *NGSS* document. In the representation are the three dimensions that taken together with the performance expectations form a "standard." As explained in the *Reader's Guide*, *NGSS* does not have standards statements that lead to instruction and assessment as are currently found in most state and district standards documents. Rather the *NGSS* indicates that every "standard" is made up of the integration of the three dimensions: disciplinary core ideas (DCIs), science and engineering practices (SEPs), and crosscutting concepts (CCCs). At the heart of the *NGSS* are the performance expectations (PEs), which are written to identify the mastery criteria that reveal what evidence students must provide to show proficiency. Because the PEs may be interpreted differently, the associated learning goals (DCIs, SEPs, and CCCs) were included to guide curriculum and assessment development.

Figure 2.1

A Sample Foundation Box From the NGSS for High School Life Science

HS.Structure and Function

HS.Structure and Function

Students who demonstrate understanding can:

HS-LS1-1. **Construct an explanation based on evidence for how the structure of DNA determines the structure of proteins which carry out the essential functions of life through systems of specialized cells.** [Assessment Boundary: Assessment does not include identification of specific cell or tissue types, whole body systems, specific protein structures and functions, or the biochemistry of protein synthesis.]

HS-LS1-2. **Develop and use a model to illustrate the hierarchical organization of interacting systems that provide specific functions within multicellular organisms.** [Clarification Statement: Emphasis is on functions at the organism system level such as nutrient uptake, water delivery, and organism movement in response to neural stimuli. An example of an interacting system could be an artery depending on the proper function of elastic tissue and smooth muscle to regulate and deliver the proper amount of blood within the circulatory system.] [Assessment Boundary: Assessment does not include interactions and functions at the molecular or chemical reaction level.]

HS-LS1-3. **Plan and conduct an investigation to provide evidence that feedback mechanisms maintain homeostasis.** [Clarification Statement: Examples of investigations could include heart rate response to exercise, stomate response to moisture and temperature, and root development in response to water levels.] [Assessment Boundary: Assessment does not include the cellular processes involved in the feedback mechanism.]

The performance expectations above were developed using the following elements from the NRC document *A Framework for K-12 Science Education*:

Science and Engineering Practices	Disciplinary Core Ideas	Crosscutting Concepts
Developing and Using Models Modeling in 9–12 builds on K–8 experiences and progresses to using, synthesizing, and developing models to predict and show relationships among variables between systems and their components in the natural and designed world. • Develop and use a model based on evidence to illustrate the relationships between systems or between components of a system. (HS-LS1-2) **Planning and Carrying Out Investigations** Planning and carrying out in 9-12 builds on K-8 experiences and progresses to include investigations that provide evidence for and test conceptual, mathematical, physical, and empirical models. • Plan and conduct an investigation individually and collaboratively to produce data to serve as the basis for evidence, and in the design: decide on types, how much, and accuracy of data needed to produce reliable measurements and consider limitations on the precision of the data (e.g., number of trials, cost, risk, time), and refine the design accordingly. (HS-LS1-3) **Constructing Explanations and Designing Solutions** Constructing explanations and designing solutions in 9–12 builds on K–8 experiences and progresses to explanations and designs that are supported by multiple and independent student-generated sources of evidence consistent with scientific ideas, principles, and theories. • Construct an explanation based on valid and reliable evidence obtained from a variety of sources (including students' own investigations, models, theories, simulations, peer review) and the assumption that theories and laws that describe the natural world operate today as they did in the past and will continue to do so in the future. (HS-LS1-1) - *Connections to Nature of Science* **Scientific Investigations Use a Variety of Methods** • Scientific inquiry is characterized by a common set of values that include: logical thinking, precision, open-mindedness, objectivity, skepticism, replicability of results, and honest and ethical reporting of findings. (HS-LS1-3)	**LS1.A: Structure and Function** • Systems of specialized cells within organisms help them perform the essential functions of life. (HS-LS1-1) • All cells contain genetic information in the form of DNA molecules. Genes are regions in the DNA that contain the instructions that code for the formation of proteins, which carry out most of the work of cells. (HS-LS1-1) *(Note: This Disciplinary Core Idea is also addressed by HS-LS3-1.)* • Multicellular organisms have a hierarchical structural organization, in which any one system is made up of numerous parts and is itself a component of the next level. (HS-LS1-2) • Feedback mechanisms maintain a living system's internal conditions within certain limits and mediate behaviors, allowing it to remain alive and functional even as external conditions change within some range. Feedback mechanisms can encourage (through positive feedback) or discourage (negative feedback) what is going on inside the living system. (HS-LS1-3)	**Systems and System Models** • Models (e.g., physical, mathematical, computer models) can be used to simulate systems and interactions—including energy, matter, and information flows—within and between systems at different scales. (HS-LS1-2) **Structure and Function** • Investigating or designing new systems or structures requires a detailed examination of the properties of different materials, the structures of different components, and connections of components to reveal its function and/or solve a problem. (HS-LS1-1) **Stability and Change** • Feedback (negative or positive) can stabilize or destabilize a system. (HS-LS1-3)

Connections to other DCIs in this grade-band: **HS.LS3.A** (HS-LS1-1)

Articulation across grade-bands: **MS.LS1.A** (HS-LS1-1),(HS-LS1-2),(HS-LS1-3); **MS.LS3.A** (HS-LS1-1); **MS.LS3.B** (HS-LS1-1)

Common Core State Standards Connections:

ELA/Literacy –

RST.11-12.1	Cite specific textual evidence to support analysis of science and technical texts, attending to important distinctions the author makes and to any gaps or inconsistencies in the account. (HS-LS1-1)
WHST.9-12.2	Write informative/explanatory texts, including the narration of historical events, scientific procedures/experiments, or technical processes. (HS-LS1-1)
WHST.9-12.7	Conduct short as well as more sustained research projects to answer a question (including a self-generated question) or solve a problem; narrow or broaden the inquiry when appropriate; synthesize multiple sources on the subject, demonstrating understanding of the subject under investigation. (HS-LS1-3)
WHST.11-12.8	Gather relevant information from multiple authoritative print and digital sources, using advanced searches effectively; assess the strengths and limitations of each source in terms of the specific task, purpose, and audience; integrate information into the text selectively to maintain the flow of ideas, avoiding plagiarism and overreliance on any one source and following a standard format for citation. *(HS-LS1-3)*
WHST.9-12.9	Draw evidence from informational texts to support analysis, reflection, and research. (HS-LS1-1)
SL.11-12.5	Make strategic use of digital media (e.g., textual, graphical, audio, visual, and interactive elements) in presentations to enhance understanding of findings, reasoning, and evidence and to add interest. *(HS-LS1-2)*

*The performance expectations marked with an asterisk integrate traditional science content with engineering through a Practice or Disciplinary Core Idea.

The section entitled "Disciplinary Core Ideas" is reproduced verbatim from A Framework for K-12 Science Education: Practices, Cross-Cutting Concepts, and Core Ideas. Integrated and reprinted with permission from the National Academy of Sciences.

May 2013   ©2013 Achieve, Inc. All rights reserved.   1 of 1

Since the *NGSS* foundation boxes (the three dimensions) represent the goals and the PEs clarify what a student should know and be able to do, it is still up to teachers to determine how to teach the ideas and decide which learning opportunities and strategies to provide in the classroom. Teachers still have discretion in how to sequence learning to meet the performance expectations and the experiences to implement during instruction.

The *NSTA Reader's Guide to the* Next Generation Science Standards provides a recommended design process. The organization of the design process begins with the end in mind, as was proposed by Wiggins and McTighe in *Understanding by Design* (1998). This design process is presented in Figure 2.2.

Figure 2.2

The NSTA Reader's Guide to the Next Generation Science Standards **Design Process**

Design Process for Planning Instruction Using the *NGSS*	
1. Select a performance expectation (PE)	Understand the PE statement and the clarifications and assessment boundaries
2. Determine the applicable Disciplinary Core Idea (DCI)	Select the DCI and read about the idea in the *Framework* document.
3. Identify the appropriate science and engineering practices (SEP) identified in the foundation box	Understand how the SEP(s) is related to the PE and read the material found in the *Framework* for these practices.
4. Identify the appropriate crosscutting concept (CCC) identified in the foundation box	Understand how the CCC is related to the PE and read the material found in the *Framework* for these concepts
5. Create one or more learning goal(s) for the instruction	Integrate the three dimensions found in the foundations box into the learning goals that will be the focus of instruction
6. Determine the acceptable evidence students would provide to demonstrate proficiency	Draft the summative assessment process for the learning goal(s)
7. Create the learning sequence	Using the BSCS 5E instructional model (or other instructional model), determine the sequence of learning that matches the learning goal(s)
8. Create the summative assessment	Based on the learning experiences, finalize the summative assessment and check its alignment with the performance expectation (PE)

Source: Adapted from Figure 4.1 in Pratt, H. 2013. *The NSTA Reader's Guide to the* Next Generation Science Standards. Arlington, VA: NSTA Press.

The *NGSS* is organized both by topic and by disciplinary core ideas. The approach we took asks you to first look at the disciplinary core ideas configuration to determine the essential understanding(s) represented by one or more of the DCIs that you would bundle into a unit (e.g., heredity and evolution *or* structure/function and heredity). The stages in the design process can then be followed, once that initial decision is made. We will return to the stages of the design process later in this chapter.

Once we identified the appropriate DCIs to "chunk" together for a unit of study, we looked at the science and engineering practices and crosscutting concepts that should be integrated with the core ideas to form the content basis for instructional planning. Depending upon your years of experience teaching biology, you already have some ideas about how to group the content located in the foundation boxes. The disciplinary core ideas themselves represent broad biological ideas, which in our experience include several subgoals or learning targets. As you begin the design process, we suggest you sequence the learning targets into content storylines that align with the SEPs and CCCs. In this way you adhere to the central thinking of the *NGSS* that integrates the three dimensions into coherent learning experiences for students. Creating content storylines is also consistent with the way that *NGSS* is organized around learning progressions.

As was explained in the *NSTA Reader's Guide to the* Next Generation Science Standards, designing with the end in mind then leads us to the stage in the process where we align the content storyline to the performance expectations. We know that with clear learning goals, students need to know the success criteria by which they will be assessed. How are they meant to show proficiency? What will they need to do, say, write, or demonstrate to show that they have mastered the learning goals? Our approach is still consistent with the backward design process (Wiggins and McTighe 1998). In our Instructional Planning Framework Phase 1, once the essential content is identified, success criteria are written. Remember that if you have bundled multiple learning goals together then you will also have several performance expectations that need to be met. Once the rest of the planning phase is completed, planning for instructional experiences can commence. We expect that for each performance expectation there would be multiple lessons that would address one PE. Students will need several opportunities to engage with and learn the science ideas. We suggest a model that is based on a conceptual change approach because our goal is to work together with our students to make what is confusing and mystifying clear as understandable biology ideas. With the *NGSS* approach and goals for science education, the learning progressions or content storylines are not meant to identify a series of facts, definitions, or ideas to teach students about; instead with the integration of the science and engineering practices and crosscutting concepts we need to plan instruction that allows students to figure out scientific ideas using evidence that they gather or are given and develop explanations about how and why phenomena occur (Passmore and Svoboda 2012).

The stages (steps) included in the *NSTA Reader's Guide* design process and in the Instructional Planning Framework in this book both feature planning approaches that maintain a focus on what matters. Both approaches include integrating the three dimensions of the standards, aligning them to the performance expectations, and developing activities into an instructional sequence to connect the learning in significant ways for students. This new focus on developing explanations about science ideas and phenomena will pose a challenge to most of us. With the Instructional Planning Process and the Instructional Tools in this book, we will provide suggestions for how to make student learning visible, how to help students develop explanations, and how to engage students in evidence-based arguments that lead to explanatory models of how and why science phenomena work the way they do. Looking at the instructional shifts will help you reflect on practices to develop our students' explanatory knowledge.

Instructional Shifts in the Context of the Standards

With the new vision for science education in this country, it is important to be aware of the conceptual shifts found in the *Framework* and *NGSS* documents. In Appendix A: Conceptual Shifts in the *Next Generation Science Standards*, seven significant shifts are included with elaborations about the changes that they represent (NGSS Lead States 2013, Volume 2 Appendixes, p. 1):

1. K–12 science education should reflect the interconnected nature of science as it is practiced and experienced in the real world.

2. The *Next Generation Science Standards* are student performance expectations—NOT curriculum.

3. The science concepts in the *Next Generation Science Standards* build coherently from K–12.

4. The *Next Generation Science Standards* focus on deeper understanding of content as well as application of content.

5. Science and engineering are integrated in the *Next Generation Science Standards* from kindergarten through twelfth grade.

6. The *Next Generation Science Standards* are designed to prepare students for college, careers, and citizenship.

7. The *Next Generation Science Standards* and *Common Core State Standards, English Language Arts and Mathematics* are aligned.

These conceptual shifts are provided to help us reflect on changes to our instructional practices. Once you get familiar with the documents, some teachers will want to start small and incorporate some of the recommendations into their curriculum. In some ways, to fully revise curriculum to align with *NGSS*, significant changes to our current pedagogical approaches and content planning are needed. From the research on change (Bridges 2003), making the transition to a new approach means ceasing to do some things while we start to do other things to implement a new approach or a new initiative. So keeping instructional tools that we know work with students is important. At the same time, changing our planning to integrate the disciplinary core ideas, science and engineering practices, and crosscutting concepts will necessitate a significant amount of change.

From the work of Dean Fixsen on implementation (2005), new initiatives or approaches require a major shift from current practices to new ways of doing things. To maintain the change, research indicates that it is frequently easier to maintain shifts when implementers adopt all of the recommended changes into practice rather than making small changes over time. The tendency with small changes is to revert to prior practices when faced with challenges. When implementers make broad sweeping changes, they are more likely to continue to work to make improvements even when first efforts may not work out as planned.

As we look at the conceptual shifts, think about the changes in practice that might result for you. For each shift we will suggest some implications for your practice—some that may represent significant shifts as you plan for teaching and learning. Let's think about these shifts as they apply to our Instructional Planning Framework.

- *Conceptual Shift #1.* This reminds us of the fundamental change in the *NGSS*: Science education should reflect the nature of science as it is practiced and experienced in the real world. *Implications for planning:* Teaching about the Nature of Science cannot be just part of a stand-alone unit but must be integrated into every unit and lesson. Student learning should focus on conceptual understanding and applications of learning. We need to focus on students thinking and behaving as scientists where they use the terminology correctly and engage regularly in science and engineering practices.

- *Conceptual Shift #2. NGSS* are performance expectations and are not a curriculum. *Implications for planning:* From the performance expectations, we still need to identify the learning goals, instruction, and assessments that form coherent instructional programs. Sequencing the learning into appropriate progressions, teaching and reinforcing the science and engineering practices, and incorporating the crosscutting concepts throughout the programs is a necessary planning step as new curriculum maps and curriculum units are created. There is a lot of work ahead to get this done, so work with a team to create curriculum drafts, pilot them, and make revisions based on the piloting experiences.

- *Conceptual Shift #3.* The science concepts in *NGSS* build coherently from K–12. *Implications for planning:* As biology teachers, we can use the *Framework* and *NGSS* to determine the expected student prior learning and understandings. From the progression documents, we also know the end points for student understanding for the DCIs; the science and engineering practices; the understandings about the Nature of Science; progressions of crosscutting concepts; and science, technology, society, and the environment connections. Students should become more skilled and sophisticated in their conceptual understanding. Planning content storylines and progressions for SEPs and CCCs means planning with the end in mind.

- *Conceptual Shift #4. NGSS* focus on deeper understanding of content as well as application of content. *Implications for teaching:* We need to look at our current curriculum and units of instruction and prune away technical vocabulary and content that is outside the core ideas. To move students beyond a novice level of understanding, they need time and opportunity to deepen their understandings. Planning learning experiences that help students learn the core ideas through engaging in scientific and engineering practices is more likely to result in the planned outcomes.

- *Conceptual Shift #5.* Science and engineering are integrated in the *NGSS* from K–12. *Implications for planning:* Even with a recent emphasis nationally on STEM learning, engineering and technology are still not being integrated significantly into science classrooms. With *NGSS*, engineering practices are now as important as the features of inquiry (scientific practices). Planning that includes problem-based approaches helps students address important societal challenges and promotes relevant learning. Ultimately experiential learning will lead to a greater depth on understanding.

- *Conceptual Shift #6. NGSS* are designed to prepare students for college, careers, and citizenship. *Implications for planning:* Scientifically literate students prepared for postsecondary experiences means we must incorporate 21st-century skills into our teaching. First and foremost this means teaching our student how to learn.

- *Conceptual Shift #7. NGSS* and *Common Core State Standards, English Language Arts and Mathematics* are aligned. *Implications for planning:* These standards overlap in meaningful ways. All three sets of standards include student practices around argumentation associated with reasoning. Teaching procedures that support all three standards increase the connections for students.

Detailed elaborations about the seven conceptual shifts are included in the *NGSS* appendixes. For more connections to the Instructional Planning Framework, we offer additional information in Chapter 4.

Learning Progressions

Educational research and science education research has turned its attention to learning progressions in recent years (Heritage 2008; NRC 2012). Embedded within the *Framework,* progressions are emphasized. Margaret Heritage describes specific learning progressions that inform K–12 learning as trajectories where sophistication of understanding and skills develop increasingly over time (Heritage 2008).

In the past, state standards were not organized this way and each performance expectation was as important as another. Some learning progressions have tried to incorporate what we know from cognitive research about students' ideas to help create classroom learning sequences. According to the *Framework,* "Developing detailed learning progressions for all of the practices, concepts, and ideas that make up the three dimensions was beyond the committee's charge; however, we do provide some guidance on how students' facility with the practices, concepts and ideas may develop over multiple grades " (NRC 2012, p. 33). Within the narratives related to the science and engineering practices and the crosscutting concepts, a discussion of how student understanding may progress across grade levels is included. Of particular importance within the *Framework* is the notion that understanding develops over time and "Building progressively more sophisticated explanations of natural phenomena is central throughout grades K–5 as opposed to focusing only on description in the early grades and leaving explanation to the later grades" (NRC 2012, p. 26). Understanding the progressions of all three dimensions is important background information so we can determine the prior knowledge expected of our students. It also elucidates what we know from the research on teaching and learning about how student thinking increases in sophistication and complexity.

During the writing of the *Framework,* the committee decided that rather than trying to identify specific disciplinary core idea learning progressions, grade band endpoints informed by research on teaching and learning would be included. The notion that learning becomes more sophisticated as students master the conceptual ideas is central to the *NGSS.* Two of the key instructional shifts identified in the introduction (NGSS Lead States 2013, p. xix) state that "Science concepts build coherently across K–12 and the *NGSS* focus on deeper understanding and application of content."

The Appendix E: Disciplinary Core Idea Progressions in the *Next Generation Science Standards* summarize the end of grade-band core ideas and reveals the increasing development of student thinking. An example is included in Figure 2.3. This is just a visual representation that is in the *NGSS* derived from the full progressions in the *Framework.* What are not included in this document are the other two dimensions, which for our planning purposes need to be integrated with the disciplinary core idea.

How do we use this information? Learning progressions inform both phases of the Instructional Planning Framework. In Phase 1 we focus on identifying the learning goals (DCIs, SEPs and CCCs) and essential understandings that are part of the unit plan. In the progression identified in Figure 2.3 we can see that the student thinking

Figure 2.3

Example Disciplinary Core Progression

DCI	K–2	3–5	6–8	9–12
LS2.B Cycles of matter and energy transfer in ecosystems	Content found in LS1.C and ESS3A	Matter cycles between the air and soil and among organisms as they live and die.	The atoms that make up the organisms in an ecosystem are cycles repeatedly between the living and non-living parts of the ecosystem. Food webs model how matter and energy are transferred among producers, consumers, and decomposers as the three groups interact within an ecosystem.	Photosynthesis and cellular respiration provide most of the energy for life processes. Only a fraction of matter consumed at the lower level of a food web is transferred up, resulting in fewer organisms at higher levels. At each link in an ecosystem elements are combined in different ways and matter and energy are conserved. Photosynthesis and cellular respiration are key components of the global carbon cycle.

Source: NGSS Lead States. 2013. Appendix E. p. 44.

gets more sophisticated and the biological ideas included as endpoints include several important concepts. In the first edition of *Hard-to-Teach Biology Concepts* we looked at the content storyline that we proposed for teachers to develop student understanding of the process of photosynthesis. Figure 2.4 (p. 30) shows the sequence of learning targets that provided one pathway to connect the learning for students to lead to the broader understanding represented by the essential understanding statement.

If the essential understanding is represented by the disciplinary core ideas integrated with the science and engineering practices and crosscutting concepts, then when planning the content sequence—the content storyline—we believe that we can help students develop their understanding both across K–12 and across a unit of study. Now, compare the core ideas in Figure 2.3 with the essential understandings in Figure 2.4. You will see that we deconstructed the disciplinary core ideas about the process of photosynthesis into subgoals or learning targets that build a DCI progression of understanding that when taken together aggregate into the broader core idea

Figure 2.4

Photosynthesis Learning Sequence

Target #1:	Target #2:	Target #3:	Target #4:
The vast majority of plants are able to convert inorganic carbon in CO_2 into organic carbon through photosynthesis. Carbon dioxide and water are used in the process to create biomass. The surrounding environment is the source of raw materials for photosynthesis	Photosynthesis captures the energy of sunlight that is used to create chemical bonds in the creation of carbohydrates. Chloroplasts in the cells of plant leaves contain compounds able to capture light energy. Photosynthesis utilizes CO_2 and the hydrogen from water to form carbohydrates, releasing oxygen.	Carbohydrates produced during photosynthesis in leaves can be used immediately for energy in the plant, stored for future use, or converted to other macromolecules that help the plant grow and function.	A variety of gases move into and out of plant leaves. Leaves use CO_2 and release O_2 during photosynthesis. When they respire, leaves use O_2 and release CO_2. Other gases enter the leaf as well, but are excreted.

Essential Understandings: Plants have the ability to capture the energy of sunlight and use it to combine low-energy molecules (carbon dioxide and water) to form higher energy molecules (glucose and starches). This is called *photosynthesis*. They can use the glucose immediately as a source of energy, convert it into other molecules that help their cells function, or store it for later use. Regardless, the total amount of matter and energy in the system stays the same.

Source: Koba, S., and A. Tweed. 2009. *Hard-to-teach biology concepts: A Framework to Deepen Student Understanding.* Arlington, VA: NSTA Press.

represented in the endpoint concept found included in *NGSS*. With the philosophy of *NGSS*, we remind you that you would still need to take your progression of DCIs and incorporate the appropriate SEPs and CCCs before planning the lessons. We recommend that you determine a reasonable chunk of learning that is represented in the foundation boxes by the DCIs, SEPs, and CCCs and the corresponding performance expectations and generate a content storyline. In Figure 2.3 the summary statements represent broad ideas once again that can only be understood when we identify the smaller learning targets that connect into a coherent science idea. Ultimately we want students to understand that photosynthesis and cellular respiration provide most of the energy for life processes. It is significant to remember that we are shifting away from students learning about the learning targets and toward learning experiences that figure out the science ideas. That can't happen unless we consider the learning targets and appropriate science and engineering practices. In the learning sequence provided in Figure 2.4, we can see that to develop a conceptual understanding of photosynthesis and the role that it plays when providing energy for life processes, students have to be able to answer a variety of questions based on evidence gathered during instruction. This is where we need to reflect on teaching shifts needed for students to develop explanations. We can start by having students ask questions related to the learning targets.

- Why is photosynthesis such an important biological process?

- Which organisms are capable of this energy generating process?

- How does energy flow though ecosystems?

- How is the energy created in photosynthesis made available for life processes?

With these and other questions we can get smarter about what smaller learning targets are needed for students, the science and engineering processes necessary to advance their learning, and crosscutting concepts to expand their mental models for them to progress from their naïve conceptions of photosynthesis to the sophisticated level where they develop coherent explanations supported by evidence.

Kathy Roth's research into content storylines (Roth et al. 2009, p. 1) explains the need for content storylines in the following way,

> Teachers may present accurate science content and engaging hands-on activities, but these content ideas and activities are often not carefully woven together to tell a coherent story. Students miss the point of the activities they are carrying out, and instead pick up random pieces of scientific terminology without fitting them together to develop rich conceptual understandings of important learning goals. To help students learn science, you can use a "science content storyline" lens. The science content storyline lens focuses attention on how the science ideas in a lesson (or unit) are sequenced and linked to one another to build a coherent "story" that makes sense to students.

Her research findings were gathered prior to release of the *Framework* and *NGSS* and were considered by the committee that included learning progressions as a fundamental component of the documents.

Learning progressions also link to Phase 2 of the Instructional Planning Framework. When planning for responsive action designed to help students make sense of the science ideas, learning activities and content representations need to connect the learning targets so that little by little, students can generate explanations using evidence gathered in the learning activities. This is why the core ideas must be linked to the practices and crosscutting concepts. In the meta-analysis within *How People Learn* (Bransford, Brown, and Cocking 1999), children try to make sense of what is confusing and mystifying to them. In the absence of evidence gathered to answer scientific questions, their explanations are based on suppositions and conjecture. When we provide well-planned learning experiences that include accurate content representations and investigations, sense-making can occur.

Interpreting an *NGSS* Standard

The *NGSS* articulate how science ideas build coherently over time and across science disciplines. What is not clear in the document is what is meant by a "standard." This is a key difference between *NGSS* and the previous *National Science Education Standards* (NRC 1996). Because the *NGSS* were written as performance expectations that depict what students should do to show proficiency with science ideas, typical one-line standards are not included. The performance expectations are, according to the *NGSS*, "the policy equivalent of what most states have used as their standards" (NGSS Lead States 2013, p. xiii).

The *NGSS* include learning goals, which are represented in the "foundation boxes" that align with the descriptions in the *Framework.* The science and engineering practices (SEPs), coupled with the disciplinary core ideas (DCIs) and crosscutting concepts (CCCs), make up the performance expectations (PEs). State standards usually represent the core science ideas separately from science practices like inquiry. In the *NGSS* the science and engineering practices are not instructional strategies for teachers but represent learning goals themselves. Because of this, the DCIs, SEPs, and CCCs are integrated and are included in the *NGSS* to help teachers determine the intent of the performance expectations. This approach reflects real-world science where content and practice are linked naturally. The *NGSS* explains it this way: "K–12 science education should reflect the interconnected nature of science as it is practiced and experienced in the real world" (NGSS Lead States 2013, Volume 2, Appendix A, p. 5).

When writing the performance expectations, the developers of the *NGSS* recognized that to be well understood, the material included in foundation boxes and the *Framework* are needed. To assist users of *NGSS*, the National Science Teachers Association has provided the following definition: "NSTA considers a standard to be the performance expectations and foundation box associated with a core idea at a given grade level or band" (NGSS@NSTA).

With these clarifications, let's spend a little time interpreting an *NGSS* "standard."

In Figure 2.5 we see a representation of a standard. What does a teacher do with the information in the performance expectation and foundation boxes? With a think-aloud strategy we will talk through the understanding and inferences that we might make based on the *NGSS* documents. Additional examples are included in the content chapters 5–8.

Traditional curriculum materials would begin a unit on ecosystems by introducing a definition or what ecosystems are and are not and the level of organization that they represent. This would be followed by opportunities to learn about interrelationships between populations within an ecosystem and the impacts of the abiotic conditions. To explain the interactions within ecosystems, students would study populations that are identified as producers and consumers and then introduce the relationships between

Figure 2.5

NGSS "Standard"

Performance Expectation
Students who demonstrate understanding can: **HS-LS2-6. Evaluate claims, evidence and reasoning that the complex interactions in ecosystems maintain relatively consistent numbers and types of organisms in stable conditions, but changing conditions may result in a new ecosystem.**
[*Clarification Statement:* Examples of changes in ecosystems conditions could include modest biological or physical changes, such as moderate hunting or a seasonal flood, and extreme changes, such as volcanic eruption or sea-level rise.]

Science and Engineering Practice (SEP) Engaging in Argument from Evidence	Disciplinary Core Ideas (DCI) LS2C: Ecosystem Dynamics, Functioning, and Resilience	Crosscutting Concepts (CC) Stability and Change
Engaging in argument from evidence in 9–12 progresses to using appropriate and sufficient evidence and scientific reasoning to defend and critique claims and explanations about the natural and designed world(s). Arguments may also come from current scientific or historical episodes in science. • Evaluate the claims, evidence, and reasoning behind currently accepted explanations or solutions to determine the merits of arguments. (HS-LS2-6)	A complex set of interactions within an ecosystem can keep its numbers and types of organisms relatively constant over long periods of time under stable conditions. If a modest biological or physical disturbance to an ecosystem occurs, it may return to its more or less original status (i.e., the ecosystem is resilient), as opposed to becoming a very different ecosystem. Extreme fluctuations in conditions or the size of any population however can challenge the functioning of ecosystems in terms of resources and habitat availability. (HS-LS2-6)	Much of science deals with constructing explanations of how things change and how they remain stable. (HS-LS2-6)

them. Students might get a chance to look at historical data and generate interpretations. In a broader unit, we might also include information about the cycling of matter and the flow of as it moves through the ecosystem. Students would be introduced to food webs and energy pyramids and asked to create them from information provided. The outcome of the learning experiences for students would probably be memorization of terms and basic explanations of types of interactions. For the most part, text-based materials feature facts and definitions and don't lead students to an understanding of the science phenomena. They are often intended to be comprehensive and provide detailed explanations of science ideas that tell the students what scientists have already learned or discovered.

If we return to the performance expectation, we can see that the traditional approach will not prepare students to meet the PE. To do that we need to be clear about the phenomena that the students should be attempting to explain. We can see that the science idea we want students to understand is the stability of ecosystems that occurs due to the complex interactions that exist between populations of organisms. If conditions remain relatively unchanged, then the numbers and kinds of populations remains stable. To fully involve students in the learning we want to know what students already understand about ecosystems and how they work and the questions that they have related to the phenomena of system stability over time, which is sometimes described as *dynamic equilibrium*. Students should have already developed the idea that different ecosystems, described by their typical plant and animal populations are dependent on the temperature and precipitation in a geographic area. They also have a basic understanding of the role that producers play in ecosystems to provide food for consumers, which in turn can eat other consumers or both producers and consumers to get their food energy and matter for growth, development, and reproduction. Based on the performance expectation and clarifications included in Figure 2.5, we need to push students to consider how and why this happens. What conditions cause an ecosystem to become unstable? What conditions would cause significant changes to an ecosystem? How do the complex interactions within ecosystems maintain stability? What are the impacts of human activity on the biodiversity and stability of ecosystems? How resistant to change are ecosystems that are subjected to natural disasters (e.g., hurricanes, tsunamis, volcanos)? Why is ecosystem stability important? Who benefits? To answer these questions, we can refer students to historical and current natural disasters (e.g., fires in Yellowstone National Park, 2013 typhoon in the Philippines, earthquake and Tsunami in Japan 2011). And what about the impact of human activity causing habitat loss?

To answer the questions, students need to work on investigations using historical data and ecosystem models. Using appropriate and significant evidence, students can develop their explanations. Through argumentation from evidence, students present their scientific reasoning and critique the claims of others. In this way we link the science and engineering practice of Argumentation from Evidence to the disciplinary core ideas. We can also integrate the crosscutting concept of Stability and Change as students add to their mental models of stability and change in biological systems.

The focus on explanations suggests that our challenge is to incorporate strategies into our lessons that use the science and engineering practices. We need to provide realistic content representations and experiences that provide options for students to gather data and determine evidence to answer the questions that arise when studying the phenomena of system stability.

When we interpret the *NGSS* standard, we can see that organizing instruction around questions that arise when studying the phenomena must replace typical instruction that sequentially moves from topic to topic. The goal then is to provide investigations that provide the evidence needed to develop explanatory models and not just test a hypothesis. With Argumentation as the identified focus of the SEPs, substantive time needs to be planned for argumentation and reaching consensus based on critical analysis of evidence and student thinking revealed in their communication within the classroom. With student learning centered on the performance expectation, it is important to build a positive classroom climate where argumentation is an everyday occurrence, where students work together cooperatively and peer interactions move toward answering the questions posed.

Lesson Design Process Aligned With the Instructional Framework and the *NGSS*

The lesson planning process for hard-to-teach biology concepts is critical to the success of the teaching and learning experiences for students. One approach described in *Designing Effective Science Instruction* (Tweed 2009) suggests using a C-U-E lesson framework. Each lesson should include all three elements of effective lesson design. The C is for Content, with the focus on setting clear learning goals and success criteria that can be communicated with students. The U is for Understanding and suggests identifying instructional strategies to develop student reasoning that leads to conceptual understanding. With *Next Generation Science Standards* this means designing lessons incorporating not just the science and engineering practices that we now integrate with the learning goals, but also including the metacognitive strategies, and the connections to the Nature of Science and *Common Core State Standards*. Finally, the E represents Environment and reminds us that promoting positive learning environments is necessary to engage students with the biology phenomena. It is also important to support cooperative learning strategies; develop peer and self-assessment opportunities; and develop a climate where intellectual rigor, constructive criticism, and challenging of ideas are evident. Teaching procedures that invite student questioning contribute to classroom environments where students can share their ideas in a safe setting.

So how do we start the lesson planning process? First, remember that a lesson is not always a single class period of 45–50 minutes. A lesson is actually a sequence of learning aligned to one of the learning targets (DCI, SEP, and CCC). Thinking about it this way, a lesson can be anywhere from a couple of days to even a week in length. Use whatever template you already have and revise it to incorporate the *NGSS* components. Figure 2.6 (pp. 36–37) provides a sample unit planning template aligned to our Instructional Planning Framework and incorporating *NGSS*.

With the unit planning framework you can determine where you are going with your lessons and how you will get there. The invited chapters 5–8 will provide examples of how the authors thought about the planning process as they tell their stories of planning for lesson implementation. Once you have established the series of lessons within the unit plan, you can plan for a lesson (sequence of learning) along with the lesson implementation strategies.

Figure 2.6

Unit Planning Template

Unit Topic—		
Phase 1. Identifying Essential Content		
Conceptual Target Development	Disciplinary Core Ideas Addressed	
	Crosscutting Concepts Addressed	
	Science and Engineering Practices Addressed	
Essential Understandings		
Criteria to Determine Understanding		
Performance Expectations Addressed		

Phase 2. Planning for Responsive Action	
Identifying Student Preconceptions	

Learning Sequence Targets	
Learning Target #1	
Research-Identified Misconceptions Addressed	
Initial Instructional Plan	*Eliciting Preconceptions:* *Confronting Preconceptions:* *Sense-Making:*
Formative Assessment Plan *(Demonstrating Understanding)*	

Figure 2.6 (continued)

Learning Target #2	
Research-Identified Misconceptions Addressed	
Initial Instructional Plan	*Eliciting Preconceptions:* *Confronting Preconceptions:* *Sense-Making:*
Formative Assessment Plan *(Demonstrating Understanding)*	

Learning Target #3	
Research-Identified Misconceptions Addressed	
Initial Instructional Plan	*Eliciting Preconceptions:* *Confronting Preconceptions:* *Sense-Making:*
Formative Assessment Plan *(Demonstrating Understanding)*	

Learning Target #4	
Research-Identified Misconceptions Addressed	
Initial Instructional Plan	*Eliciting Preconceptions:* *Confronting Preconceptions:* *Sense-Making:*
Formative Assessment Plan *(Demonstrating Understanding)*	

Course Scope and Sequence Implications

In Appendix K of the *NGSS* you will find model course maps (*www.nextgenscience.org/sites/ngss/files/Appendix%20K_Revised%208.30.13.pdf*). This provides a starting point for curriculum development. There are multiple ways to organize the performance expectations so that all disciplinary core ideas, all science and engineering practices, and all crosscutting concepts are included in the proposed curriculum maps. It is still up to districts, states, or teams of teachers to figure out the organizational sequences that work with students. Three options are provided as examples to guide development of scope and sequence documents across grades 6–12 that lead to curricula. One approach is to organize by conceptual progressions where the courses are based on the concepts and not by science discipline. The second approach is to organize by science domains. This model is similar to current practices, and the grade-band PEs are grouped into content-specific courses in physical, life, and Earth and space sciences. The third model is for grades 9–12 and organized into a modified science domain model. In this structure the PEs are clustered into typical high school course sequences that include biology, chemistry, and physics. The Earth and space science concepts are spread across the course sequences.

Each of these approaches has its advantages and disadvantages. What is clarified in the examples (NGSS Lead States 2013, Appendix K, pp. 117–120) is that core ideas are written as broad statements and are at too large a grain size (big idea) to be useful for curriculum mapping. To get to a grain size that was useable, the core ideas were split into each of the component ideas identified in the *Framework*. With this information the core ideas could be sorted in a variety of useful ways. A series of recommendations are provided if you are going to do this work. To find out more about the process, select the model that you find most useful and read the descriptive process provided for that model. With 50 sets of state standards, there is no expectation that all states and districts would use the same model!

An Introduction to the Design Process for Implementing the Instructional Planning Framework

The design process is organized into two phases. Phase 1: *Identifying Essential Content*, details the planning stages when preparing a unit of study that incorporates the learning goals from *NGSS*. Phase 2: *Planning for Responsive Action*, elaborates on the steps you would take developing lessons for implementation with students. This part of the Instructional Planning Framework includes the Instructional Tool selection process.

With the second edition of this book, some of the stages in the planning are still the same but the outcomes of the planning will likely change. With a focus on student outcomes rather than decontextualized materials (e.g., textbook resources), incorporating the science and engineering practices and linking them to the instructional strategies may look quite different from the lessons you have used in the past. When it comes to implementing the Instructional Planning Framework, everyone will approach the decision-making based on prior experiences with students and what works for you. We encourage you to try some of the Instructional Tools that you haven't used before to increase your options for developing student reasoning. For now, let's look at the design process shown in Table 2.1.

Table 2.1

Design Process for Implementing the Instructional Planning Framework

Phase 1: Identifying Essential Content	
Stage I	Identify the disciplinary core idea(s) (key concept or learning goal) that are the focus for unit of instruction and the related crosscutting concepts and practices.
Stage II	Deconstruct the selected disciplinary core idea(s) (key concept or learning goal) into its sub-goals (learning targets), sequence into a content storyline, and align with appropriate practices and crosscutting concepts.
Stage III	Determine the success criteria for each learning goal and align NSGG performance expectation(s) to determine student expectations for understanding.
Stage IV	Determine appropriate inclusion of Nature of Science connections.
Stage V	Identify metacognitive goals and strategies.
Phase 2: Planning for Responsive Action	
Stage VI	Research student misconceptions common to this topic that are documented in the research literature.
Stage VII	Determine appropriate strategies to identify your students' preconceptions.
Stage VIII	Determine appropriate strategies to elicit and confront your students' preconceptions.
Stage IX	Determine appropriate sense-making strategies.
Stage X	Determine responsive actions based on formative assessment evidence.

Note: This table will be referred to throughout the book. You might want to flag it to make it easy to find.

Phase 1: Identifying Essential Content

Stage I: For your selected unit of study, determine the disciplinary core ideas and related crosscutting concepts that together will form a broader statement of essential science understanding. Identify the practices from the *NGSS* that align to your selected core ideas and concepts. Taken together, these ideas will constitute the essential learning that will form the basis for student learning. Refer to the *Framework* (NRC 2012) to help with this. This is an important step for you the teacher. Without determining how the learning links together into a bigger idea or essential understanding will mean that the students will also not be able to connect the ideas into a broader framework of understanding.

Stage II: The disciplinary core ideas and the crosscutting concepts are generally broad statements themselves that will still need to be deconstructed into subconcepts that most teachers and educational literature refer to as learning targets. The learning target(s) become the focus for daily instruction. You will need to include learning targets that incorporate the crosscutting concepts (CCCs) and also the science and engineering practices (SEPs). You can use the learning progressions for the CCCs and the SEPs found in Appendixes F and G of *NGSS*. All of the identified targets should then be sequenced into a storyline where the learning targets are connected through instruction in ways that promote student thinking and sense-making. There are numerous pathways that a teacher can use when sequencing the learning targets. When we describe the decision-making process in subsequent chapters, each chapter will highlight one proposed pathway determined by the invited authors and the rationale that resulted in the coherent content storyline. This process is easier when the teacher has a well-developed understanding of the disciplinary core ideas and crosscutting concepts themselves and integrates them with the science and engineering practices. This part of the process should not be rushed and pedagogical content knowledge about when and how to sequence the learning will be useful when planning for responsive action in Phase 2.

Stage III: For each learning goal, crosscutting concept, and set of practices previously identified, write success criteria that identify the evidence that students need to provide to show they have mastered the learning goal. Success criteria provide guidance for students. If we want them to learn the science ideas, then we need to be sure they know the "look-fors" for effective work. Next, look at the performance expectations in the *NGSS* to determine performance tasks that can be used to gather evidence of student learning. Remember that students will need to use evidence gathered through learning to develop explanations of their conceptual understanding. This is really at the core of *NGSS*—learning through development of explanations based on reasoning and critical discussions. Rubrics are useful tools to determine different levels of student performance and if done well, will provide the "look-fors" that students need to move their learning forward as they strive to meet the performance expectations.

Stage IV: The *NGSS* include connections to the Nature of Science. Review the information in the foundation boxes and from Appendix H to determine any additional opportunities that you should include in your Phase 1 planning to help students think and act like scientists. Connections to the Nature of Science should link to learning experiences throughout the unit. Obvious links can be made between the Nature of Science connections and the science and engineering practices. Many of these procedures or practices need to be taught to the students, modeled, and reinforced. Planning for those experiences then is part of Phase 1. Appendix H of the *NGSS* includes the basic understandings about the Nature of Science and their association by grade level with the practice and crosscutting concepts.

Stage V: Metacognition is an important skill to teach students. Helping them to identify what they need to learn, where they are in the learning sequence and what they need to work on next are all pieces necessary to develop metacognition in students. Ultimately we would like students to be self-regulated learners. To support positive learning environments, planning for student self-reflection is an important aspect of the planning in Phase 1. Essentially, metacognition concerns a learner's ability to monitor his or her own thinking, which results in bringing one's learning to a conscious level. We know from the cognitive science literature that being able to monitor one's thinking is a characteristic of effective learning. If we teach students to constantly monitor their thinking and what makes sense, then when it doesn't, they have strategies to do something about it. When they engage in metacognition and become conscious of the process of learning, they are able to take control over their own learning.

Phase 2: Planning for Responsive Action

Now, refer back to Table 2.1 and consider Phase 2: *Planning for Responsive Action*. While Phase 1: *Identifying Essential Content* is research-based, Phase 2 implements a research-based plan and *responds* to our students' ideas. We will now explore this phase and its stages.

Stage VI–VII: Identify preconceptions. We recognize that, in the ideal, teachers should elicit their students' preconceptions and use these preconceptions to plan instruction. However, we also understand the importance of long-term planning. This step is designed to help you study the research base as well as identify your students' initial ideas so you can outline instruction focused on the conceptual target. Operationally, the first part of Phase 2 is to identify preconceptions, including common, research-identified misconceptions and the preconceptions of your own students.

Stage VIII: Elicit and confront preconceptions. We begin instruction by eliciting our students' preconceptions and discussing those preconceptions with our students. Once you know what conceptual understandings your students already have related to the concepts, you can use the tools in Chapter 3 to select strategies that address the conceptual change process.

Stage IX: Use sense-making strategies to address preconceptions. This step in the process is at the core of the Instructional Planning Framework. Teachers need to provide opportunities for students to make sense of the ideas they learn because it is unlikely that students will draw the appropriate conclusions on their own. Teachers must select strategies that engage students through questioning, discussion, and other methods, so students can make connections between their ideas and what they are meant to learn. With the *NGSS* goal of helping students figure out the science ideas rather than learning about them, getting students to share their ideas and reasoning should be part of everyday activity in our classrooms. Later, you will ask students to think about their initial ideas, relate them to their current learning, and then determine how their thinking has changed.

Stage X: Determine responsive actions based on formative assessment evidence. To complete the framework, student learning must match the criteria for developing student conceptual understanding. If it does not, you will need to revisit the final steps in Phase 1 and select additional tools to help students understand the concepts. Formative assessment is part of best practices in our classrooms and when this process is implemented, there is a regular opportunity for feedback about the learning that occurs between teachers and students and among students when engaging in peer assessment. More about this process is included in Chapter 4. It is important to remember that formative assessment is a process and not actually an assessment test, so should not be confused with interim or benchmark assessments. Rather it is assessment for learning where at regular intervals within the learning sequence, students reveal their progress toward the performance expectations.

The design process may seem like an overwhelming number of tasks to prepare for teaching and learning. But as we begin to make shifts to the learning goals included in *NGSS*, spending additional time planning will benefit both you and your students. You will have a clear vision of where the learning is going and students will also know what is expected of them. As always, this planning and implementation work is easier for us as teachers when we can collaborate with colleagues.

Chapter 3 will go through a detailed example of the design process using a unit previously discussed in the first edition of this book, now revised to include information from *NGSS* and the *Framework*. Take time to reflect on your current approach to teaching and what instructional shifts you can make so that the learning is accessible to all your students.

Chapter 3

Implementation of the Hard-to-Teach Instructional Planning Framework

"Science teachers have always used multiple strategies, so we need not make a decision about the one best strategy for teaching science. There isn't one; there are many strategies that can be applied to achieve different outcomes. Science teachers should try to sequence them in coherent and focused ways. This is how inquiry can contribute to the prepared mind."

—Bybee 2006, p. 456

Overview

As a teacher, you know a lot of different strategies to use and activities to implement in the biology classroom. However, strategies themselves are not the total solution to learning for your students. Nor are activities. You must purposefully select strategies—and activities that target specific science phenomena—that best fit your instructional goals and then sequence classroom experiences to support students' conceptual understandings.

A well-developed science storyline where the learning targets are connected through instruction in ways that promote student thinking and sense making is important, but it is not a simple endeavor. This chapter provides an example of how to design instruction around a subset of life science ideas in one of *Next Generation Science Standards* (*NGSS*), using the Instructional Planning Framework (Figure 1.1, p. 7) and design process (Table 2.1, p. 39). It also introduces a well-researched set of Instructional Tools (pp. 57–139) to help you select strategies that have been shown to develop students' conceptual understanding.

Those of you familiar with the first edition of this book know that the design process was outlined in a series of chapters. That approach was helpful in some ways since the process was gently paced, with each chapter focusing on a different topic and a different aspect of the design process. A negative side effect of that approach was that only the final topic was developed using the entire process. We decided for this second edition to present the process in one chapter focusing on one topic. As a result, this chapter includes a lot of information, but the process is completely explored. The second half of the book presents four additional topics from invited authors, and again, each topic reflects the full design process.

We chose the topic "proteins and genes" to model use of the Instructional Planning Framework, design process, and Instructional Tools, in particular how segments of DNA (genes) guide the function or dysfunction of proteins at the various levels of biological organization (cells, tissues, organs, and systems)—that is, the connection between genotype and phenotype. We purposefully chose to repeat the same topic from the first edition to demonstrate that an *NGSS*-aligned set of lessons need not be totally created from scratch.

Why This Topic?

The media are filled with commentary about stem cell research, gene therapy, and genetic modification of foods, yet few people understand the relationship between the function of genes, the production of proteins, and phenotype—a relationship essential to understanding each of those issues. Molecular genetics is a topic likely to have an immediate effect on and direct relevance to students' lives in the 21st century, yet traditional teaching methods are not adequate to motivate and educate students about that topic (Eklund, Rogat, Alozie, and Krajcik 2007; Sinan, Aydin, and Gezer 2007).

Genetics is a difficult topic to understand (Mertens and Walker 1992). Students must integrate information from multiple levels of biological organization (Bahar, Johnstone, and Hansell 1999; Duncan and Reiser 2003; Kindfield 1994; Lewis and Wood-Robinson 2000). This is also the problem when it comes to the molecular basis of heredity, because students must connect what happens at the molecular level to the resulting cellular proteins and they must understand how that impacts the appearance and function of cells, tissues, and organisms.

These difficulties are compounded by the common approach to teaching genetics—that is, Mendelian inheritance is taught before DNA, and DNA is unconnected to the actions of proteins (Roseman, Caldwell, Gogos, and Kurth 2006). Also, passive instruction using lecture and textbooks, typical for this topic, relies on abstract, complex figures and chemical formulas, posing difficulties for typical high school students (Rotbain, Marbach-Ad, and Stavy 2005). Finally, "This area of science is moving so quickly that many of us lack appropriate content training" (Texley and Wild 2004, p. 89).

Application of Phase I to "Proteins and Genes" Topic

We want you to meet Mrs. Hernandez, a teacher who has tried in many ways to improve her instruction yet has met with less than adequate student understanding. We'll use her context (Figure 3.1) as we delve into the biological content and the pedagogical implications for our chosen topic.

Figure 3.1

Case Study About Protein Synthesis Instruction

Mrs. Hernandez worked for years to improve her students' understandings about protein synthesis. She took a course on technologies appropriate to molecular genetics, attended a workshop on multiple intelligences, and learned through district professional development about cooperative learning and reading in the content area. She even completed a graduate degree with a focus on molecular biology. In her classroom, she used a good combination of interactive lectures and hands-on experiences. Her students completed activities that included laboratory-based activities and simulations. She thought she understood what her students knew and the things that were most difficult for them. But regardless of her attempts, they still struggled to understand the central idea and importance of protein synthesis. What might be the problem?

Perhaps the answer to her dilemma is best addressed through the design process as we implement the Instructional Planning Framework (Figure 1.1, p. x). The remainder of this chapter (1) introduces the design process and how to use the instructional tools provided as we make some decisions about a unit of study for Mrs. Hernandez; (2) thoroughly describes the resulting lesson designed for learning target #1; and (3) briefly outlines the resulting lessons for learning targets #2 through #4. We will model the use of the entire design process. Appendix 2 (pg. xx) provides a summary of hints and resources you can use as you complete this process yourself.

Note again that the framework is made up of two phases: *Identifying Essential Content* (Phase 1) and *Planning for Responsive Action* (Phase 2). Phase 1 occurs before instruction. As a classroom teacher, you might participate in that process; however, in many school districts, instructional leaders for all the schools and subjects in the district—rather than individual teachers—complete this work, deconstructing the science standards in the process. However, there is a lot to learn from going through the process yourself, especially because it provides a deep look into the content you teach. If you decide you want to tackle this work for a science topic, it is best done with a team or your professional learning community. Whether or not you participate in the process, your knowledge of it is essential for your deep understanding of the designed instructional sequence.

The heart of Phase 1 is the identification of learning goals and subgoals and the design of the best instructional sequence to reach these goals. Clarification of the content focus is important to both you and your students. It is equally important that all of you know and understand the criteria by which student understanding will be determined (Stage III in Phase 1). Finally, you cannot assume that just because proteins and genes are covered in your textbook that this coverage truly addresses the core ideas in science that you want your students to walk away with. So, the question becomes, "How might you, during Phase 1 planning, determine the content focus, the learning sequence, and the criteria for determining student understanding of the topic "proteins and genes"?

Stage I: Identify Disciplinary Core Ideas, Practices, and Crosscutting Concepts

We begin this process with a review of *Next Generation Science Standards*, finding the appropriate Disciplinary Core Ideas (DCIs) that relate to the planned topic of instruction. There are two arrangements of the standards: Topic Arrangements and Disciplinary Core Idea (DCI) Arrangements. We chose in this book to use the DCI arrangement of the standards, focusing on high school life science.

First identify the DCIs that most closely align with the chosen topic—proteins and genes—realizing that you might actually find DCIs from more than one standard that can be bundled together. Indeed, depending on how your state/district organizes the high school science course map, you might even bundle DCIs from the various science disciplines.

In this case, the relationship between proteins and genes is part of HS-LS1: From Molecules to Organisms: Structures and Processes (See Figure 3.2). Refer to the foundation boxes, the second component of the standard, and begin with the center of the foundation boxes that describes the Disciplinary Core Ideas (DCIs) for this standard. Notice that the circled first, second, and third bullets under DCI LS1A focus on the content we want to address. The second bullet specifically targets the relationship between DNA and protein synthesis, while the first and third bullets address the multiple levels of organization that are part of the system. We include the content in these bullets to help students connect what happens at the molecular level to the resulting cellular

Figure 3.2

HS-LS1: From Molecules to Organisms Foundation Boxes

The performance expectations above were developed using the following elements from the NRC document *A Framework for K-12 Science Education*:

Science and Engineering Practices

Developing and Using Models
Modeling in 9–12 builds on K–8 experiences and progresses to using, synthesizing, and developing models to predict and show relationships among variables between systems and their components in the natural and designed worlds.
- Develop and use a model based on evidence to illustrate the relationships between systems or between components of a system. (HS-LS1-2)
- Use a model based on evidence to illustrate the relationships between systems or between components of a system. (HS-LS1-4),(HS-LS1-5),(HS-LS1-7)

Planning and Carrying Out Investigations
Planning and carrying out in 9-12 builds on K-8 experiences and progresses to include investigations that provide evidence for and test conceptual, mathematical, physical, and empirical models.
- Plan and conduct an investigation individually and collaboratively to produce data to serve as the basis for evidence, and in the design: decide on types, how much, and accuracy of data needed to produce reliable measurements and consider limitations on the precision of the data (e.g., number of trials, cost, risk, time), and refine the design accordingly. (HS-LS1-3)

Constructing Explanations and Designing Solutions
Constructing explanations and designing solutions in 9–12 builds on K–8 experiences and progresses to explanations and designs that are supported by multiple and independent student-generated sources of evidence consistent with scientific ideas, principles, and theories.
- Construct an explanation based on valid and reliable evidence obtained from a variety of sources (including students' own investigations, models, theories, simulations, peer review) and the assumption that theories and laws that describe the natural world operate today as they did in the past and will continue to do so in the future. (HS-LS1-1)
- Construct and revise an explanation based on valid and reliable evidence obtained from a variety of sources (including students' own investigations, models, theories, simulations, peer review) and the assumption that theories and laws that describe the natural world operate today as they did in the past and will continue to do so in the future. (HS-LS1-6)

- -

Connections to Nature of Science

Scientific Investigations Use a Variety of Methods
- Scientific inquiry is characterized by a common set of values that include: logical thinking, precision, open-mindedness, objectivity, skepticism, replicability of results, and honest and ethical reporting of findings. (HS-LS1-3)

Disciplinary Core Ideas

LS1.A: Structure and Function
- Systems of specialized cells within organisms help them perform the essential functions of life. (HS-LS1-1)
- All cells contain genetic information in the form of DNA molecules. Genes are regions in the DNA that contain the instructions that code for the formation of proteins, which carry out most of the work of cells. (HS-LS1-1) *(Note: This Disciplinary Core Idea is also addressed by HS-LS3-1.)*
- Multicellular organisms have a hierarchical structural organization, in which any one system is made up of numerous parts and is itself a component of the next level. (HS-LS1-2)
- Feedback mechanisms maintain a living system's internal conditions within certain limits and mediate behaviors, allowing it to remain alive and functional even as external conditions change within some range. Feedback mechanisms can encourage (through positive feedback) or discourage (negative feedback) what is going on inside the living system. (HS-LS1-3)

LS1.B: Growth and Development of Organisms
- In multicellular organisms individual cells grow and then divide via a process called mitosis, thereby allowing the organism to grow. The organism begins as a single cell (fertilized egg) that divides successively to produce many cells, with each parent cell passing identical genetic material (two variants of each chromosome pair) to both daughter cells. Cellular division and differentiation produce and maintain a complex organism, composed of systems of tissues and organs that work together to meet the needs of the whole organism. (HS-LS1-4)

LS1.C: Organization for Matter and Energy Flow in Organisms
- The process of photosynthesis converts light energy to stored chemical energy by converting carbon dioxide plus water into sugars plus released oxygen. (HS-LS1-5)
- The sugar molecules thus formed contain carbon, hydrogen, and oxygen: their hydrocarbon backbones are used to make amino acids and other carbon-based molecules that can be assembled into larger molecules (such as proteins or DNA), used for example to form new cells. (HS-LS1-6)
- As matter and energy flow through different organizational levels of living systems, chemical elements are recombined in different ways to form different products. (HS-LS1-6),(HS-LS1-7)
- As a result of these chemical reactions, energy is transferred from one system of interacting molecules to another. Cellular respiration is a chemical process in which the bonds of food molecules and oxygen molecules are broken and new compounds are formed that can transport energy to muscles. Cellular respiration also releases the energy needed to maintain body temperature despite ongoing energy transfer to the surrounding environment. (HS-LS1-7)

Crosscutting Concepts

Systems and System Models
- Models (e.g., physical, mathematical, computer models) can be used to simulate systems and interactions—including energy, matter, and information flows—within and between systems at different scales. (HS-LS1-2), (HS-LS1-4)

Energy and Matter
- Changes of energy and matter in a system can be described in terms of energy and matter flows into, out of, and within that system. (HS-LS1-5), (HS-LS1-6)
- Energy cannot be created or destroyed—it only moves between one place and another place, between objects and/or fields, or between systems. (HS-LS1-7)

Structure and Function
- Investigating or designing new systems or structures requires a detailed examination of the properties of different materials, the structures of different components, and connections of components to reveal its function and/or solve a problem. (HS-LS1-1)

Stability and Change
- Feedback (negative or positive) can stabilize or destabilize a system. (HS-LS1-3)

proteins and the impact on appearance and function of cells, tissues, and organisms. Several bullets address the content we plan to teach, but when we design instruction for other topics we might find that fewer or more bullets address the selected content.

Now determine the best crosscutting concept(s) (See the last of the foundation boxes in Figure 3.2.) for this content. A review of the crosscutting concepts in *A Framework for Science Education* (NRC 2012) is appropriate at this time to help you decide. Again, a single or multiple concepts might be addressed. For the targeted DCIs in this chapter, the crosscutting concept Structure and Function is clearly appropriate. The structure of DNA and its impact on protein structure and thus its function or dysfunction is at the heart of understanding the topic proteins and genes.

We should also consider system and system models. According to the *Framework,*

> The parts of a system are interdependent, and each one depends on or supports the functioning of the system's other parts. Yet the properties and behavior of the whole system can be very different from those of any of its parts, and large systems may have emergent properties, such as the shape of a tree, that cannot be predicted in detail from knowledge about the components and their interactions. (NRC 2012, p. 92)

To fully understand the topic proteins and genes, students must understand the connections among the various components of the living system across the various levels of organization. Also, students "recognize that often the first step in how a system works is to examine in detail what it is made of and the shapes of its parts" (NRC 2012, p. 98). Thus, to fully understand the system on which we are focusing, a student must also understand "structure and function."

Finally, select the science and engineering practices (see the first of the foundation boxes in Figure 3.2.) most appropriate for this content. It is unlikely that this is a topic for investigation since it requires equipment rarely available to a high school classroom. However, Developing and Using Models is an appropriate choice since an "explicit model of a system under study can be a useful tool not only for gaining understanding of the system but also for conveying it to others" (NRC 2012, p. 92), and models are "valuable in predicting a system's behaviors or in diagnosing problems or failures in its functioning, regardless of what type of system is being examined" (NCR 2012, p. 93). The latter point is important if we want students to understand the role genes play in protein function and dysfunction.

Constructing explanations is also a good practice to include for this topic. Indeed, modeling and explanations often work hand-in-hand. For this standard, the specific aspect of the practice is to "construct and revise an explanation based on valid and reliable evidence obtained from a variety of sources" (NGSS Lead States 2013). As we said, it is unlikely that students will directly gather evidence through experimentation to support explanations about genes and proteins. However, models and simulations are appropriate sources of evidence to develop explanations. Modeling and explanations are a good blend of practices to use. Our challenge is to use the practices most effectively at the appropriate times in the instructional sequence.

Stage II: Deconstruct DCIs, Create a Storyline, and Align Practices and Crosscutting Concepts

We suggest that you have the Stage II conversation with other biology teachers in collaboration. Discourse around the DCIs helps clarify the content and establish a common instructional storyline. First review the information about LS1.A: Structure and Function in *A Framework for K–12 Science Education: Practices, Crosscutting Concepts, and Core Ideas* (NRC 2012, pp. 143–145) to thoroughly familiarize yourself with the content focus as outlined there since it is the document that guided the development of *NGSS*. Also review the disciplinary core idea progressions as outlined in the Standards and then study the selected DCI bullets and crosscutting concepts to translate them into the *essential understanding* for the unit of study (Table 3.1, Column 2, p. 50).

Table 3.1

Deconstructing DCIs Into Learning Targets

Bullets Selected from LS1.A	Essential Understanding	Learning Targets	Pruning
Systems of specialized cells within organisms help them perform the essential functions of life.	Organisms are systems made of dynamic and complex subsystems of interacting molecules in cells, tissues, and organs (levels of organization). Proteins are molecules that carry out the major biological processes of cells and impact the functioning of the tissues and organs they build. If they are not made properly, the entire organisms can be affected. Changes (mutations) in the gene/DNA can impact not only the genetic code but also the protein, cells, tissues and organs, since various components of the system interact and depend on each other. Though some mutations can be helpful, they might also stop or limit the protein's ability to function and potentially lead to physiological disorder in the entire organism.	**#1.** Proteins carry out the major work of cells and are responsible for both the structure and function of organisms. These proteins are made based on the code found in the organism's DNA (genotype) and result in the organism's traits (phenotype). How well the cellular system functions and interacts among cells and at the various levels of organization impacts the entire organism.	The biochemistry of protein synthesis is beyond the scope of this unit. Specific descriptions of how proteins function can also be pruned.
All cells contain genetic information in the form of DNA molecules. Genes are regions in the DNA that contain the instructions that code for the formation of proteins, which carry out most of the work of cells.		**#2.** Genetic information (genes) coded in DNA provides the information necessary to assemble proteins. The sequence of subunits (nucleotides) in DNA determines the sequence of amino acid in proteins.	The biochemistry of protein synthesis is beyond the scope of this unit, as is RNA structure, and details of transcription and translation. Names and structures of nucleotides and peptides and peptide bonds are not necessary.
		#3. The sequence of amino acids determines not only the kind, but also the shape of the protein, and thus its function.	Interactions and functions at the molecular or chemical reaction level are beyond the scope of this unit. Prune RNA and its structure, as well as primary, secondary, and tertiary protein structure.
Multicellular organisms have a hierarchical structural organization, in which any one system is made up of numerous parts and is itself a component of the next level.		**#4.** Mutations, changes in the DNA, impact protein production. Errors in the DNA (mutation) can result in missing proteins or ones that function inadequately. This results in a change in phenotype/trait.	Interactions and functions at the molecular or chemical reaction level are beyond the scope of this unit.

Next, break down the essential understanding into small "chunks" of specific content to reduce the gap between students' current understandings and scientific explanations and then sequence the chunks of content into *learning targets*, creating an instructional storyline (Table 3.1, Column 3). We need to make certain that we not only build on ideas taught in earlier grades and during the school year (the *prerequisite knowledge/skills for the lesson*) but also teach first the ideas that are foundational to understanding our targeted concepts. The learning sequence we developed using this process for a unit on proteins and genes is shown in Figure 3.3.

Figure 3.3

Proteins and Genes Learning Sequence

Target #1:
Proteins carry out the major work of cells and are responsible for both the structure and function of organisms. These proteins are made based on the code found in the organism's DNA (genotype) and result in the organism's traits (phenotype). How well the cellular system functions and interacts among cells and at the various levels of organization impacts the entire organism.

Target #2:
Genetic information (genes) coded in DNA provides the information necessary to assemble proteins. The sequence of subunits (nucleotides) in DNA determines the sequence of amino acid in proteins.

Target #3:
The sequence of amino acids determines not only the kind, but also the shape of the protein, and thus its function.

Target #4:
Mutations, changes in the DNA, impact protein production. Errors in the DNA (mutation) can result in missing proteins or ones that function inadequately. Since actions in the cell impact the other levels of organization in the organism, changes in the DNA and proteins can result in a change in phenotype/trait.

Essential Understandings: Organisms are systems made of dynamic and complex subsystems of interacting molecules in cells, tissues, and organs (levels of organization). Proteins are molecules that carry out the major biological processes of cells and impact the functioning of the tissues and organs they build. If they are not made properly, the entire organisms can be affected. Changes (mutations) in the gene/DNA can impact not only the genetic code but also the protein, cells, tissues, and organs, since various components of the system interact and depend on each other. Though some mutations can be helpful, they might also stop or limit the protein's ability to function and potentially lead to physiological disorder in the entire organism.

Now identify the most important concepts and vocabulary to teach for students' conceptual understandings, trying not to burden lessons with unnecessary language and instead *prune* what is unnecessary (Table 3.1, Column 4). It is important to use only language that is essential for scientific literacy and vocabulary that will not require an inordinate amount of time to learn (AAAS 2001a). Consider unburdening curricula of some major topics, subtopics, and technical vocabulary. Wasteful repetition can also be eliminated. The American Association for the Advancement of Science (AAAS 2001a) used the following criteria when deciding what to eliminate:

- No compelling argument that it would be essential for science literacy, or

- The amount of time and effort that would be needed for all students to learn was out of proportion to its importance.

Think about the topic *proteins and genes*. What do you think is necessary? Must students know about the detailed process of transcription and translation, the names and structures of nucleotides, RNA and its structure, or primary, secondary and tertiary structures of proteins? Are these ideas necessary to grasp the essential understandings? What does *NGSS* say? The standards focus on the disciplinary core ideas and do not include all of the additional subtopics. We don't need to share everything we know with our students. Don't lose sight of the essential understandings and don't start with the details. Focus on the conceptual core and then add vocabulary when needed.

How does all of this help develop a science content storyline? And how can we best help students understand the content using practices? To help clarify content storylines for those of you who might not be familiar, consider the description in a paper from the NARST 2009 Symposium on science content storylines:

> There are two key aspects of a science content storyline. First, the science ideas and terms in the lesson and unit are carefully chosen and sequenced to develop a story about one main learning goal. Second, the activities that students carry out in the lesson and unit help develop this content story, with the science ideas and terms explicitly linked to the activities. Thus, each activity that students engage in – whether it is hands-on observations of phenomena, analyzing data, arguments, and conclusion, reading, creating and analyzing representations, predicting new observations, writing a summary, participating in a class discussion, or something else—helps develop a key part of the science content storyline. (Roth et al. 2009)

To align with *NGSS*, the three dimensions (DCIs, crosscutting ideas, and practices) must be included in our thinking. Planning to this point (see Table 3.1, p. 50) includes the DCIs and crosscutting ideas, but the practices are not yet included. What is the best way to infuse practices in support of the content storyline? Table 3.2 outlines possible connections.

Table 3.2

Developing a Storyline

Learning Targets	Crosscutting Concept Addressed in Learning Target	Science and Engineering Practices to Include	Possible Aligned Representations or Activities to Include
#1. Proteins carry out the major work of cells and are responsible for both the structure and function of organisms. These proteins are made based on the code found in the organism's DNA (genotype) and result in the organism's traits (phenotype). How well the cellular system functions and interacts among cells and at the various levels of organization impacts the entire organism.	Models (e.g., physical, mathematical, computer models) can be used to simulate systems and interactions—including energy, matter, and information flows—with and between systems at different scales.	Develop and use a model based on evidence to illustrate the relationships between systems or between components of a system.	Research a genetic disorder and trace impact from genes to phenotype. Develop a model that represents the interactions among the various levels of organization and results in the disorder.
#2. Genetic information (genes) coded in DNA provides the information necessary to assemble proteins. The sequence of subunits (nucleotides) in DNA determines the sequence of amino acid in proteins.	Investigating or designing new systems or structures requires a detailed examination of the properties of different materials, the structures of different components, and connections of components to reveal its function and/or solve a problem.	Use a model based on evidence to illustrate the relationship between systems or between components of a system. Construct an explanation based on valid and reliable evidence obtained from a variety of sources…	Animations and simulations about transcription Develop an initial explanation of transcription
#3. The sequence of amino acids determines not only the kind, but also the shape of the protein, and thus its function.	Investigating or designing new systems or structures requires a detailed examination of the properties of different materials, the structures of different components, and connections of components to reveal its function and/or solve a problem.	Use a model based on evidence to illustrate the relationship between systems or between components of a system. Construct an explanation based on valid and reliable evidence obtained from a variety of sources…	Animations and simulations about protein synthesis Revise and expand the explanation about protein synthesis
#4. Mutations, changes in the DNA, impact protein production. Errors in the DNA (mutation) can result in missing proteins or ones that function inadequately. This results in a change in phenotype/trait.	Models (e.g., physical, mathematical, computer models) can be used to simulate systems and interactions—including energy, matter, and information flows—with and between systems at different scales.	Develop and use a model based on evidence to illustrate the relationships between systems or between components of a system.	Revise the model developed in Learning Target #1 and use it to further revise the explanation developed in Targets #1 and #2.

The science ideas are now chosen and sequenced to develop a story about our learning goal, and some activities are identified that should support that learning. This completes Stage II of the design process.

Stage III: Determine Performance Expectations and Identify Criteria to Determine Student Understanding

Establish criteria by which to measure student understanding, with one criterion for each learning target. Criteria are what you should look for when you examine student products and performances to determine student success or acceptability of work. In other words, "[criteria are] the qualities that must be met for work to measure up to a standard" (McTighe and Wiggins 1999, p. 275). It is very important that you identify criteria *before* you fully develop activities or performance tasks. Consider using the "backwards design" process as described in *Understanding by Design* (Wiggins and McTighe 1998) if your school district has provided professional development on this model.

Develop criteria that reflect understanding of both the content of the learning target and the intersection of the target with the practices and the crosscutting concepts. We established the following set of criteria, one for each learning target:

- Learning Target #1: Develop and explain a model that illustrates the production of a protein and its action across levels of organization (cell, tissue, organ, organism), resulting in a particular phenotype.

- Learning Target #2: Construct an explanation for how DNA coding determines the structure of a protein.

- Learning Target #3: Construct and refine an explanation of how amino acid sequence determines the shape of a protein and thus its function.

- Learning Target #4: Refine the model developed in Learning Target #1 to demonstrate the impact of a mutation on a phenotype/trait, including a discussion of the protein's role in the process.

As teachers, we use two basic forms of student assessment: summative and formative. Summative assessments are cumulative assessments that try to capture what a student has learned and can be used for grading, placement, promotion, or accountability (NRC 2001a). However, this type of assessment is not what is intended in Stage III of the design process. Instead, at this stage we establish criteria by which students' understandings are formatively assessed so that we can use students' current understandings to modify instruction. The form of assessment used depends on the context. It is ideal to focus on big ideas and conceptual understanding, but if assessments in your school district focus instead on facts and vocabulary, you'll need to teach both the big ideas *and* facts and vocabulary. You'll have to use both formative and summative assessment to ensure that your students understand both the big ideas and the facts and vocabulary. Formative assessment strategies will be selected later in this chapter and specific examples of formative assessment and feedback will be described in Chapter 4.

Now that criteria are established, connect the learning targets and criteria that measure understanding to the Performance Expectations (PEs) outlined in *NGSS*. Which of the PEs best align with the identified targets and criteria? And is there anything in these PEs not addressed in the identified learning targets and established criteria?

For standard HS-LS1, a component of which is the focus of the proteins and genes unit of study, performance expectations HS-LS1-1 (Construct an explanation based on evidence for how the structure of DNA determines the structure of proteins, which carry out the essential functions of life through systems of specialized cells) and HS-LS1-2 (Develop and use a model to illustrate the hierarchical organization of interacting systems that provide specific functions within multicellular organisms) align with the criteria we outlined. So students who successfully complete this unit of study should show proficiency on each of the two performance expectations. Additional instruction would be required prior to assessment using the PE as the assessment criterion if essential information from any of the three dimensions were missing. In this case, that is not necessary.

Stage IV: Determine Nature of Science (NOS) Connections

NGSS include an NOS reference for each life science standard. Make certain to note this reference while planning a unit of study and include it appropriately in lessons. Refer back again to the standards table for HS-LS1 (Figure 3.2, p. 47). Notice that the NOS connection for this standard is related to the practices foundation box and is stated, "Scientific inquiry is characterized by a common set of values that include: logical thinking, precision, open-mindedness, objectivity, skepticism, replicability of results, and honest and ethical reporting of findings." This ties beautifully to our selected practices of modeling and explanations, both of which require logical thinking, precision, and open-mindedness. However, there is no specific tie of the NOS target to a performance expectation. How well students understand the NOS target will only be measured if formative or summative assessments are included that specifically infuse NOS.

Stage V: Identify Metacognitive Goals and Strategies

Cognition is the thought process—processing information and applying knowledge— and *metacognition* is thinking about thinking—reflecting on your own thinking. Metacognition requires that students are aware of and have control over their learning. Metacognitive goals (in addition to cognitive goals) should be addressed in each lesson because teaching metacognitive strategies is second only to classroom management in influencing student learning (Wang, Haertel, and Walberg 1993/1994).

Thinking is invisible, but there are ways that it can be made visible to students, helping them to become more metacognitively aware and to see school as more about exploring ideas than memorizing content. The following three metacognitive approaches (Marzano 1992), when used in your classroom, will help students develop expertise in this area at the same time that they are learning science content. (*Note:* These three approaches are explored at greater length in Instructional Tool 3.1, p. 57)

- Critical Thinking and Learning: "Critical thinking and learning include being accurate and seeking accuracy, being clear and seeking clarity, being open-minded, restraining impulsivity, taking a position when the situation warrants it, and being sensitive to others' feelings and level of knowledge" (Marzano 1992, pp. 133–134).

- Creative Thinking and Learning: "Creative thinking and learning include engaging intensely in tasks even when answers or solutions are not immediately apparent, pushing the limits of your knowledge and abilities, generating, trusting, and maintaining your own standards of evaluation, and generating new ways of viewing a situation outside the boundaries of standard conventions" (Marzano 1992, p. 134).

- Self-Regulated Thinking and Learning: Self-regulated thinking includes "being aware of your own thinking, planning, being aware of necessary resources, being sensitive to feedback, and evaluating the effectiveness of your actions" (Marzano 1992, p. 133).

Step 1

First determine which of these three metacognitive approaches will be the lesson focus, keeping in mind that during the year abilities in each area should be addressed. The sequence is not critical, but it might be appropriate to work on "self-regulated thinking" early in the school year. This will quickly establish student-centered instruction in the classroom. Also, self-regulated thinking is an appropriate metacognitive focus when students are expected to plan and organize experiments and fieldwork.

We selected critical thinking and learning for this model lesson since it requires being accurate and seeking accuracy as well as being clear and seeking clarity. Notice that this ties well with the NOS target for this standard, which stresses the common set of values that characterizes inquiry. The practices on which this unit of study focuses are modeling and explanations where accuracy and clarity are extremely important. It is essential that students understand the importance of accuracy and precision in measurement and the best instruments to achieve that accuracy. Even if they do not engage in their own experimentation, they should look for accuracy and precision in the research they read about and reference. Clarity is also essential in student explanations. In addition, if student groups are to develop models and explanations then being open-minded is essential.

Instructional Tool 3.1

Three Approaches That Support Metacognition

1. Critical Thinking and Learning	
Visible Thinking: Truth Routines Truth Routines help students identify truth claims and explore strategies to uncover the truth, think more deeply about the truth of something, clarify claims and sources, explore truth claims from various perspectives, and determine the various factors relevant to a question of truth and see beyond an either/or approach to truth. These routines promote critical thinking because they encourage students to seek accuracy and clarity, be open-minded, and restrain impulsivity. Explore the Visible Thinking website (*www.visiblethinkingpz.org*) to learn more.	
The Research	The Visible Thinking website is based on years of research about thinking and learning as well as research and development in classrooms. All strategies shared on the site were developed in classrooms and were revised multiple times to ensure that they were applicable in the classroom and that they promoted student thinking and engagement. These strategies allow students with disabilities and general education students to show what they know, provide gifted and talented students a way to explore ideas at a deeper level, and give teachers a window into their students' thinking, allowing them to provide targeted and differentiated instruction. Visible thinking changes the nature of classroom discussions, making them more student-directed (i.e., increasing students' active participation) and inclusive (i.e., making sure everyone feels safe about participating) (Ritchart and Perkins n.d.).
Classroom Implications	Some of these strategies work very well during inquiry when you want your students to formulate explanations from evidence and connect their explanations to scientific knowledge. They also work well when exploring controversial issues related to science. You can use the various strategies with individuals and small groups and during whole-class discussion.
Application Example	One strategy at the website is Claim/Support/Question. The student makes a claim, identifies support for the claim, and then questions the claim. You can use this strategy during initial explorations of specific content. For example, set up a series of stations with equipment to explore osmosis—one station might look at the impact of temperature and another at sodium concentrations. After visiting the stations, each student makes a claim related to his or her observations and identifies support for the claim that is based on things that can be seen, felt, and known. Then other students ask questions about the original student's claim—for example, about what isn't explained or what's left hanging. These questions point students to areas of experimentation.

Instructional Tool 3.1 (continued)

1. Critical Thinking and Learning	
Visible Thinking: Fairness Routines Fairness Routines help students explore diverse perspectives, consider attitudes and judgments, separate fact and feeling, and explore the complexity of dilemmas. They promote critical thinking because they relate to open-mindedness and sensitivity to others' feelings. Explore the Visible Thinking website (*www.visiblethinkingpz.org*) to learn more.	
The Research	See Fairness Routines on the Visible Thinking website.
Classroom Implications	These routines are easy to use and can be infused into all kinds of content instruction. The Circle of Viewpoints strategy is useful to begin discussions about controversial issues and dilemmas. Reporter's Notebook helps students separate fact from feeling and might be useful when teaching science concepts that conflict with students' beliefs and feelings.
Application Example	The Circle of Viewpoints strategy can be used in an ecology unit that looks at water pollution. Students brainstorm a list of perspectives related to the issue. Then each student selects a viewpoint, describes the topic from that viewpoint, and asks a question from that viewpoint. Finally, students consider new ideas and questions they now have about the topic as a result of having done the activity.
Resources	See graphic organizers in Instructional Tool 3.7 (pp. 127–133).

2. Self-Regulated Thinking	
Identify What You Know and What You Don't Know At the beginning of an activity, students identify both what they know and what they don't know about a topic. As they research and learn about the topic, they verify, clarify, expand, or replace their original thinking (Blakey and Spence 1990).	
The Research	• Personal goals are important because we all need to plan and manage resources (Marzano 1992). • Concept mapping has been shown to benefit students' metacognitive abilities (Stow 1997). • Concept cartoons increase students' awareness of their ideas (Keogh and Naylor 1999).
Classroom Implications	• Concept mapping coupled with interviews help students analyze their thinking, identify their strengths and weaknesses, and set learning targets (Stow 1997). • 3-2-1 Bridge and Generate, Sort, Connect, Elaborate are visible thinking routines that help activate prior knowledge and make connections (Ritchart and Perkins 2008) (Visible Thinking website: *www.visiblethinkingpz.org*). • Many graphic organizers require students to identify what they know and what questions they have about a topic. (KWL is probably the most common example.)
Application Example	Use a probing question to start a lesson on just about any concept. Have students respond to that question using a mapping tool. Students then reflect on their individual charts or maps and identify personal learning goals for the lesson.

Instructional Tool 3.1 (continued)

2. Self-Regulated Thinking	
Technology Applications	You can use the computer program Inspiration to generate initial ideas and questions. Then flag and annotate areas of digital documents to mark questions and capture unfamiliar concepts and terms. Helpful tools include digital sticky notes (Google "sticky notes" for multiple links), highlighting and commenting tools in Word or Acrobat, and smart highlighters (available for sale at *http://firedoodle.com*).
Resources	The *Uncovering Student Ideas in Science* series by Page Keeley and her colleagues includes numerous probes that help you determine and explore your students' ideas about many science concepts. This series is available through NSTA Press (*www.nsta.org/publications/press/uncovering.aspx*).

Talk About Thinking	
Students externalize their thinking by using a variety of strategies to think out loud. By verbalizing their thinking, students gain awareness of and control over their problem-solving abilities and a fresh perspective on their own thoughts (Hartman and Glasgow 2002).	
The Research	Talking about their thinking helps students develop the language of thinking (Blakey and Spence 1990).
Classroom Implications	• You can model and discuss thinking, labeling your thinking as you talk. For example, you can analyze a set of data that students have gathered about growth of a population of mice, talking through what the data indicate. You might say, "I notice the line graph has an upward slope but then it levels off. That tells me that the number of individuals increased for a while but then the population stopped growing. I wonder why that happened?" Use of words like *slope* and *increased* are good insights into your thinking process. Hearing you talk in this way helps students realize just what thinking processes are (Blakey and Spence 1990). • Paired problem solving formalizes this process. Using this technique, a student talks through a problem, describing his or her thinking, while a partner listens and asks questions to clarify thinking (Blakey and Spence 1990). Paired problem solving makes problems more engaging, promotes self-monitoring and self-evaluation, and gives students feedback on their thinking. It also improves collaboration and communication. It is necessary, however, for you to monitor progress of each pair of students and provide them feedback (Hartman and Glasgow 2002). • Reciprocal teaching (Palincsar and Brown 1985) is another activity that formalizes the process of labeling one's thinking. In this case, the teacher and the students take turns assuming the role of the teacher in leading a dialogue regarding segments of a text. The dialogue uses four strategies: summarizing, question generating, clarifying, and predicting (see Resources on the next page for more information on this strategy).

Instructional Tool 3.1 (continued)

Application Example	Imagine that a student pair has just completed working through a simulation about protein synthesis. They are trying to develop an explanation of the simulation, based on their current understandings. One student verbalizes all the thoughts she has as she thinks through what happened in the simulation. The other student actively listens. He also points out what he thinks are errors, examines the accuracy of the statements, and probes the "thinker" to continue voicing her thinking. They can then change roles, and the second student voices his thoughts about the simulation and the first student actively listens and probes. At the end of this activity, they prepare a written summary of their explanation.
Technology Applications	There are very simple voice annotations in PowerPoint, as well as in VoiceThread, which is a much more powerful and collaborative tool: *http://voicethread.com*.
Resources	Reciprocal teaching: *www.ncrel.org/sdrs/areas/issues/students/atrisk/at6lk38.htm*

Plan and Self-Regulate	
This strategy involves estimating time requirements, organizing materials, scheduling procedures needed to complete an activity, and developing evaluation criteria (Blakey and Spence 1990).	
The Research	• For students to become self-directed, they must take on increasing responsibility for planning and regulating their learning (Blakey and Spence 1990). • Contractual agreements with students in which they set goals and subgoals (within the context of teacher goals) have a positive impact on planning and self-regulating (Marzano, Pickering, and Pollock 2001). • Graphic organizers can act as thinking tools and memory support systems that scaffold self-regulation. They help students to uncover their prior knowledge and can be used to mark progress and to contrast what was known to what is now known (Lipton and Wellman 1998).
Classroom Implications	• Peel the Fruit is a tool to make thinking visible and to plan and track over time the exploration of a topic (Ritchart and Perkins 2008). This routine allows the entire class, small groups, or individuals to track progress on a long-term project. Further information can be found at the Visible Thinking website (*www.visiblethinkingpz.org*). • Encourage students to personalize the goals you establish and adapt them to their personal needs. Your goals should be general enough to allow this flexibility (Marzano, Pickering, and Pollock 2001). • Use of rubrics is an effective way for students to track their efforts and the impact of those efforts on their achievement (Marzano, Pickering, and Pollock 2001).
Application Example	Students are working on a long-term research project—for example, the study of a local habitat. They can use a planning map (e.g., Peel the Fruit [*www.visiblethinkingpz.org*]) initially to map and plan their work, add tasks, establish research routines, and track progress. They periodically revisit the map, choose next steps, and monitor their progress. "'Peeling the fruit' is a metaphor for getting familiar with the surface of something, seeing puzzles and mysteries to investigate, and pursuing these in various ways to arrive at core understandings" (*www.visiblethinkingpz.org*).

Instructional Tool 3.1 (continued)

Technology Applications	NoteStar, an online tool, helps students develop research papers. It lets them create subtopics to research, assign topics to group members (if the research paper is being developed by more than one student), take notes, track source information, and organize notes and sources. It is found at *http://notestar.4teachers.org* and can be used alone or together with ThinkTank (*http://thinktank.4teachers.org*), another tool that helps students manage a topic for online research.
Resources	Monitoring and self-management site: *www.muskingum.edu/~cal/database/general.*

Debrief the Thinking Process

This strategy includes activities that help students bring closure to a lesson and focus discussion on the thinking process itself. It shows students how to develop their awareness of the ways in which various strategies might be used in other situations (Blakey and Spence 1990).

The Research	• When students talk about learning, they can check their thinking and performance, gain deeper understandings of their learning, and use better strategies for planning and monitoring their work. It also prepares students for the risk-taking required of learning (Davies 2003). It increases their confidence in their thinking and willingness to share ideas in a school environment where typically answers are "right" or "wrong." • Sharing their work and organizing evidence of their learning helps students see clearly what they have learned, what they still need to learn, and what kinds of support are required for them to learn. The presence of others students encourages reflection (Davies 2003), which improves the metacognitive skills of the students who are sharing their work.
Classroom Implications	Blakey and Spence (1990) recommend a three-step process. First, you facilitate a review of an activity, eliciting student responses on thinking processes and feelings. Next, your students classify related ideas and identify the thinking strategies they used. Third, they evaluate their success, identify helpful strategies, and eliminate unproductive strategies.
Application Example	You can use the Blakey and Spence (1990) three-step process to have student groups complete an inquiry activity about, for example, variables that impact photosynthesis. Review the activity with the class and ask students to summarize their thinking and feelings about the process they have gone through. Then have them complete an entry in their thinking journals about what challenged—and what clarified—their thinking; why they think they were challenged as they were; and how certain ideas clarified their thinking. Finally, ask each group to talk about the thinking processes they used and whether or not they were effective. This discussion is likely to lead the group to revisit the activity and to modify their initial approaches.

Self-Evaluate

Students are self-evaluating when they judge the quality of their work based on evidence and explicit criteria. The purpose of self-evaluation is to do better work in the future.

The Research	• Self-evaluation enhances self-efficacy and increases intrinsic motivation (Rolheiser and Ross n.d.) • When students set criteria, self-assess, and reset criteria, they better understand assessment and the language of assessment. They acquire a clearer image of what they need to learn, where they are in relationship to that, and how they might take steps to get there (Davies 2003).

Instructional Tool 3.1 (continued)

Classroom Implications	• We can move students toward self-evaluation by first providing guided self-evaluation—for example, individual conferences with the teacher and use of checklists that focus on thinking processes. We can then slowly allow students to complete this process more independently (Blakey and Spence 1990).
	• We should constantly encourage students to compare their current thinking to their original thinking and try to determine what helped them achieve their current understandings. Various graphic organizers support this before-and-after process.
	• Rolheiser and Ross (n.d.) recommend a four-stage process: (1) involve students in defining the criteria used to evaluate performance, (2) teach students how to apply the criteria to their work, (3) give students feedback on their self-evaluations, and (4) help students develop goals and action plans.
Application Example	Students are given time in class to keep a journal in parallel with other learning processes as a means to practice self-evaluation.
Technology Applications	Once again, use of VoiceThread (*http://voicethread.com*) is appropriate.
Resources	*Science Formative Assessment: 75 Practical Strategies for Linking Assessment, Instruction, and Learning* (Keeley 2008) includes numerous strategies for self- and peer-assessment.

Keep a Thinking Journal or Learning Log
See Instructional Tool 2.4 for more information on these strategies.

---

3. Creative Thinking and Learning

Visible Thinking: Creativity Routines
Creativity Routines look at purposes and audiences, generating creative questions, creative thinking about options, creative decision making, and ways to consider various perspectives. Explore the Visible Thinking website (*www.pz.harvard.edu/vt*) to learn more.

The Research	See Creativity Routines on the Visible Thinking website.
Classroom Implications	These easy-to-learn routines can be incorporated into any classroom and can address various content areas. Simply select a routine that aligns well with the targeted content goals.
Application Example	A great routine to use in a decision-making situation is called Options Diamond (*www. visiblethinkingpz.org*). Perhaps your students are engaged in a case involving an environmental impact decision. They draw a large diamond. In the center of the diamond they write the decision that has to be made. At the left and right corners of the diamond, they write the one or two main trade-offs of making a particular decision. (As noted on the website, "Usually there are trade-offs or tensions between [two or more options] that make the decision hard: Choose one and you get X but lose Y; choose the other option and you lose X but get Y.") Students brainstorm (1) solutions for each trade-off; (2) compromises between the trade-offs that they write at the bottom point of the diamond; and (3) clever solutions that combine what seem to be the opposites from the right and left corners and write these at the diamond's apex. They then reflect on the diamond to determine what they learned.

Note: Instructional Tool 3.2, "Instructional Strategy Selection Tool," is on pages 67–68.

Step 2

Refer to Instructional Tool 3.1, Three Approaches That Support Metacognition (pp. 57–62), to learn about approach #1: Critical Thinking and Learning. We chose this approach for the genes and proteins unit because students are required to identify, clarify, and explore claims that align well with the development of models and explanations.

Step 3

Develop the criterion to determine students' understanding of this metacognitive goal. The metacognitive criterion we developed is, "Accurately and carefully develop explanations and models, providing evidence for each."

Application of Phase 2 to Proteins and Genes Unit

During Phase 1, we determined the content focus, established a learning sequence and storyline, developed criteria to demonstrate understanding, and identified metacognitive goals. In Phase 2, we design the actual classroom learning experiences, specifically selecting strategies that uncover students' ideas and promote conceptual understanding. While this instruction is planned ahead of time, the plan may change *in response to* students' learning needs—thus the phase is titled "planning for responsive action."

First, build on the thinking during storyline development and identify activities that address the learning target content about proteins and genes. Ideas can be drawn from textbooks, NSTA resources, the internet, and other sources of teaching ideas. A good curriculum will sequence activities appropriately to build students' content understanding. Once activities are identified, they might need modification or altered sequencing to best support the storyline. Of course, even the best activities do not ensure conceptual understanding. That is why the instructional strategies are so important. The plan that results from Phase 2 is a prediction, based on research, of what will work in the classroom for a given topic. Stages VI through X outline a process you can use to develop such a plan.

Stage VI: Research Student Misconceptions Common to This Topic That Are Documented in the Research Literature

During this stage, we study the misconception research and compile the misconceptions about proteins and genes found to be common among students. There are many resources available to assist in identifying common misconceptions (see Appendix 2, p. 311). We used the referenced resources and began the misconception search for you. Our findings—together with instructional ideas to confront the misconceptions—are summarized in Table 3.3 (pp. 64–66). Consider the implications of these misconceptions for instruction.

Table 3.3

Misconceptions and Instructional Ideas for Proteins and Genes

Learning Target #1: Proteins carry out the major work of cells and are responsible for both the structures and functions of organisms. An organism's traits (phenotype) are a reflection of the work of proteins.

Misconceptions	Instructional Ideas
• Students demonstrate confusion over levels of organization, particularly with cells and molecules. They tend to think of molecules as related to the physical sciences and cells to life science. Some students even think that proteins are made of cells and that molecules of protein are bigger than cells (Driver et al. 1994). • The majority of upper division biology students and future science teachers recognize the physical constitution of an organism as its phenotype, yet do not understand the role of genes and proteins in producing the phenotype (Elrod n.d.). Because some students do not connect genes to proteins to phenotypes (Lewis and Wood-Robinson 2000), they assume that genes directly express traits in organisms (Lewis and Kattmann 2004). • Though students usually equate genes with traits, they do not understand that genes code for specific proteins and that the production of these proteins results in the traits (Friedrichsen and Stone 2004).	• Student understanding of proteins, genes, the connection between them, and genomes all increased when teachers did the following: integrated proteins into the same context as genes; introduced the importance of proteins before introducing genes; and scaffolded students' written explanations of a trait at the levels of gene, protein, cell, tissue, and organisms (Eklund et al. 2007). • Students' limited understanding of the specific contexts at the cellular, tissue, and whole organism levels may be a reason students struggle to make a connection between protein and trait. The examples teachers use should be simple and familiar to students. In addition, activities should be spread throughout the lesson not just used at the end of the lesson (Eklund et al. 2007).

Learning Target #2: Genetic information (genes) coded in DNA provides the information necessary to assemble proteins. The sequence of subunits (Nucleotides) in DNA determines the sequence of amino acids in proteins.

Table 3.3 (continued)

Misconceptions	Instructional Ideas
• Some students think a gene is a trait or that the DNA makes proteins (Elrod n.d.). • Less than half of upper-division biology students and future science teachers understand the nature of the genetic code (Elrod nd). Only 22% of undergraduate students with some biological science course work defined the gene in terms of nucleotide sequences involved in protein synthesis (Chattopadhyay and Mahajan 2006). • Students often think that genes code for more than proteins. They also often think that the genes code for information at multiple levels of organization (e.g., the gene "tells" a tissue or organ to malfunction), which bypasses the need for students to provide a mechanistic explanation of molecular genetics phenomena (Duncan and Reiser 2005). • 30–52% of upper-division biology students and future science teachers do not recognize RNA as the product of transcription. 50–75% of introductory biology and genetics students in college and future science teachers do not identify proteins as the product of translation (Elrod nd; Fisher 1985).	Attention should be paid to confusing terms (*proteins* and *amino acids, gene, mutation*) but not to unnecessary words (*codon* and *anticodon*) (Eklund et al. 2007).

Learning Target #3: The sequence of amino acids determines not only the kind, but also the shape of the protein, and thus its function.

Misconceptions	Instructional Ideas
• Because students are not aware that proteins play a role that is central to living things (most/all genetic phenomena are mediated by proteins) and robust (many functions), it hampers their ability to provide mechanistic explanations of genetic phenomena (Duncan and Reiser 2005).	Attention should be paid to confusing terms (*proteins* and *amino acids, gene, mutation*) but not to unnecessary words (*codon* and *anticodon*) (Eklund et al. 2007).

Learning Target #4: Mutations, changes in the DNA, impact protein production. Errors in the DNA (mutation) can result in missing proteins or ones that function inadequately. This results in a change in phenotype/trait.

Table 3.3 (continued)

Misconceptions	Instructional Ideas
• Though 80% of undergraduates knew that a disease could be linked to a gene, only 35% correctly represented a flow diagram between the genes and disease. Even if they could explain the concept of the central dogma, they could not extrapolate their understanding to a real-life situation (Chattopadhyay and Majahan 2006). • Most students are unable to explain a situation where a change to the DNA sequence does not change the protein sequence (Eklund et al. 2007).	Lewis and Kattmann (2004) recommend using sickle-cell anemia as an example that links the gene, the structure/function of the gene product, and the resulting phenotype. Other research recommends that the teacher cover a number of different disease traits as well as nondisease traits to provide context (Eklund et al. 2007).

The learning sequence for this topic in the first edition of our book did not include Learning Target #1. As we reviewed the misconception research for this edition, we realized how important it is for initial instruction to integrate proteins into the same context as genes; introduce the importance of proteins before introducing genes; and scaffold students' written explanations of a trait at the levels of gene, protein, cell, tissue, and organisms (Eklund, Rogat, Alozie, and Krajcik 2007). This is just one example of the reflexive nature of the design process. The process is outlined and demonstrated as a linear process, but the plan often changes as we learn more about the research related to an instructional topic.

Instructional Strategic Selection Tool: Identifying Strategies for Phase 2 Planning

Notice that Stages VII through X (see Table 2.1, p. 39 for a list of the stages) require strategy selection. We need to find strategies that help *identify*, *elicit*, and *confront* students' preconceptions, as well as strategies that help them *make sense* of those experiences. Coupling these strategies with effective activities (which address the targeted phenomenon and link with the learning targets) helps ensure that selected activities will build conceptual understanding—as long as we act upon the information about student understanding that the strategies yield, responding to student learning needs.

This section of the chapter is devoted to Instructional Tool 3.2, the Instructional Strategy Selection Tool (pp. 67–68). The terms *instructional approach* and *instructional strategy* are used frequently in this part of the chapter. An *approach* is defined as a broad method used in instruction (e.g., reading as a linguistic approach or modeling as a nonlinguistic approach to learning science) and *strategies* are specific ways in which to achieve a specific goal (e.g., use of analogies).

Instructional Tool 3.2

Instructional Strategy Selection Tool

			Framework Components							
			Identifying Preconceptions	Eliciting and Confronting Preconceptions		Sense Making			Demonstrating Understanding	
			Bringing (by the teacher) students' preconceptions to the surface and determining prior knowledge	Explicitly eliciting preconceptions	Confronting preconceptions	Perceiving, interpreting, and organizing information	Connecting information	Retrieving, extending, and applying information; using knowledge in relevant ways	Formative assessment	Peer and self-assessment
Linguistic Representations of Knowledge										
Approaches	**Strategies**	*NGSS* **Practices**								
Writing to Learn (Instructional Tool 3.3, p. 93)	Learning Logs	QP, I, OEC	√	♥		♥	√	√	√	♥
	Probes	E/S		♥	√				♥	♥
	Science Notebooks	QP, I, D, E/S		√	√	♥	♥	√	♥	√
	Scientific Explanations	M, E/S, AE, OEC		√	♥	♥	√	√	♥	√
	Science Writing Heuristic	E/S, AE, OEC		♥	♥	♥	♥	√	♥	√
Approaches	**Strategies**	*NGSS* **Practices**	Identifying Preconceptions	Eliciting and Confronting Preconceptions		Sense Making			Demonstrating Understanding	
Reading to Learn (Instructional Tool 3.4, p. 102)	Vocabulary Development Strategies	E/S, OEC		√		♥	♥		√	
	Informational Text Strategies	OEC, I, QP, AE	♥	√	♥	♥	♥	√	√	
	Narrative Text Strategies	D, E/S, OEC			√	♥	√	√		
	Reflection Strategies	D, E/S, OEC		√	√	√	♥	♥	√	♥
Speaking to Learn (Instructional Tool 3.5, p. 110)	Large- and Small-Group Discourse	QP, E/S, AE, OEC	√	♥	♥	√	√	√	♥	♥
	Student Questioning	QP, I, E/S, AE, M		♥	♥				√	♥
	Communication	AE, OEC	√		♥	♥	√	√	♥	♥

Note: See legend on bottom of page 68.

Instructional Tool 3.2 (continued)

Nonlinguistic Representations of Knowledge										
Approaches	**Strategies**	*NGSS* **Practices**	Identifying Preconceptions	Eliciting and Confronting Preconceptions		Sense Making			Demonstrating Understanding	
Models (Instructional Tool 3.6, p. 119)	Mathematical Models	M, MCT, E/S				♥	♥		√	
	Physical Models	M, E/S			√	♥	♥	√	√	♥
	Verbal Models: Analogies	M, E/S				♥	√	♥	√	√
	Verbal Models: Metaphors	M, E/S				♥	√	♥	√	√
	Visual Models	M, E/S	♥	♥	√	♥	√	√	√	√
	Dynamic Models	M, I, D, E/S	♥	√		√	♥	♥	√	√
Approaches	**Strategies**	*NGSS* **Practices**	Identifying Preconceptions	Eliciting and Confronting Preconceptions		Sense Making			Demonstrating Understanding	
Visual Tools (Instructional Tool 3.7, p. 127)	Brainstorming Webs	QP, M	♥	√		√	♥		√	
	Task-Specific Organizers	I, E/S		√		♥	♥	√	√	♥
	Thinking-Process Maps	M, D, MCT, E/S, OEC	√	♥	♥	♥	♥	√	♥	♥
Drawing Out Thinking (Instructional Tool 3.8, p. 134)	Drawings and Annotated Drawings	M, D, E/S, OEC	♥	♥	√	√	√		♥	♥
	Concept Cartoons	QP, E/S		♥	♥				√	
Kinesthetic Activities (Instructional Tool 3.9, p. 137)	Hands-on Experiments and Activities and Manipulatives	QP, I, M, OEC			♥	√	♥	♥		
	Physical Movements/ Gestures	M, OEC		♥	√	√	♥	♥	♥	

Note: A check mark (√) means that that strategy will promote student understanding, if effectively implemented. A heart symbol (♥) means that that strategy is very strongly supported by research (and is also a favorite of the authors!).

Note: The abbreviations for practices referenced in the tool are:

- Asking questions and defining problems (QP)
- Developing and using models (M)
- Planning and carrying out investigations (I)
- Analyzing and interpreting data (D)
- Using mathematics and computational thinking (MCT)
- Constructing explanations and designing solutions (ES)
- Engaging in argument from evidence (AE)
- Obtaining, evaluating, and communicating information (OEC)

Notice in Instructional Tool 3.2 that the Framework Components run across the top of the tool and that the row below the components further defines each one. Specific strategies in Instructional Tool 3.2 are listed according to one of two types of representations—linguistic representations (Instructional Tools 3.3–3.5, pp. 93–118) and nonlinguistic representations (Instructional Tools 3.6–3.9, pp. 119–139)—and then specific strategies within each approach are shown in first and second columns. (Instructional Tools 3.3–3.9 are at the end of this chapter.)

Notice that Tool 3.2 includes check marks and hearts. A check mark (√) means that that strategy will promote student understanding, if effectively implemented, and is a useful strategy for that particular framework component. A heart symbol (♥) means that that strategy is very strongly supported by research and is also one of our favorites.

Linguistic Representations of Knowledge

The linguistic representational tools in this book examine writing (Instructional Tool 3.3, pp. 93–101), reading (Instructional Tool 3.4, pp. 102–109), and speaking to learn (Instructional Tool 3.5, pp. 110–118) as ways to linguistically represent knowledge. These tools provide great strategies and resources to support the struggling reader and should be carefully studied to determine how you might support the improvement of your students' linguistic skills. However, we chose to focus on strategies that support the development of not only literacy but also science literacy and that are foundational to meaning making through inquiry. Speaking and listening, writing, and reading are critical parts of inquiry and essential to implementing the practices of science. These strategies can be used during engagement and exploration, designing and conducting investigations, analyzing and interpreting data, and presenting findings and understandings (Century et al. 2002). Reading, writing, speaking, and listening are each critical for explanations and argumentation, our practice foci for this particular unit of study.

Nonlinguistic Representations of Knowledge

The nonlinguistic representational tools in this book are also important for sense making. Whereas linguistic representations use language in learning, nonlinguistic representational tools store knowledge in the form of visual images. The more that students use both systems of representation, the more they will think, learn, and recall information. Instructional Tools 3.6–3.9 (pp. 119–139) will help you select appropriate strategies to promote nonlinguistic representations. Table 3.4 (p. 70) provides rationales for the use of the four groups of nonlinguistic representations for which we provide Instructional Tools: models; maps and graphic organizers; drawing; and kinesthetic activities.

Table 3.4

Four Nonlinguistic Representations of Learning (Models; Maps and Graphic Organizers; Drawing; and Kinesthetic Activities) and the Rationales for Their Selection

Nonlinguistic Representation of Learning	Rationale	Supportive Instructional Tool in This Book
Models	We want our students to build nonlinguistic representations in their minds (Marzano 1992). Such "mental models" represent ideas, objects, events, processes, and systems. To help build and communicate mental models, people use expressed models that are simplified from those held in their brains (Hipkins et al. 2002). Types of expressed models are mathematical, physical, verbal (metaphors and analogies), visual (graphs, pictures, and diagrams), and dynamic (simulations, computer simulations, virtual manipulatives, and animations).	Instructional Tool 3.6 (p. 119)
Visual Tools: Maps and Graphic Organizers	Maps and graphic organizers are among the most commonly used strategies to help construct nonlinguistic representations. They combine linguistic and nonlinguistic modes because they call for words and phrases as well as symbols. When used as advanced organizers, they can help students retrieve what they already know, thus activating their prior knowledge (Dean, Hubbell, Pitler and Stone 2012). Furthermore, these visual thinking tools are used for storing ideas already developed and for construction of content knowledge.	Instructional Tool 3.7 (p. 127)
Drawing	Student drawings can be used to determine students' levels of conceptual understanding, observational skills, and abilities to reason, as well as their beliefs. Teachers can use drawings as learning experiences or assessments (McNair and Stein 2001).	Instructional Tool 3.8 (p. 134)
Kinesthetic Activities	Kinesthetic activities involve movement. In such activities, there is a constant interplay between movement and learning (this interplay can even occur in our adult lives). The association of movement with specific knowledge produces a mental image of that knowledge in the learner's mind (Dean, Hubbell, Pitler and Stone 2012). Teachers need to purposefully integrate movement into their everyday instruction.	Instructional Tool 3.9 (p. 137)

The following section on Stages VII through X provides a guide for strategy selection as we plan for Learning Target #1. We work through the process using only this learning target, but a completed instructional template for all learning targets is found in Appendix 3 (p. 319). As we select strategies, we also complete the Strategy Selection Template (Table 3.5) for the topic of proteins and genes. The selected metacognition strategy, "Critical Thinking and Learning: Truth Routines," is already filled in the template.

Table 3.5

Strategy Selection Template for Proteins and Genes Unit

	Learning Target #1	Learning Target #2	Learning Target #3	Learning Target #4
	Proteins carry out the major work of cells and are responsible for both the structures and functions of organisms. An organism's traits (phenotype) are a reflection of the work of proteins.	Genetic information (genes) coded in DNA provides the information necessary to assemble proteins. The sequence of subunits (Nucleotides) in DNA determines the sequence of amino acids in proteins.	The sequence of amino acids determines not only the kind, but also the shape of the protein, and thus its function.	Mutations, changes in the DNA, impact protein production. Errors in the DNA (mutation) can result in missing proteins or ones that function inadequately. This results in a change in phenotype/trait.
Possible Strategies for:				
Identifying Preconceptions				
Eliciting and Confronting Preconceptions				
Sense Making				
Demonstrating Understanding				
Selected Metacognitive Strategy	Critical Thinking and Learning: Truth Routines			

Stage VII: Determine Strategies to Identify Students' Preconceptions

We need to consider students' preconceptions because learning occurs when they make connections and construct patterns—something that depends on their prior knowledge (Lowery 1990). Although it is true that we should gear lessons to our students' developmental levels and provide them multiple pathways to understanding (Weiss et al. 2003), we must also know their preconceptions in order to design instruction that helps them examine their misconceptions.

We already shared the misconceptions that students in general bring to the topic of proteins and genes (refer back to Table 3.3). It is equally important to determine *our own* students' preconceptions because misconceptions can vary by age, sex, geography, and student motivation or interest (Westcott and Cunningham 2005). The following process helps us identify our students' preconceptions well ahead of actual instructional time so we can take their thinking into account as we plan lessons.

Step 1

Return to Instructional Tool 3.2: Instructional Strategy Selection Tool (pp. 67–68). See the first column of the framework components, "Identifying Preconceptions." Scan this column on all the pages. There are three great nonlinguistic strategies (♥) to identify preconceptions.

Step 2

Learn more about each of these strategies by finding the appropriate Instructional Tool at the end of the chapter. Select one or two strategies that might work well for this topic. We focused on three possible strategies. *Brainstorming webs*, which include clustering, mind mapping and circle maps, allow students to generate ideas and to brainstorm without restriction while connecting ideas. Brainstorming webs can be used to determine prior knowledge and can be revised during the course of a unit of study. Some *informational text strategies* (e.g., concept diagrams, anticipation guides, KWLs) can easily be used to identify student preconceptions. An extra advantage is that students might actually be used to some of these strategies if they have been used for reading instruction. *Drawings and annotated drawings* are less biased ways than verbal literacy activities to determine understanding.

Any one of these strategies might be used. Even if we do not use all of the strategies at this point, we can use any one of them later to elicit preconceptions during instruction of individual learning targets. We selected anticipation guides to identify preconceptions.

An anticipation guide is a quick, easy tool that activates and assesses students' prior knowledge. It is made up of carefully selected questions that can be used as a pre- and post-instruction inventory. Notice that the statements in the sample anticipation guide (Figure 3.4) are drawn directly from the list of common misconceptions

Figure 3.4

Proteins and Genes Anticipation Guide

Anticipation Guide
Proteins and Genes

Directions: In the first column, "Before Lessons," place a check mark (√) when you agree with the statement on the right. After the end of our unit on proteins and genes, place a check mark (√) in the second column, "After Lessons." Then compare what your opinions about these 10 statements were before the lessons on proteins and genes with your opinions after the lessons.

Before Lessons (Date_____)	**After Lessons** (Date_____)	
I Agree	I Agree	
_____	_____	1. Cells are larger than proteins.
		2. Proteins are made of cells
_____	_____	3. Genes are directly responsible for an organism's traits.
_____	_____	4. Genes code for proteins, which then determine traits.
_____	_____	5. Genes are a section of DNA involved in protein synthesis.
_____	_____	6. Genes "tell" a tissue or organ to function as they do.
_____	_____	7. Genetic phenomena are mediated by proteins.
_____	_____	8. Changes in genes always cause changes in proteins.
_____	_____	9. Genes directly cause diseases.
_____	_____	10. Genes can code for proteins, fats, and carbohydrates.

about proteins and genes (Table 3.3, pp. 64–66). This instrument can be administered well before instruction and used to guide planning of the entire unit of study since the statements in the guide include misconceptions that cross the four learning targets. (*Note:* Have students write the date in the first column when you first administer the guides. Then, when you hand them back to students after the unit, have students write in *that* date. It can be interesting to students to see how they have learned new information and changed their opinions over time.)

For the remaining stages in the Phase 2, identify specific strategies for each aspect of the Instructional Planning Framework (Figure 1.1, p. 7) and for each learning target. That process is modeled in this chapter for Learning Target #1 only. However, possible strategies and lesson ideas for all learning targets are found in the planning template (Appendix 3, p. 317).

Stage VIII: Determine Strategies to Elicit and Confront Students' Preconceptions

It is important to focus on students' ideas throughout a lesson. Begin by *eliciting* their preconceptions, making them public so that they, their peers, and you can grapple with thinking about the concepts. Provide opportunities for students to express their preconceptions as well as design experiences that confront these preconceptions. That means that students' alternative explanations are made public and that students have experiences that make them question their existing ideas. Use the following three steps and the Instructional Tools to select strategies to elicit and confront student preconceptions for Learning Target #1.

Step 1

Review Learning Target #1 (Figure 3.3, p. 51), as well as the related misconceptions and instructional ideas in Table 3.3 (pp. 64–66). This will help focus on the content details of the learning target; it will also help anticipate what learning difficulties might arise regarding these concepts.

Step 2

Refer to the Instructional Tool 3.2, Instructional Strategy Selection Tool (pp. 67–68). Go to the column headed Eliciting and Confronting Preconceptions for the strategies you need. Remember to focus on the best (♥) strategies if your time is limited.

Step 3

Find the specific Instructional Tools that provide information about the specific selected strategies (e.g., if learning logs are selected, go to Instructional Tool 3.3, Writing to Learn, pp. 93–101). Identify strategies you think will work best. Then refer to Table 3.6 to see the strategies we identified as well as our comments about using them.

As seen in Table 3.6, there are many effective linguistic and nonlinguistic strategies, and it is good to use a mix of them. Annotated drawings used in conjunction with student discourse are especially effective. Refer back to Instructional Tool 3.2 and notice that the linguistic approach, reading to learn (specifically Informational Text Strategies), holds potential to confront students' conceptions since appropriate text offers alternative explanations to students. Students need to establish a systemic understanding of the connection between genotype and phenotype—in other words the impact of coded genetic information across the various levels of organization. Reading for information and then processing that information, perhaps using thinking process maps, might be an effective instructional strategy to achieve this.

Table 3.6

Nine Strategies to Elicit and Confront Students' Preconceptions

Strategy	Elicit or Confront Preconceptions	Both Elicit and Confront Preconceptions
1. Probes	Probes help teachers know more about what students understand, since they tend to listen more carefully to what students say and thus determine next instructional steps.	
2. Scientific Explanations	As compared to writing that simply transmits information, scientific explanations help students understand the nature of science, inquiry, and science content.	
3. Science Writing Heuristic (SWH)		This is a great strategy, but it is complex. You might spend some time learning about it before using it (see Instructional Tool 3.3, p. 93, for more information). If you are already familiar with SWH, its use with the topic of the flow of matter and energy in ecosystems is appropriate.
4. Small- and Large-Group Discourse		Discourse, used with a strategy such as concept cartoons to establish the purpose of the discourse, promotes learning about concepts, metacognition, and the nature of science. It can also serve as a formative assessment. It helps students build on one another's ideas to develop explanations and plan investigations.
5. Thinking-Process Maps		Thinking-process maps promote cognitive and metacognitive learning. They help students "learn how to learn" by having them distinguish between accurate conceptions and misconceptions. In addition, the maps tend to reduce student anxiety and increase self-confidence and motivation. They also help learners see linkages among ideas and connectivity in systems (the latter an important concept when studying the proteins and genes) better than written text. In particular, systems diagrams can be understood using the thinking-process maps. For this unit of study, systems diagrams are perfect.
6. Visual Models		With our focus on modeling, visual models may be a good choice. If the visual model is a two-dimensional annotated drawing, it has the benefits outlined in Table 3.9 (p. 82).

Table 3.6 (continued)

Strategy	Elicit or Confront Preconceptions	Both Elicit and Confront Preconceptions
7. Dynamic Models	This strategy has potential with this content. However, dynamic models are better used with content that is best understood through visualization. (*Note:* This is a strong strategy for Learning Targets #2 and #3.)	
8. Drawing	This is a great strategy, as discussed in Instructional Tool 3.9 (p. 82).	
9. Concept Cartoons		Concept cartoons work very well because they draw from different "intelligences" and are a safe way for students to express alternative explanations.

Stage IX: Determine Sense-Making Strategies

What do we mean by *sense making?* Eliciting and confronting students' preconceptions will not, by itself, promote conceptual understanding. Effective instruction requires that students make sense of the ideas with which they grapple, connecting what they already understand with the learning intent of the lesson, linking the ideas to the larger scientific body of knowledge, organizing that knowledge, and applying the ideas to new contexts. It is not likely that students will make all these connections by themselves, so it is important that teachers facilitate this process (Banilower, Cohen, Pasley, and Weiss 2010). Sense making should occur throughout the lesson, even as teachers elicit and confront student preconceptions.

Step 1

Use the same approaches as in Stage VIII, but this time with a focus on sense making. Go to the column headed Sense Making in the Instructional Tool 3.2, p. 67) to look for possible strategies. Under "Sense Making," find the following four categories:

1. Perceiving, interpreting, and organizing information
2. Connecting information
3. Retrieving, extending, and applying information
4. Using knowledge in relevant ways

These categories support sense making. Strategies can be selected from among these categories. Since we are working on the first lesson in the unit of study, we might focus on the first and second of these subcategories, knowing that later lessons will build on this lesson and apply what students learned to further learning experiences.

Because students benefit from multiple exposures to content, select three or four sense-making strategies. Having extra strategies on tap is useful if the formative assessment results require that students have additional learning experiences. When formative assessments indicate that students are still struggling with the lesson content, these strategies can be used to teach certain aspects of the lesson differently. Furthermore, a variety of strategies help differentiate instruction.

A quick look at the strategies under Perceiving, Interpreting, and Organizing Information indicates that writing-to-learn strategies and some reading-to-learn strategies are especially effective (♥) as are some models and visual tools.

Step 2

Now refer to the Instructional Tools (pp. 93–139), which tell more about the strategies identified in Step 1 as effective for sense making. After reading the information there, decide on three or four strategies that might work in the initial lesson or for additional exposures to the content after instruction, if necessary.

We found multiple strategies among the linguistic representations in Instructional Tool 3.3 that work well. Both science notebooks and learning logs are possibilities. Science Writing Heuristic (SWH) should be used only if you are familiar with the process or are willing to learn about it before implementation. Scientific explanations are also appropriate, especially since one of the practices on which we focus is "construct and revise an explanation." Also, some of the reading tools (Instructional Tool 3.4), especially informational text strategies, might be useful. It might be useful to teach strategies for note taking like two-column notes that allow for note taking (record key information during reading or lecture) and note making (reflecting on notes and making notes on them—through annotation, putting in own words, summarizing, highlighting, and so on).

Several nonlinguistic representations can also be used. Task-specific organizers (e.g., network tree maps) and thinking-process maps (specifically systems diagrams) show quite a bit of promise and so are on our short list. Take a look at Instructional Tool 3.7 (p. 127) and you can see why. The section for Systems Diagrams, under thinking-process maps, says that if used when studying a system, they are considered a model, which helps with both our aligned practices and crosscutting concepts. The research section says they help understand connectivity in a system, central to this unit of study.

Specific Graphic Organizers (Categorical Organizers) says that since they are highly structured they help develop important habits of mind that include persistence, accuracy, and precise language and thinking. Recall that this aligns with both our metacognitive and NOS foci. The research section also says, "Their step-by-step nature provides concrete models, thereby providing scaffolds for students who might otherwise give up. They also result in a written display of students' ideas, which students can reflect on and perhaps use to modify their thinking." Network trees, in particular, demonstrate causal information and hierarchies, and they can illustrate a system of thinking, visualizing interrelated parts of a whole, and study systems. This should help students build an understanding of the system, one of our targeted crosscutting concepts. We included some of these strategies in the Strategy Selection Template completed for Learning Target #1 (See Table 3.8 on p. 80).

Stage X: Determine Responsive Actions Based on Formative Assessment Evidence

Many of the strategies found in the Instructional Tools can serve as formative assessments during a course of instruction. Assessment results should be used to modify instruction and support student learning throughout a unit. The steps to identify assessment strategies are the same basic steps as used in Stages VII through IX.

Step 1

Refer once again to the Instructional Tool 3.2 (p. 67). Look at the column Demonstrating Understanding. Note the two sub-columns. Strategies listed in the Formative Assessment sub-column are intended for use directly by the teacher. Strategies in the Peer and Self-Assessment column are for students, and both should be included in the lesson.

A quick scan of both subcolumns indicates that almost any of the strategies listed on the left can serve as assessments. Use a variety. Keep in mind that it makes sense to use as formative assessments the student work that will be produced as a result of the strategies already built into your lesson. Why add extra assessments if the students' work will serve the purpose?

Step 2

Proceed to the specific Instructional Tools at the end of the chapter and review the information about the strategies selected from the Instructional Strategy Selection Tool. Then refer to Table 3.7 to compare your choices to the potential strategies we selected and our rationales for the selections.

That completes the process for strategy selection (see the summary in Table 3.8). Normally, we would complete the entire table, but because the process has been modeled in this chapter for only Learning Target #1, that column alone is filled in. The selected strategies can now be used in conjunction with activities typically employed when teaching a unit on proteins and genes.

Consider working with other science teachers in your department or in other schools to generate a strategy selection template for another hard-to-teach science topic from *NGSS*. The work of lesson planning is greatly facilitated by completion of this template. Furthermore, you will end up with valuable research-based strategies

Table 3.7

Formative Assessment Strategies and Rationales for the Selections

Assessment Strategy	Rationale for Selection
Learning Logs	They can be used for metacognitive reflection for students about their understandings of concepts, procedures, and the NOS, thus serving as a good self-assessment. If you ask open-ended questions as writing prompts, you can assess students' understandings of content as well as their thinking skills, especially their abilities to analyze, evaluate, and solve problems (a good tie to our metacognitive and NOS targets).
Graphic Organizers, Two-Column Notes and Tree-Maps	These result in written displays of students' ideas, which they can then reflect on and possibly modify their thinking – another good self- or peer-assessment. The display also provides you with a concrete model of their thinking.
Thinking-Process Maps	Thinking-process maps can serve as evidence for the thinking of individual students or of groups; they can be revised during a lesson or unit. Perhaps you can have students generate a systems diagram in this unit and revise the map as their thinking changes. The map could be expanded to include content related to the other learning targets as the unit progresses.
Large- and Small-Group Discourse	• Discourse is effective for both teacher assessment and for peer and self-assessment. • Discourse gives the teacher information about students' thinking, which can be used to modify instruction. • Student-to-student discourse, focused on developing explanations, allows peers to assess their own and one another's thinking.
Concept Cartoons	Concept cartoons can inspire discussions and written explanations that can serve as formative assessments.
Physical Movements and Gestures	A simple gesture such as a thumbs-up or thumbs-down can provide a quick sense of where students' thinking is during a lesson.

Table 3.8

Completed Strategy Selection Template for Learning Target #1

	Learning Target #1	**Learning Target #2**	**Learning Target #3**	**Learning Target #4**
	Proteins carry out the major work of cells and are responsible for both the structures and functions of organisms. An organism's traits (phenotype) are a reflection of the work of proteins.	Genetic information (genes) coded in DNA provides the information necessary to assemble proteins. The sequence of subunits (Nucleotides) in DNA determines the sequence of amino acids in proteins.	The sequence of amino acids determines not only the kind but also the shape of the protein, and thus its function.	Mutations, changes in the DNA, impact protein production. Errors in the DNA (mutation) can result in missing proteins or ones that function inadequately. This results in a change in phenotype/trait.
Strategies for:				
Identifying Preconceptions	Anticipation Guide (See Figure 3.4, p. 73)			
Eliciting and Confronting Preconceptions	Brainstorming Webs, Student Discourse, Informational Text Strategies, Visual Models, Annotated Drawings, Thinking-Process Maps			
Sense Making	Learning logs, Informational Text Strategies (two-column notes), Thinking Process Maps (Systems Diagrams), Categorical Organizers (network tree maps)			
Demonstrating Understanding	Learning logs (with two-column notes and system diagrams), questions and probes during discussions, presentation of developed models and explanations			
Selected Metacognitive Strategy	Critical Thinking and Learning: Truth Routines			

shown to be effective for helping students learn difficult science topics. The lessons themselves flow quite easily once the strategies are selected. After you go through this process once or twice, it will become second nature. Remember to refer to Appendix 2 (p. 313) as you begin this process on your own or with your peers, since it includes valuable hints and resources to support you during the process.

Resulting Lessons for Learning Targets #1–4

What might the lessons look like for this topic? Table 3.9 (pp. 82–84) provides a brief summary of the lesson for the first learning target and the following segment of the chapter more thoroughly describes the lesson, based on the planning process just completed. A description of all four lessons in the unit of study is found in Appendix 3 (p. 317).

Assume that Mrs. Hernandez (see Figure 3.1, p. 45) attended a professional development workshop we facilitated and worked with us to co-develop this lesson.

Lesson for Learning Target #1

Mrs. Hernandez's students complete the anticipation guide (Figure 3.4, p. 73), and she finds out that they hold many of the misconceptions identified in the research. She naturally wants her students to have a better understanding of what they think a protein is and does.

Elicit and Confront Student Preconceptions

To elicit student preconceptions, she prepares for a brainstorming session using the probe, "What are proteins and why are they important?" She considers two of the options we identified earlier: The "Think Puzzle Explore" strategy (Visible Thinking website at *www.visiblethinkingpz.org/VisibleThinking_html_files/03_ThinkingRoutines/03c_CoreRoutines.html*) or brainstorming webs. She wants students to research various genetic disorders, so decides on one of the brainstorming webs, in particular Cluster Maps (See example in Figure 3.5, p. 85).

She places students in groups of four and gives each a whiteboard on which to record their thinking. She models the use of clustering on her interactive whiteboard. Student groups then proceed to draw a circle in the center of their board and write "What are proteins and why are they important?" in the circle. They draw branches out from the central circle to other circles and brainstorm ideas to include in them. They then add new details or ideas in even further branches. Elicited student ideas are recorded and, at the same time, individual student's ideas are confronted when students in the group have alternative ideas. Each group shares their whiteboard with the class and the teacher generates a list of categories shared among the groups. She facilitates a class discussion during which students summarize their thinking and identify questions they have about proteins. They also discuss the use of the process itself, how

Table 3.9

Planning Template for Proteins and Genes Learning Target #1

Unit Topic—Proteins and Genes		
Phase 1. Identifying Essential Content		
Conceptual Target Development	Disciplinary Core Ideas Addressed	• Systems of specialized cells within organisms help them perform the essential functions of life • All cells contain genetic information in the form of DNA molecules. Genes are regions in the DNA that contain the instructions that code for the formation of proteins, which carry out most of the work of cells. • Multicellular organisms have a hierarchical structural organization, in which any one system is made of numerous parts and is itself a component of the next level.
	Crosscutting Concepts Addressed	• Models (e.g., physical, mathematical, computer models) can be used to simulate systems and interactions—including energy, matter, and information flows—within and between systems at different scales. • Investigating or designing new systems or structures requires a detailed examination of the properties of materials, the structures of different components, and connections of components to reveal the function and/or solve a problem.
	Science and Engineering Practices Addressed	• Develop and use a model based on evidence to illustrate the relationship between systems or between components of a system. • Construct and revise an explanation based on valid and reliable evidence obtained from a variety of sources (including students' own investigations, models, theories, simulations, peer review) and the assumption that theories and laws that describe the natural world operate today as they did in the past and will continue to do so in the future.
Essential Understandings	Organisms are systems made of dynamic and complex subsystems of interacting molecules in cells, tissues, and organs (levels of organization). Proteins are molecules that carry out the major biological processes of cells and impact the functioning of the tissues and organs they build. If they are not made properly, the entire organism can be affected. Changes (mutations) in the gene/DNA can impact not only the genetic code but also the protein, cells, tissues, and organs, since various components of the system interact and depend on each other. Though some mutations can be helpful, they might also stop or limit the protein's ability to function and potentially lead to physiological disorder in the entire organism.		

Table 3.9 (continued)

Phase 1. Identifying Essential Content	
Criteria to Determine Understanding	• Develop and explain a model that illustrates the production of a protein and its action across levels of organization (cell, tissue, organ, organism), resulting in a particular phenotype. • Construct an explanation for how DNA coding determines the structure of a protein. • Construct and refine an explanation of how amino acid sequence determines the shape of a protein and thus its function. • Refine the model developed in learning target #1 to demonstrate the impact of a mutation on a phenotype/trait, including a discussion of the protein's role in the process. • Accurately and carefully develop explanations and models, providing evidence for each.
Performance Expectations Addressed	• Construct an explanation based on evidence for how the structure of DNA determines the structure of proteins, which carry out the essential functions of life through systems of specialized cells. • Develop and use a model to illustrate the hierarchical organization of interacting systems that provide specific functions within multicellular organisms.
Phase 2. Planning for Responsive Action	
Identifying Student Preconceptions	Use an anticipation guide (See Figure 3.4, p. 73) to identify students' preconceptions about all of the learning targets.
Learning Target #1	Proteins carry out the major work of cells and are responsible for both the structure and function of organisms. These proteins are made based on the code found in the organism's DNA (genotype) and result in the organism's traits (phenotype). How well the cellular system functions and interacts among cells and at the various levels of organization impacts the entire organism.
Research-Identified Misconceptions Addressed	• Students demonstrate confusion over levels of organization, particularly with cells and molecules. They tend to think of molecules as related to the physical sciences and cells to life science. Some students even think that proteins are made of cells and that molecules of protein are bigger than cells (Driver et al. 1994). • The majority of upper division biology students and future science teachers recognize the physical constitution of an organism as its phenotype, yet do not understand the role of genes and proteins in producing the phenotype (Elrod n.d.). Because some students do not connect genes to proteins to phenotypes (Lewis and Wood-Robinson 2000), they assume that genes directly express traits in organisms (Lewis and Kattmann 2004). • Though students usually equate genes with traits, they do not understand that genes code for specific proteins and that the production of these proteins results in the traits (Friedrichsen and Stone 2004).

Table 3.9 (continued)

Phase 2. Planning for Responsive Action	
Initial Instructional Plan	• *Eliciting Preconceptions:* Facilitate a brainstorming session using the probe, "What are proteins and why are they important?" This can be as a whole-class or in small groups (but share as a whole class after small-group work). You can use one of the brainstorming webs or use the understanding routine, Think/Puzzle/Explore, found at the Visible Thinking website. During the class discussion, elicit ideas about various proteins and introduce the concept that missing or ill-functioning genes can significantly impact protein function and, in many cases, cause disease. The goal is to ensure that students understand the robust and important functions of proteins. End the discussion by generating student questions. • *Confronting Preconceptions:* Show the YouTube video "Protein Functions in the Body" (see Recommended Resources at the end of this chapter). After a brief class discussion, have students revisit and revise their brainstorming webs. Facilitate another class discussion about the question, "What would happen if there was a missing or ill-functioning protein?" show the YouTube video "Sickle Cell" (see Recommended Resources at the end of this chapter) and then have students again modify their brainstorming webs. • *Sense-Making:* Provide student groups a list of various proteins and the associated disorders that can arise if the protein is missing or malfunctioning. Have each group choose a disorder. Their task is to research the protein and disorder to determine the impact of the missing or malfunctioning protein on the various levels of organization (cells, tissues, organs, organisms). A resource that students can use as a starting point is a summary of disorders at the Your Genes, Your Health website (*www.ygyh.org*). Review with students the use of two-column notes and system diagrams before they begin research. Facilitate their use during small-group work, probing for student understanding of both concepts and use of the tools. Require students to develop individual explanations in their learning logs, share and critique their explanations (as a group), and prepare a presentation to share with the entire class. Support students during individual learning log summary development and during presentation preparation. After student presentations, facilitate a whole-class discussion, summarizing key ideas, and generating questions. Finally, ask groups to develop a single systems map that generalizes what they learned about proteins and genes from the various presentations.
Formative Assessment Plan *(Demonstrating Understanding)*	• Review of brainstorming webs • Review learning logs, including two-column notes, and systems diagram • Analysis of systems diagram for inclusion of key ideas • Questions and probes during small-group work and class discussion • Final presentation critique using pre-established rubrics *Note:* The formative assessment section in Chapter 4 provides specific suggestions for questions, probes, and feedback (pp. 142–149).

Figure 3.5

Sample Cluster Map

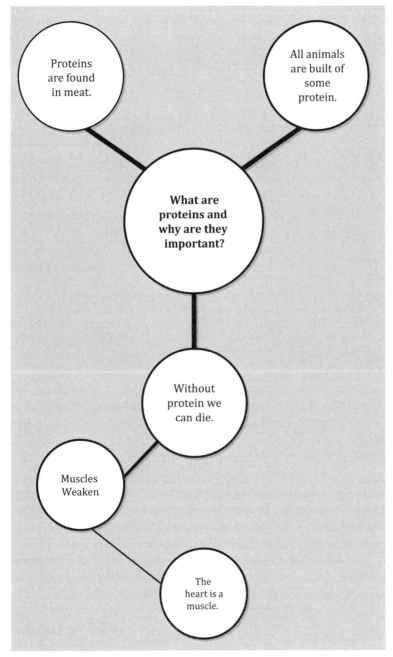

well it worked and what might be done to make it work better so that during its next use it would be a more effective strategy.

Mrs. Hernandez reflects on the work after school that day, using the brainstorming webs and student discussion of them as formative assessments, and realizes that her students understand that proteins are an important part of food and that they help build the body. But it is quite clear that they have little conception of the importance of proteins at the cellular level and even, in some cases, at the tissue level. She uses the summary of their thinking and her reflection on it to modify the upcoming portion of the lesson.

The next day she provides feedback to students and starts with her reflections on the previous day's work, suggesting that proteins are actually important at the cellular level—an idea that did not show up in their cluster maps. She shows the YouTube video "Protein Functions in the Body" (see Recommended Resources at the end of the chapter) to further confront student preconceptions. This video not only provides students with an overview of proteins and their importance but also stresses the importance of protein structure and functioning at all levels of organization and builds a nice tie to one of the targeted crosscutting concepts. After the four-minute video, the class briefly discusses it. Mrs. Hernandez asks students to revisit their cluster maps and incorporate any changes they think are important. She reviews their maps as they work, asks probing questions, and uses their responses and questions to further assess student understanding and inform lesson development. She uses what she has learned to further plan for the upcoming activities, designed for student sense-making. She learned during the workshop we facilitated that this planning is very important and that many of her peers neglect it— some simply ask students questions, have them write responses, and then grade their written work. She wants her classroom to be about grappling with the ideas instead of right answers because she wants to help her students resolve any conflicts between their thinking and a scientific explanation. So the strategies she chooses are critical.

Mrs. Hernandez learned that "what would happen if" questions can be very productive, so she asks the students what might happen if there was a missing or ill-functioning protein in a cell? A full-class discussion results in the idea that it might impact the cell but also tissues, organs, and possibly the entire organism. She uses sickle-cell anemia as a content representation and shows another very brief YouTube video "Sickle Cell" (see Recommended Resources at the end of the chapter). She chose this particular video because many of her students are familiar with sickle-cell anemia and she knows that it is important to start with what students know about already. After the video, she asks each student group to identify the various levels of organization, from molecules to organism, impacted by the disorder and further modify their cluster map based on what they learned. They then spend some time as a whole class talking about the modifications they made.

Sense Making

Mrs. Hernandez reflects that evening on the cluster maps that students modified and the discussions during the day. It is clear to her that students are beginning to connect DNA and proteins, realizing that what happens at the molecular level might impact the other levels of organization in an organism. But she wants them to make better sense of this and more fully understand the connections among system components, so decides to use simple systems diagrams to serve as models. She also determines that two-column notes (an informational text strategy) will help them in note taking as they research and note making (reflection on their notes) to help make sense of what they read. She decides to have students maintain both their systems diagrams and two-column notes in their learning logs so she can easily assess their understanding and make appropriate adjustments to her questioning and lesson. Her last task of the evening is to finalize a resource she started early in her planning that lists various genetic disorders and some initial resources related to each disorder. She will distribute these to students when they begin their research. (Note: A great initial resource that provides summaries of disorders is at the Your Genes, Your Health website: *www.ygyh.org*.)

As the lesson progresses the next day, she tells them they will further explore the connection between genes and proteins by researching some common genetic disorders and finding out what happens at the cellular level and how it impacts the body. The intent of this activity is to help them in sense-making about the idea that an organism's traits (phenotype) are a reflection of the work of proteins and that the components of a system impact how well the entire system works.

Student groups identify their top two choices of disorders to research from a list the teacher provides. She clearly explains what she expects from their research, including a clear description of the goals, timeline, and products (their two-column notes and systems diagrams, a group systems diagram that results from consensus after discussion, and a group presentation using the interactive whiteboard). Her students have already used two-column notes, so she simply refreshes their understanding of this strategy by sharing an example of student work from a previous project. She then models the use of a systems diagram, since this strategy is new to them. She provides a brief practice assignment to assess their understanding of the strategy. While they work on the practice diagram, she reviews their lists of preferred research topics and assigns topics to groups. When students complete their work the class discusses the process—what worked, what didn't, what questions they had, and how they might improve their process to complete a diagram.

Now the students are ready to begin their research. Mrs. Hernandez assigns group topic assignments and displays the document that shares initial resources for student research. She also assigns partner groups so that during the research process they can periodically share their learning with one other group. She also cautions them to use some of the planning strategies (See Plan and Self-Regulate in Instructional Tool 3.1,

p. 60) they learned earlier in the year to ensure that during their research project they map and plan the work, add tasks as necessary, establish research routines, and track their progress. The students are used to using these strategies. Students complete these initial plans and begin their research for the remainder of the class period.

Over the next two days students work individually on research, share their thinking with their group as they begin to map their systems diagram, and occasionally bounce ideas off their partner group. Mrs. Hernandez observes each group and periodically joins them, asking probing questions and checking progress on their systems maps. She also listens carefully to the groups and determines which groups demonstrate the most sophisticated understanding of the topic. She wants to use this information as she orchestrates the order of presentations.

Students have one additional day during which to prepare their presentation. Mrs. Hernandez lets them know the order in which they will present the following day and stresses that their presentation should focus on the model (systems diagram) they developed and their explanation of it.

On the next day of this lesson, Mrs. Hernandez asks the group with the least sophisticated understanding to present first. This lets her recognize the aspects of the system that they understand and ask probing questions about areas in which their thinking is less developed. She provides a verbal summary of their work at the end of the presentation—this is foundational for the class understanding but other groups' presentations can build further upon it. As each group presents, she asks questions, as do the students. The final group to present is the group she deemed to have the most sophisticated understanding of the topic. By the end of their presentation, Mrs. Hernandez is quite pleased that the students seem to have a pretty solid understanding of their assigned disorder and how it impacts the system from DNA to protein.

A final step in the sense-making process is to see if they can generalize the learning across disorders. She asks each group to build a general systems diagram on poster paper that reflects the interactions of a system to describe how genotype leads to phenotype. They discuss their diagrams with their partner group and revise accordingly. These posters are displayed and quickly reviewed in a round-robin fashion. Mrs. Hernandez reviews the posters as well, and closes the class period with a summary of their work and tells them that it lays the foundation for their upcoming series of lessons where they will learn more about what happens at the molecular and cellular level in this system. For detailed information about assessing students' understandings of this lesson, refer to the section in Chapter 4 titled "Formative Assessment for Learning Target #1" (pp. 142–149).

Thoughts About Learning Targets #2-4

The lesson template (see Appendix 3, pp. 317–324) outlines the lessons for Learning Targets #2 through #4. We did not include fully developed lessons for these learning targets beyond what is outlined in the appendix, but some of the thinking we used as we completed the design process follows.

Learning Targets #2 through #4 have several things in common. They all benefit from the use of dynamic modeling, primarily because dynamic models are effective for teaching abstract concepts and complex systems and connecting what happens at the submicroscopic and macroscopic levels (Calwetti 1999; Trunfio et al. 2003).

Here are some important research findings about models that were considered as we selected strategies and designed instruction (see Instructional Tool 3.6 on p. 119 for a complete list of ideas):

- Conceptual development is enhanced if students are able to construct and critique their own models (Hipkins et al. 2002).

- Students who receive instructions and guiding questions along with a model better understand molecular genetics (Rotbain, Marbach-Ad, and Stavy 2005).

- An interactive dynamic model combined with scaffolding improves students' abilities to explain human inheritance and evolutionary phenomena, connecting their ideas about phenotypes, chromosomes, and gametes (Schwendimann 2008).

- Simulations can promote misconceptions unless teachers work explicitly with students to identify limitations of the models (Calwetti 1999).

- Teachers should model scientific modeling to their students, encourage the use of multiple models in science lessons, and encourage negotiation of model meanings. Systematic presentation of in-class models using the Focus, Action, and Reflection (FAR) Guide to socially negotiate model meanings is effective (Harrison and Treagust 2000).

We reflected on this research and decided on the following strategies for these learning targets:

- Written explanations, which interpret dynamic models and text

- Group mapping that translates student experiences with the dynamic model into students' own mental and written models

- Large- and small-group discourse to interpret and apply interactions with the models

- Comparison of interactive models and student maps with text (scientific explanations)

- Creating and critiquing analogies to extend and apply mental models, using the FAR Guide.[1]

In addition, Learning Target #4 requires students to make predictions (claims) about mutations, gather evidence using a protein folding interactive developed by the Concord Consortium, and use this evidence to further explain the impact of their chosen disorders. The development, critique, and support of these explanations address both "retrieving, extending, and applying information" and "using knowledge in relevant ways" (as cited in Instructional Strategy Selection Tool 3.2).

Consider the following as you review the listed formative assessments for Learning Target #4. Critique of explanations is most effective when clear targets and rubrics that include the criteria for success are made available before instruction. Sutherland, McNeill, and Krajcik (2006) propose that the components of the rubric include a critique of the claim (that it responds to the question asked or problem posed), the evidence (that scientific data are used to provide support for the claim), and the reasoning (that students use scientific principles to explain why the data provide evidence to support the claim). Keeley (2008) describes a formative assessment strategy, "Explanation Analysis," that builds on this thinking. This information is helpful not only in your review of the appendix but also as you work with your students on explanations in the future.

Time for Reflection

Take a moment and reflect on the design process and then complete these two tasks:

1. Review Figure 3.1, the case study of Mrs. Hernandez, and the redesigned lesson just described. Based on what you learned in this chapter, what major changes did she make to improve her students' understandings about the relationship of genes and proteins? Do you think they were effective changes or not? Why?

2. Consider your current level of understanding about your understanding of the Instructional Planning Framework, the design process, and the Instructional Tools. What parts are clear to you and what parts are still a bit confusing? What questions do you have? As you study and use the remainder of this book, focus on the aspects that will help you learn more about your remaining questions.

Recommended Resources

Technology Applications and Websites

- Concord Consortium's Molecular Workbench (*http://mw.concord.org/modeler*) offers a rich set of interactive simulations, many of which apply to the biological sciences.

- The YouTube video "Protein Functions in the Body" (*www.youtube.com/watch?v=T500B5yTy58*) presents the wide array of proteins that make up living organisms.

- The YouTube video "Sickle Cell" (*www.youtube.com/watch?v=9UpwV1tdxcs*) is a good starting place for students to gain an overview of a genetic disorder.

- A resource that serves as a starting point as students research different disorders is the Your Genes, Your Health website (*www.ygyh.org*).

- Teachers Domain (*www.teachersdomain.org*) offers a flash interactive—Cell Transcription and Translation—that includes an overview of protein synthesis as well as more detailed interactions with transcription and translation (*www.teachersdomain.org/resource/lsps07.sci.life.stru.celltrans*); at this point, we recommend sharing only the overview, not the details. If you do not have a Teachers Domain account, you can register at no cost.

- The DNA Workshop site (*www.pbs.org/wgbh/aso/tryit/dna/#*) has a protein synthesis shockwave. Also, more extensive explorations are available at the Concord Consortium's Molecular Logic Project (*http://molo.concord.org*). Many of the explorations are linked to specific biology textbooks, perhaps including one your district uses.

- For your personal content learning, the NSTA SciPack "Cell Structure and Function" is a good option. In particular, the last Science Object called "The Most Important Molecule" relates directly to this chapter. Visit the NSTA Science Store (*www.nsta.org/store*) and search for "protein."

Build Your Library

- A great resource for formative assessments, which we used to construct the Instructional Planning Framework, is *Science Formative Assessment: 75 Practical Strategies for Linking Assessment, Instruction, and Learning* (Keeley 2008). You might also consider some of the *Uncovering Student Ideas in Science* volumes available through NSTA Press (*www.nsta.org/publications/press/uncovering.aspx*).

- An excellent resource to learn more about the use of analogies in secondary science classrooms is *Using Analogies in Middle and Secondary Science Classrooms: The FAR Guide—An Interesting Way to Teach With Analogies* (Harrison and Coll 2008).

- Chapter 11, "How Nature Builds Itself: Self-Assembly," in *Nanoscale Science: Activities for Grades 6–12* (Jones et al. 2007), ties in nicely with some of the Concord Consortium dynamic models used in this lesson.

- Black and Harrison (2004) provide examples and suggestions for descriptive feedback for students in *Science Inside the Black Box: Assessment for Learning in the Science Classroom*.

Endnote

1. The FAR Guide has three main aspects:

   - Consider the focus. Look at the concept itself (difficult? unfamiliar? abstract?), the ideas the students bring to the table, and whether the analog is familiar to students.

   - What is the action? Discuss how the concept is like the analog (similarities) and unlike the analog (differences).

   - Take time to reflect. After using the analog, determine if it was clear or confusing and if it achieved the desired outcomes. Also consider any changes you would make in future use.

We urge you to learn more about the entire FAR Guide process prior to implementation. This summary, in itself, is not adequate for implementation.

Instructional Tool 3.3

Sense-Making Approaches: Linguistic Representations—Writing to Learn

Learning Logs	
The terms *learning log* and *science notebook* are often used interchangeably. However, learning logs are a less-structured way for students to maintain written records of their observations and thinking. They permit a free-form type of writing that can express feelings as well as observations. They are a type of reflective writing structure in which students create a narrative and participate in his/her own learning story (Nathanson 2006).	
The Research	• Most writing in science classes has involved writing to answer test questions rather than as a form of dialogue between the student and the teacher. In addition, writing is rarely used in science classes as a way for students to express their thinking, either for their own benefit or for sharing with their peers. Thus, students lose potential opportunities to learn more about science and about the writing process (Rangahau 2002).
	• Writing can be used to clarify ideas in and about science (Hipkins et al. 2002; Rangahau 2002), and journaling and micro-themes can be used to promote students' personalization and understanding of science (Ambron 1987). Learning logs integrate content, process, and personal feelings and help students learn *from* their writing as opposed to writing *about* what they learn (Costa 2008). Extended writing sequences can be used for metacognitive reflection for students about their understandings of concepts, procedures, and the nature of science (NOS) (Baker 2004; Rangahau 2002). If you use open-ended questions as writing prompts, you can assess your students' understanding of content as well as their thinking skills, especially their abilities to analyze, evaluate, and solve problems (Freedman 1994).
	• The use of tablets, smartphones and mobile apps foster more collaboration and better communication, as well as better engaging students with each other and the information being taught. The use of them would be appropriate for e-learning logs.
Connections to *NGSS* Practices	• **QP** – Since learning logs allow students a flexible format in which to record observations and thinking, they lend themselves to translating these observations into questions about what they observe.
	• **I** – The written record of observations in a learning log can lead to the more formalized organization of scientific investigation.
	• **OEC** – A learning log provides a format for students to reflect upon the information they obtain, evaluate it, and consider ways in which to communicate what they've learned.

Note: See legend at bottom of page 101.

Instructional Tool 3.3 (continued)

Classroom Implications (cont'd)	• Learning logs can engage student thinking about a topic because they require students to focus on content in their writing (Barton and Jordan 2001). A journal or learning log allows your students to reflect on their thinking in a diary format. It lets them make notes about their awareness of difficulties and comment on steps they've taken to deal with these difficulties (Blakey and Spence 1990). You can use open-ended questions developed and used as probes to assess students' thinking and content understanding. Consider using the wonderful probes that make up the *Uncovering Student Ideas in Science* series (see Resources on the next page).
	• Be sure to respond carefully to your students' writing. When they know that you act on what they write and support them in the process, they are better able to identify their learning difficulties and overcome them (Rangahau 2002). Be careful in how you provide feedback. Rather than giving feedback in the form of a grade, indicate how students might improve their work as you provide suggestions for what they should work on next. Follow the guidelines or effective descriptive feedback that is part of formative assessment processes. A constructive comment is more effective than praise. It is helpful to provide suggestions for what to do next or ask questions that prompt students to connect what they learn to other experiences (Black and Wiliam 1998, Harlen 2000, Heritage 2007). The feedback should include both what was correct and what was not correct, in which case feedback should include information that helps them understand why something was incorrect (Dean et al. 2012).
Application Example	A teacher asks an open-ended question about how students think a tree could possibly come from a seed. Students enter their responses to the prompt with an annotated drawing in their learning logs. This, of course, involves annotated drawings as well as written descriptions. The teacher reviews the logs and provides written responses that encourage students to think further—both about the content and their own thinking about the content. Her responses include questions such as "What leads you to think that...?" and "How would you test your ideas?" and "How will you find out more about this topic to make a stronger response?" The following day, students review their logs and respond in writing to any questions the teacher has asked. Small groups read one another's responses and discuss the various explanations, reaching some consensus on their thinking.
Technology Applications	• Students can keep electronic learning logs and share them with other students and the teacher, allowing them to reflect on their own thinking as well as defend their viewpoints and react to those of others (Baker 2004). For additional recommendations for instructional strategies and use of e-learning logs in the classroom, see Urquhart and McIver (2005, pp. 140–141).
	• Blogs and vlogs are very appropriate to use here.
Resources	• *Open-Ended Questioning: A Handbook for Educators* (Freedman 1994)
	• *Teaching, Learning & Assessing Science 5–12* (Harlen 2000)
	• Various books to probe student thinking have been written by Keeley and colleagues. Read more about the *Uncovering Student Ideas in Science* series at *www.nsta.org/publications/press/uncovering.aspx*.
	• *Teaching Writing in the Content Areas* (Urquhart and McIver 2005)

Instructional Tool 3.3 (continued)

Probes
Probes are formative assessments used for both diagnostic and monitoring purposes, and they are considered assessments *for* learning. The purpose of the probes is to reveal to you and your students the conceptions they have about various science concepts. We focus here specifically on the probes developed by Page Keeley and her colleagues (see Resources listed below).

The Research	Information gained from use of the probes has generated much feedback and general research has been conducted. Here are some key findings: <ul><li>Probes help teachers know more about what their students understand.</li><li>Probes significantly change instruction in the classroom.</li><li>Teachers tend to listen very closely to their students' reactions to the probe, which helps the teachers determine their next instructional steps.</li><li>Probes are used successfully in professional development, helping teachers learn more about content (teachers confront their own misconceptions) and instruction (they discover that some traditional science activities actually reinforce common misconceptions).</li></ul>
Connections to *NGSS* Practices	<ul><li>**E/S** – Probes provide students the opportunity to provide their initial and ongoing "explanatory accounts of the features of the world" (NRC 2012, p. 52).</li></ul>
Classroom Implications	<ul><li>It is important to understand what each probe is designed to reveal so that you are sure to select probes that align with your learning target.</li><li>You might use multiple probes that target a single concept, which will give you even more information about students' understandings and about how to modify instruction.</li><li>In the Page Keeley et al. books, be sure to read the Explanation section in the Teacher Notes for each probe. It provides clarification about the content.</li><li>Carefully consider the Curricular and Instructional Considerations section to assure yourself that you are using a probe that is appropriate for your grade level. Also use the Administering the Probe section for ideas about using the probe appropriately at your grade level.</li><li>Carefully read the Related Research section for each probe. You will better understand the purpose of the probe and anticipate possible responses from your students.</li></ul>
Application Example	As outlined in Chapter 3, students often misunderstand the relationship among DNA, genes, and chromosomes. An additional way, in addition to the way described in the chapter, to elicit student understanding about these misconceptions is to use Page Keeley's "DNA, Genes, and Chromosomes" probe (Keeley 2011, pp. 129–133). This probe offers several possible explanations and then asks students to choose the explanation with which they most agree and to explain why they agree.
Technology Applications	<ul><li>Students could complete a probe at home and then blog or vlog with their peers.</li><li>Students' probe responses can be kept in their electronic journals, which can be made available for your feedback.</li></ul>
Resources	Use the *Uncovering Student Ideas in Science* series by Page Keeley and her colleagues. Go to *www.nsta.org/publications/press/uncovering.aspx* for more information.

Instructional Tool 3.3 (continued)

Science Notebooks
Science notebooks are a stable location for students' work, a record of the information they value, and a window into their mental activities. They are a link between science and literacy and include writing, graphs, tables, and/or drawings that help students make meaning of science learning experiences. Science notebooks focus on the more structured writing of science while science journals or learning logs are more free form. Although both types of writing are important, science journals and science notebooks should be separately maintained (Hargrove and Nesbit 2003). Interactive science notebooks help students to communicate with each other, they support differentiated instruction and they develop student thinking in ways that prepare them with 21st-century skills (Marcarelli 2010)

The Research	• Most writing in science classes has involved writing to answer test questions rather than as a form of dialogue between the student and the teacher. In addition, writing is rarely used in science classes as a way for students to express their thinking, either for their own benefit or for sharing with their peers. Thus, students lose potential opportunities to learn more about science and about the writing process (Rangahau 2002). • Writing can be used to clarify ideas in and about science (Hipkins et al. 2002; Rangahau 2002). The first goal of writing in science is to understand something because writing is a tool with which to think. When students use their science notebooks during classroom discussions, it helps them construct meaning from the science phenomena they observe (Harlen 2001). Science notebooks focus on making sense of phenomena under investigation and help students provide evidence when they are answering the questions of their teacher or their peers. • Science notebooks are useful for assessment when (1) most of the work in the notebook is narrative and centered around authentic science tasks, (2) the notebook work is purposeful, with students investigating their own questions, (3) "right" answers or conclusions are uncommon, and (4) the notebook provides information not only to the teacher, but also to the student and possibly to parents (Hargrove and Nesbit 2003). • Science notebooks support active learning by students where they are able to pursue their own interests and tackle questions that are part of authentic problems (Hargrove and Nesbit 2003; Gilbert and Kotelman 2005) • Well-structured and thoughtfully organized science notebooks incorporate reflective writing and include think-aloud opportunities that are common in the notebooks of actual scientists as they answer questions and study science phenomena firsthand (Magnusson and Palincsar 2003). • The use of tablets, smartphones, and mobile apps foster more collaboration and better communication, as well as better engaging students with each other and the information being taught. The use of them would be appropriate for e-notebooks. Online science notebooks are one approach to storing student data and maintaining a record of their ideas and explanations.

Instructional Tool 3.3 (continued)

Connections to *NGSS* Practices	• **QP** – Science notebooks should include question posing and problem definition. Their notebooks are ideal places to develop and refine questions that drive investigation as well as ask questions that define an engineering problem. • **I** – Notebooks are the ideal place for students to design investigations (both laboratory and field) and record collected data, using the information to refine their explanations. They can also be used to gain data necessary for specifying design criteria and to test their designs. • **D** – While large data sets are more easily managed technologically, the science notebook is still a good place for students to reflect on patterns in the data, whether during an investigation or in the midst of an engineering problem. • **E/S** – The notebook is also a logical place for students to record and modify explanations developed during investigations. It is likewise a good place to record possible solutions and choose among the various solutions during the engineering design process.
Classroom Implications	• There are various ways to structure a science notebook. Consider breaking it down into these seven parts: focus question, prediction/hypothesis, planning, data, claims and evidence, making meaning conference, conclusion/reflection. Another format might include the following components: recording and organizing data, technical drawings, and students' questions (Campbell and Fulton 2003). • Science notebooks are valuable in connecting what should be taught (the standards) to classroom instruction and to the actual content students learn. Make sure to respond carefully to your students' writing. When they know that you act on what they write and support them in the process, they are better able to identify their learning difficulties and overcome them (Rangahau 2002). Be careful in how you provide feedback. Rather than giving feedback in the form of a grade, indicate to students how they might improve their work. A constructive comment is more effective than praise. It is helpful to provide suggestions for what to do next or ask questions that prompt students to connect what they learn to other experiences (Harlen 2000). This can be done through written or oral responses. Timeliness of feedback is essential. If you delay providing feedback, student misconceptions may persist.
Application Example	Students are engaged in the study of two different prairies—one maintained by mowing and one maintained by fire. Each student group will develop a question to investigate. They use their notebooks the first day in the field, recording observations and generating questions. Back in the classroom they meet with their small group and discuss their observations and questions. The group develops an investigable question they would like to answer. They spend the next several weeks gathering data in the field, reflecting on those data in the classroom, identifying patterns and possible explanations for what they have observed.
Technology Applications	Blogs, wikis, and online science notebooks are good applications.
Resources	• *www.sciencenotebooks.org* • *Science Notebooks: Writing About Inquiry* (Campbell and Fulton 2003) • *Teaching Science with Interactive Notebooks* (Marcarelli 2010) • *Teaching Writing in the Content Areas* (Urquhart and McIver 2005)

Instructional Tool 3.3 (continued)

Writing Scientific Explanations

"Scientific explanations are accounts that link scientific theory with specific observations or phenomena—for example, they explain observed relationships between variables and describe the mechanisms that support cause and effect inferences about them (NRC 2012, p. 67)." Scientific explanations address a combination of the goals of explanation (how and why something happens) and argumentation (written or oral social activity that justifies or defends a standpoint for an audience). Students justify their explanations of phenomena by making claims and supporting those claims with appropriate evidence and reasoning (McNeill and Krajcik 2006).

The Research	• "Transmissive" writing is a type of writing that transmits information and is often seen in science classrooms. However, if transmissive writing is used exclusively, students will not have the opportunity to use writing to develop and understand the nature of science; in particular, they will not be able to use writing to outline evidence for an argument and to explain the reasoning behind planning decisions (Rowell 1997). The single most powerful conceptual tool for teachers to move past this is to distinguish between descriptive and explanatory experiences in science; in other words, to shift the emphasis from *what* happens to *how* and *why* events happen (Braaten and Windschitl 2010).
	• Supporting students in the construction of explanations results in improved student understandings about inquiry, science content, and science literacy (McNeill and Krajcik 2008, Sandoval 2003). In particular, student adoption of explanatory goals (focus on explanation) promotes conceptual understanding as well as better understanding of the nature of science (Sandoval 2003).
	• It is difficult for students to learn how to construct explanations while they are in the midst of learning content in a particular area, so teacher support is essential. Furthermore, context-specific scaffolds are more effective than generic explanation scaffolds in promoting students' abilities to write scientific explanations and explain their understandings of science content (McNeill and Krajcik 2006).
Connections to NGSS Practices	• **M** – Theories are often first represented by a model and then a model-based explanation is developed (NRC 2012). Models and explanations thus work hand-in-hand and students should become familiar with this.
	• **E/S** – This strategy *is* about scientific explanations so this is a clear connection.
	• **AE** – Deciding the best explanation among alternatives requires scientific argumentation. "Deciding on the best explanation is that of argument that is resolved by how well any given explanation fits the available data, how much it simplifies what would seem to be complex, and whether it produces a sense of understanding (NRC 2012, p. 68)."
	• **OEC** – Developing explanations requires that students obtain information and evaluate information. If students are to determine the best explanation among those in consideration, then communicating information with their peers is also essential.

Instructional Tool 3.3 (continued)

Classroom Implications	<ul><li>Explanation and argumentation depend upon each other since for students to develop explanations they must be engaged in argumentation (Reiser, Berland and Kenyon 2012). Consider this interplay as you develop lessons and units of study.</li><li>Learning to construct explanations is hard for students, so science teachers should provide them with a framework such as the following: (1) make a claim, (2) provide evidence for the claim, and (3) provide reasoning that links the claim to the evidence (Sutherland, McNeill, and Krajcik 2006).</li><li>Actions the teacher can take to help students construct scientific explanations include the following: (1) uncover students' prior knowledge about what *explanation* means, (2) generate criteria for explanations, (3) make the framework explicit, (4) model the construction of explanations, (5) provide students with practice opportunities, (6) practice the critique of explanations, (7) give feedback to students, and (8) provide opportunities for students to revise their work (Sutherland, McNeill, and Krajcik 2006).</li><li>Students can first practice writing scientific explanations for phenomena with which they are already familiar and for which their content knowledge is well formed. They can then move to writing scientific explanations related to their own investigations.</li><li>You can support students' practice in writing explanations by using a "summary frame" with them before they start to develop explanations on their own. Summary frames are a series of questions you provide students, centered on a specific type of information. The questions highlight the important elements of that type of information and are intended to help students write summaries of the information. When writing scientific explanations, you employ the "argumentation pattern," using text that tries to support a claim. The argumentation pattern contains these elements: evidence, claim, support, and qualifier (restriction on the claim or evidence that counters the claim). Specific questions might be "(1) What is the basic claim or focus of the information? (2) What information is presented that leads to a claim? (3) What examples or explanations support the claim? (4) What restricts the claim?" (Dean et al. 2012, pp. 86–87).</li><li>Practices to enhance explanations through model-based inquiry include:<ul><li>Focus on unobservable events/entities and tie them to observable natural phenomena to develop an explanatory model that students study.</li><li>Use engaging tasks to elicit students' initial ideas and then use those ideas to shape instruction.</li><li>Share partial explanatory models around which students investigate and gather data; have students use the model before, during and after inquiry; and build background knowledge without reasoning for the students.</li><li>Scaffold learning about scientific explanations and what counts as evidence (Braaten and Windschitl 2010).</li></ul></li></ul>
Application Example	Have students describe everyday instances where they might need to explain something—how to use their Wii, how to play a new game they got for their birthday, how to do a particular algebra problem, or what to tell their parents if they get home late from school. This activity relies on their everyday use of the word *explain*. Share explanations with the class and analyze these everyday explanations. Which require facts or other evidence? How does the explanation usually begin? Which are "good" explanations and which are not? During this conversation, make sure students distinguish between opinion-based and evidence-based explanations. Notice that this activity addresses supporting actions #1 and #2 described in the second paragraph in Classroom Implications, above.

Instructional Tool 3.3 (continued)

Technology Applications	• There are many argument diagramming tools that can be used for building explanations (e.g., Convince Me and ExplanationConstructor) • Blogs, wikis, and online lab notebooks are good applications.
Resources	• "Inquiry and Scientific Explanations: Helping Students Use Evidence and Reasoning" (McNeill and Krajcik 2008) • *Questions, Claims, and Evidence: The Important Place of Argument in Children's Science Writing* (Hand et al. 2008) • Various books to probe student thinking have been written by Keeley and colleagues. Read more about the *Uncovering Student Ideas in Science* series at *www.nsta.org/publications/press/uncovering.aspx*. • *Everyday Science Mysteries* series by Konicek-Moran (see the NSTA Store at *www.nsta.org/store/default.aspx*) • The article, "Engaging Students in the Scientific Practices of Explanation and Argumentation: Understanding A Framework for K–12 Science Education" in *The Science Teacher* (Reiser, Berland, and Kenyon 2012) gives good examples of the interplay between explanation and argumentation. • Focused on middle school but usable at any grade level, download a document from an NSTA presentation that outlines teaching strategies for scientific explanations (McNeill and Krajcik 2011) found at *www.katherinelmcneill.com/uploads/1/6/8/7/1687518/mcneillkrajcik_nsta2011_long.pdf*

Science Writing Heuristic

The science writing heuristic (SWH) is a structured approach that combines guided inquiry, collaborative group work, and writing-to-learn activities. "The SWH provides an alternate format for students to guide their peer discussions and their thinking and writing about how hands-on guided inquiry activities relate to their own prior knowledge via beginning questions, claims and evidence, and final reflections" (Burke et al. 2005, p. 2).

The Research	Use of SWH both independently and in connection with textbook strategies had a positive impact on student conceptual understanding and metacognition (Wallace, Hand, and Yang 2004). The SWH process has been used in a variety of classrooms across grade levels and science disciplines. Productive integration of SWH has produced student-learning gains (Burke, Greenbowe, and Hand 2005).
Connections to *NGSS* Practices	• **E/S** – SWH uses language and focuses on questions, claims, and evidence, essential to developing scientific explanations • **AE** – Deciding the best explanation among alternatives requires scientific argumentation. "Deciding on the best explanation is that of argument that is resolved by how well any given explanation fits the available data, how much it simplifies what would seem to be complex, and whether it produces a sense of understanding (NRC 2012, p. 68)." • **OEC** – Developing explanations requires that students obtain information and evaluate information. If students are to determine the best explanation among those in consideration, then communicating information with their peers is also essential.

Instructional Tool 3.3 (continued)

Classroom Implications	The SWH process is highly effective, but also complex. The process engages students in activities to identify pre-instructional ideas, pre-laboratory activities to begin thinking about the concepts, the laboratory activity, small-group sharing and comparing of data, comparing ideas to those in textbooks or other print resources, individual reflection and writing, and exploring post-instruction understanding. Each component is essential and requires attention by the teacher. Note that this is a process that entails multiple strategies from across the Instructional Tools in this chapter. Instructor preparation is essential for the successful integration of SWH into a science course and for positive impacts on student learning (Burke, Greenbowe, and Hand 2005). This strategy is included here primarily to encourage you to consider training in SWH. We encourage you, if you have the opportunity, to attend a professional development session in your school district or state or at an NSTA conference. Once you are familiar with SWH, you can use it in your lesson development.
Application Example	Because of the complexity of SWH, it would take several pages to thoroughly describe an application example here.
Technology Applications	Blogs, wikis, and online lab notebooks are good applications.
Resources	A brief overview of SWH is found at *http://edutechwiki.unige.ch/en/ Science_writing_heuristic.* You can download a template at *www.aea11.k12.ia.us/science/Heuristic.html.*

Note: The abbreviations for practices referenced in the tool are:

- Asking questions and defining problems (QP)
- Developing and using models (M)
- Planning and carrying out investigations (I)
- Analyzing and interpreting data (D)
- Using mathematics and computational thinking (MCT)
- Constructing explanations and designing solutions (ES)
- Engaging in argument from evidence (AE)
- Obtaining, evaluating, and communicating information (OEC)

Instructional Tool 3.4

Sense-Making Approaches: Linguistic Representations—Reading to Learn

Vocabulary Development	
These strategies help students understand words in a lesson or unit that are unique to science. The strategies also identify certain words that have different meanings in science than they do in everyday use (e.g., *replicate, translate, class*).	
The Research	• Teachers should use strategy instruction to help students understand content-specific vocabulary. This is particularly difficult in biology because the terminology often represents ideas and concepts (Urquhart and Frazee 2012). • Pre-teaching vocabulary using effective strategies at the beginning of a unit of study can improve student comprehension. However, students learn science vocabulary best through purposeful interaction with the relevant concepts (Urquhart and Frazee 2012). • Students' gradual understanding of new vocabulary improves comprehension of text, regardless of age or population (Magnusson and Palinscar 2003). Furthermore, coaching in the language of science (both vocabulary and grammar) simultaneously supports reading and science literacy (Hipkins et al. 2002; Rangahau 2002). However, it is important to make sure students understand the conceptual foundation for a topic before introducing specialist vocabulary (Hipkins et al. 2002). • Words with dual meanings (e.g., *fault, force, food*) cause more problems in science lessons than words with single meanings. Often in science, nouns can be substituted for verbs or an entire sequence of events (e.g., *evaporation*), and nouns can be used as adjectives (e.g., *elephant population*). Explicit coaching about these grammatical features of science enhances reading literacy and science literacy (Rangahau 2002). • Learning science vocabulary is important so that students can learn the discourse of science. Embedding the language of science in guided-inquiry experiences allows you to model how science language is used (Magnusson and Palinscar 2003). Words used for hypothesizing, comparing, and other aspects of science reasoning (e.g., *frequently* or *simultaneously*) cause difficulty. Giving your students practice with argumentation (written and oral) helps them build understandings of these words (Rangahau 2002). • A good portion of vocabulary learning should occur as students learn subject matter because the context in which vocabulary is learned is important (Magnusson and Palinscar 2003). Use of models and analogies builds shared meanings that facilitate the communication of individual understandings of science technical vocabulary (Rangahau 2002).
Connections to NGSS Practices	• **E/S** – To develop explanations, students need to understand core science vocabulary, which is quite conceptually loaded. • **OEC** – Students need to read and use words (as well as tables, diagrams, graphs, and mathematical expressions) to obtain information and communicate understandings about science concepts, both in written and oral form. Even with grade level–appropriate reading skills, science is challenging to read for many reasons (NRC 2012). Precise meanings of vocabulary words are a piece of this challenge.

Note: See legend at bottom of page 109.

Instructional Tool 3.4 (continued)

Classroom Implications	• Vocabulary instruction not only prepares students to read and understand science text but also develops students' conceptual understanding so they can communicate new ideas. You might find it a challenge to prepare for vocabulary instruction. Four steps that help in this process are (1) identify learning goals, (2) develop a vocabulary list based on the goals, (3) determine required levels of understanding for the terms, and (4) select appropriate strategies (Barton and Jordan 2001, Urquhart and Frazee 2012). • One strategy that helps students with terms that don't require in-depth understanding is the Student VOC (Vocabulary) Strategy. Strategies that work better for terms that require deeper exploration include concept definition mapping, the Frayer model, and semantic feature analysis. These strategies can be used throughout a unit, with modifications as understanding grows (Barton and Jordan 2001, Urquhart and Frazee 2012). The Reading Educator website explores these strategies (see Resources below). • It is important to offer multiple opportunities to learn new terms in context, provide some instruction on the concepts prior to reading, help students connect an image to a term, and focus only on those terms critical to learning the new content (Barton and Jordan 2001, Urquhart and Frazee 2012).
Application Example	In a lesson on protein synthesis, a teacher found that her students did not fully understand the difference between DNA and RNA and some students did not understand the differences among DNA, genes and chromosomes. She used semantic feature analysis (i.e., using a grid to explore how sets of things are related to one another) with these sets of essential terms. However, there were also terms required by her school district that were not essential to students' conceptual understanding of the topic. She decided to use the Student VOC (Vocabulary) Strategy for these terms because in-depth understanding was not essential.
Technology Applications	• Inspiration software can be used for semantic mapping. • Visit *www.wordle.net* to generate word clouds from text that you provide. • VocabAhead (*www.vocabahead.com*) includes videos that explain words. • Visuwords (*www.visuwords.com*) is an online graphic dictionary and thesaurus that maps words and associated concepts.
Resources	• Teaching Reading in the Content Areas, 3rd edition (Urquhart and Frazee 2012) • Science portion of the Reading Educator website: *www.readingeducator.com/content/science/index.htm* • Vocabulary strategies portion of the Reading Education website: *www.readingeducator.com/strategies/vocabulary.htm* • AdLit. Org website (*www.adlit.org/strategy_library/*) describes classroom literacy strategies for grades 4–12. • Teaching Reading in Science: A Supplement to the Second Edition of Teaching • Reading in the Content Areas Teacher's Manual. (Barton and Jordan 2001) • *Overcoming Textbook Fatigue: 21st Century Tools to Revitalize Teaching and Learning* (Lent 2012)

Instructional Tool 3.4 (continued)

Informational Text Strategies
Informational text strategies help students better understand text structure, text coherence, and appropriateness for a particular audience and, as a result, have a positive impact on the learning of science.

The Research	• Reading and comprehension skills are important in science because of specialist language and grammatical features. This may require significant teacher support (Rangahau 2002). You should help students recognize the role of their prior knowledge and teach them how to use that knowledge when they learn science through reading (Barton and Jordan 2001, Urquhart and Frazee 2012). • Use of text in inquiry-based instruction is important because learning from text is a standard science practice and a way to promote both text comprehension and science instruction (Magnusson and Palinscar 2003). Hands-on activities stimulate student questioning, and the search for answers in textbooks or other science materials provides frames of reference to construct meaning from text (Nelson-Herber 1986). • Reading and inquiry both require that learners are aware of and use the appropriate discourse structures, can coordinate information across texts, and interpret multiple representations in texts (Magnusson and Palinscar 2003). Explicit instruction with students to both recognize and represent common text structures significantly improves student learning, as does familiarizing students with the different ways that information is presented in various text forms (Barton and Jordan 2001, Urquhart and Frazee 2012).
Connections to *NGSS* Practices	**QP** – In some cases, questions can arise as text is read. As questions are formulated, text is often referenced. Informational text strategies can help students clarify their thinking and formulation of questions. **I** – An investigation can arise from a student's reading of informational text. In some cases review of a current theory or model, described in text, can initiate thinking about further experimental design, or fieldwork. **AE** – Informational text is important to argumentation in several ways. A key way in which it is important is in making critical judgments about their own work and work of their peers, which requires the reading and evaluation of informational text. In addition, it is important to make evaluative judgments of science-related media reports. Likewise, analysis of text is important in engineering when competing ideas are compared in developing an initial design and later in the design process when analyzing strengths and weaknesses in a design (NRC 2012). **OEC** – Students need to read and use words (as well as tables, diagrams, graphs, and mathematical expressions) to obtain information and communicate understandings about science concepts, both in written and oral form. Even with grade level–appropriate reading skills, science is challenging to read for many reasons (NRC 2012). Informational text strategies improve students' abilities to obtain and evaluate information gained from text.

Instructional Tool 3.4 (continued)

Classroom Implications	• At the beginning of the year, walk students through a section of the textbook. They need to understand text presentation—that is, how the material is laid out; visual clues, such as illustrations and graphs; and textual clues such as headings, lists, and titles of visual elements (Vasquez, Comer, and Troutman 2010). Model for students how to predict text content based on these presentation clues. • Before the school year starts, you should assess your textbook for coherence if you are working with it for the first time. If there are limitations in the presentation, you need to pre-identify main ideas and then, as you work through the textbook with your class, make certain that students understand how concepts in paragraphs are related. Many science textbooks are not "user friendly" for students in elementary school. If that is the case, you may need to use alternative materials until appropriate textbooks are adopted (Barton and Jordan 2001, Urquhart and Frazee 2012). • The language used in books and other materials about science (i.e., science "text") is specific to science. Ways that science texts differ include different kinds of nouns and referent items (*generic* rather than *particular*), verb tense (use of the present tense instead of the past tense, which young children are used to from storybooks), and vocabulary (see Vocabulary Development, pp. 102–103). Science texts also include a range of images and designs (Varelas and Pappas 2006). • Text structures often used in science with which students should become familiar include the following: comparison/contrast, concept definition, description, generalization/principle, and process/cause and effect. Teachers should scaffold students' learning about each structure when students begin to learn a skill or concept and use words that will be part of the content focus (Barton and Jordan 2001, Urquhart and Frazee 2012). • Reading strategies that can be applied across disciplines in a classroom include discussion strategies (e.g., creative debate discussion webs [see *www.readingeducator. com/strategies/debate.htm*], Question-Answer Relationships (QAR), and Think Pair Share); active reading strategies (e.g., anticipation guides, KWL, and reciprocal teaching), and organization strategies (concept diagrams, graphic organizers, and two-column notes). Several of these strategies also help identify student preconceptions, which helps in teacher planning and allows students to grapple with their own thinking. Anticipation guides, the Directed Reading Thinking Activity (DR/TA), various graphic organizers, Group Summarizing, KWL, PLAN, Problematic Situation, Proposition/Support Outline, reciprocal teaching, and Think-Aloud are all effective strategies to help students make meaning from text. (Barton and Jordan 2001, Urquhart and Frazee 2012). Online resources outline many of these strategies (see Resources on p. 106).
Application Example	Students have completed an activity that shows effects of several variables on enzyme action, and they have developed explanations for what happened in the activity. The teacher then uses Pairs Read as a strategy. She provides pairs of students with a passage from the textbook or other resources accessible to readers in ninth grade. Each pair uses a different resource. She tells them that one student in each pair will be the coach and the other will be the reader. The reader reads the first paragraph aloud to the coach, and then the coach summarizes the paragraph. They then reverse roles. If the selection is more than two paragraphs, they continue to rotate being the reader and the coach. Each pair of students then summarizes what they read, and various pairs share their summaries with the class.

Instructional Tool 3.4 (continued)

Technology Applications	• Digital texts and tools are increasingly available.
	• Watch the YouTube video, "Using Screenchomp, Dropbox and the iPads for Think Alouds (*www.youtube.com/watch?v=w0Ja6rolsh0*)
Resources	• Teaching Reading in the Content Areas, 3rd edition (Urquhart and Frazee 2012)
	• AdLit. Org website (*www.adlit.org/strategy_library/*) describes classroom literacy strategies for grades 4–12.
	• Reading as a Strategic Activity on the Reading Literacy website: *www.readingeducator.com/strategies/index.htm*
	• NSTA annually publishes "Outstanding Science Trade Books for Students, K–12": *www.nsta.org/publications/ostb.*
	• *Teaching Reading in Science: A Supplement to the Second Edition of Teaching Reading in the Content Areas Teacher's Manual* (Barton and Jordan 2001, Urquhart and Frazee 2012)
	• *Strategies That Work: Teaching Comprehension for Understanding and Engagement* (Harvey and Goudvis 2007)
	• *The Comprehension ToolKit: Language and Lessons for Active Literacy* (Harvey and Goudvis 2011)
	• *Reading, Writing, & Inquiry in the Science Classroom Grades 6–12: Strategies to Improve Content Learning* (Chamberlain and Crane 2009) includes a chapter, Textbooks in the Science Classroom, that outlines strategies and activities to support science learning through informational text.

Narrative Text Strategies

Narrative text relates a series of events and includes both fiction and nonfiction. In science, narrative text is used to form *narratives of science* where scientists develop a claim supported by a series of data and *popularizing narratives of nature*, which are presented in story form rather than claim-data form. *Narrative text in which expository text is embedded* is typical of popular science to stimulate and hold the interest of the reader (Avraamidou and Osborne 2009). Narrative pedagogy includes teachers' traditional use of stories related to science concepts. Stories used in science can include stories in everyday contexts, those related to the history of science, and those drawn from different cultural views of the world. They can be used to develop understandings about the history of science, the nature of science, and science conceptual understandings. To help students learn about the history of science, teachers can use narratives such as biographies and manuscripts from historical studies. Narrative text strategies help students better understand narrative text structure and, as a result, have a positive impact on the learning of science.

Instructional Tool 3.4 (continued)

The Research	• Narrative is an engaging way to represent and communicate scientific information and can help make science approachable and meaningful to students.
	• Teaching and learning can both be improved through the use of narrative text in the content areas. Students demonstrate better retention and greater subject matter understanding since narratives help focus the readers' attention and build connections for students (Nathanson 2006).
	• Including science history and biography in class activities provides images of science and scientists in action. It also promotes interest, provides role models, and portrays a more accurate image of the NOS. It also draws students into the underlying science concepts (McKinney and Michalovic 2004).
	• Historical theories about photosynthesis in conjunction with the generative teaching model have been used to help students wrestle with the ideas of contemporary photosynthesis theory. That approach was found to be more effective than guided discovery experiences in which students participated in teacher-contrived experiments (Barker 1985, 1986).
Connections to *NGSS* Practices	• **E/S –** Narratives of science often include explanations where scientists develop a claim supported by data. Students might also be called upon to read narrative text and identify gaps or weaknesses in the explanatory accounts.
	• **OEC –** Students need to read and use words (as well as tables, diagrams, graphs, and mathematical expressions) to obtain information and communicate understandings about science concepts, both in written and oral form. Even with grade level–appropriate reading skills, science is challenging to read for many reasons (NRC 2012). Narrative text strategies improve students' abilities to obtain and evaluate information gained from text.
Classroom Implications	• Students need to learn the purpose, framework and methods to eliminate frustration when they read. They can then better follow the storyline and make predictions (Sejnost and Thiese 2010).
	• Narrative text comprehension strategies should be explicitly taught and students should practice a strategy until they understand and can apply it (Block and Pressely 2002). These comprehension strategies can be taught one at a time, but eventually students should develop a repertoire of strategies to use as appropriate (Pressley 2002, 2006).
	• Use graphic or semantic organizers to help students organize their concepts both during and after reading narrative text. It helps them see the relationships among facts, ideas, and concepts in the text and can also can help illustrate text structure (Eunice Kennedy Shriver National Institute of Child Health and Human Development 2010).
	• HOS infusion into lessons reinforces the multicultural nature of science and provides role models for students who might otherwise not consider the option of science as a career.
	• Narratives (telling stories) about the history of science help students understand the role of creativity in science as well as the complexity of science ideas.
Application Example	Students use MendelWeb for guided studies. They conduct genetic experiments and use the website as a guided review of Mendel's work. They then compare their results and reasoning with those of Mendel.
Technology Applications	Digital libraries are good sources for archived source documents. Examples include the National Science Digital Library (NSDL), the Library of Congress, the Chemical Heritage Foundation, the U.S. Patent Office, and the Alsos Digital Library for Nuclear Issues.

Instructional Tool 3.4 (continued)

Resources	• NSDL has an area called Primary Articles Learning Environment that includes Classic Articles in Context that has scientists comment on the impact of seminal scholarly works and Timely Teaching that features current articles that investigate high-profile contemporary and emerging problems. It also includes suggestions for incorporating primary literature into the classroom (*http://nsdl.org/sites/classic_articles/index3.html*). • AdLit. Org website (*www.adlit.org/strategy_library/*) describes classroom literacy strategies for grades 4–12. In particular, strategies to scaffold students' reading of narrative text can be found at *www.adlit.org/article/39884/*. • *Reading, Writing, & Inquiry in the Science Classroom Grades 6–12: Strategies to Improve Content Learning* (Chamberlain and Crane 2009) includes a chapter, Beyond the Textbook, that outlines ideas for the use of various genres in science instruction and lists numerous text and online resources. • See the history of science resources at the SciEd.ca website: *http://sci-ed.org/HOS.htm*. • See the history of science on the web: *www.ou.edu/cas/hsci/rel-site.htm*. • Go to MendelWeb at *www.mendelweb.org*.

Reflection Strategies

Reflection strategies develop students' metacognitive abilities and are important if students are to become effective readers. Specific strategies can be used to promote reflection on reading in science.

The Research	Reflection strategies that enhance metacognition related to reading in science help students (1) better understand science text, science reading, and science reading strategies; (2) enhance skills required to read science text and use science reading strategies; and (3) understand why and when to use particular strategies. These abilities help students (1) better understand science text, science reading, and science reading strategies; (2) improve the skills required to read science text and use science reading strategies; and (3) understand why and when to use particular strategies (Barton and Jordan 2001, Urquhart and Frazee 2012).
Connections to *NGSS* Practices	• **D** – Reflection strategies are important when gathering and looking for patterns and relationships among data, both by scientists and engineers. • **E/S** – Reflection is core to developing explanations since these explanations link observations or phenomena to scientific theory. Students must use reflection when developing explanations based on observations or models. Students must also use reflection when engaged in an engineering problem that requires them to develop and refine a design. • **OEC** – Students need to read and use words (as well as tables, diagrams, graphs, and mathematical expressions) to obtain information and communicate understandings about science concepts, both in written and oral form. Even with grade level–appropriate reading skills, science is challenging to read for many reasons (NRC 2012). Enhanced metacognitive abilities improve students' reading so they better understand science text.

Instructional Tool 3.4 (continued)

Classroom Implications	Reflective questioning, reflective writing, and discussion enhance students' metacognition as related to learning science (including reading science) and enhance students' understanding of science content. Various reflection strategies include creative debate; discussion webs; learning logs; Question-Answer Relationships (QAR); Questioning the Author (QtA); Role/Audience, Format/Topic (RAFT); and Scored Discussion (Barton and Jordan 2001, Urquhart and Frazee 2012). See Resources below for learning more about these strategies.
Application Example	Students read about global warming. The teacher uses a Discussion Web to facilitate class discussion. Students are asked, "Can we really control global warming?" Small groups record this question in the middle of a discussion web and then record as many "yes" or "no" reasons as they can, in a given amount of time, related to the question. These go on the right and left of the web. They discuss their responses and reach a consensus "yes" or "no" response, recording it in the conclusion box. Groups share responses and the teacher facilitates a discussion related to the question.
Technology Applications	Online journals and discussions can be implemented. Consider using a classroom blog where scientific discussions can occur.
Resources	• Teaching Reading in the Content Areas, 3rd edition (Urquhart and Frazee 2012) • *Teaching Reading in Science: A Supplement to the Second Edition of Teaching Reading in the Content Areas Teacher's Manual* (Barton and Jordan 2001)

Note: The abbreviations for practices referenced in the tool are:
- Asking questions and defining problems (QP)
- Developing and using models (M)
- Planning and carrying out investigations (I)
- Analyzing and interpreting data (D)
- Using mathematics and computational thinking (MCT)
- Constructing explanations and designing solutions (ES)
- Engaging in argument from evidence (AE)
- Obtaining, evaluating, and communicating information (OEC)

Instructional Tool 3.5

Sense-Making Approaches: Linguistic Representations—Speaking to Learn

Large- and Small-Group Discourse
Discourse is used to make sense of science learning experiences. The teacher and students explore ideas, pose questions, and listen to multiple points of view to establish understanding (Mortimer and Scott 2003). Discourse benefits student learning, teacher-student rapport, equity, expectations, and formative assessment (ASCD 2008).

The Research	• Scientific practices are language intensive so require *all* students to engage in discourse. Effective engagement in discourse can enhance both the understanding of science and the development of language proficiency for English language learners, students with language processing disabilities, students with limited literacy development, and students who commonly use non-Standard English (NRC 2012).
	• Students can engage freely in focused talk in science without being taught rules, conventions, structures, or vocabulary, especially if realistic and authentic problems are the focus of the discussions. You can develop focus using strategies that present authentic problems, focus on scientific issues, and lead to productive follow-up activities. Better discussions occur when students see several plausible viewpoints, resulting in students' own theory determination. When only one viewpoint is presented, there is no reason to have a discussion. Alternative viewpoints are effectively presented via concept cartoons or puppets (Keogh and Naylor 2007). Discussion based on reading materials such as magazine articles, online commentary, and books also enriches the discourse. Examination and discussion of text promotes skills of analytical reading, careful listening, citing evidence, respectful disagreement, and open-mindedness (Hale and City 2006).
	• "Accountable talk" deepens conversation and understanding of the topic studied (ASCD 2008). Norms and skills for "accountable talk" must be explicitly taught.
	• Large- and small-group discourse promotes learning about concepts, metacognition, and the nature of science; it also develops positive attitudes. Whole-class discussion works best in the context of an activity because it provides shared experiences for students (Hipkins et al. 2002). Small-group discourse lets students build on one another's ideas and generate explanations. Whole-class discourse demands more clarity and explanatory power on the part of students. If students understand the purposes of the two types of discourse, they will better develop shared understandings (Woodruff and Meyer 1997). There should be a balance between whole-class and small-group discussions. Notice that they reflect the two types of scientific discourse, one within the laboratory and one among laboratories working on similar research (Cartier 2000; Hipkins et al. 2002).

Note: See legend at bottom of page 118.

Instructional Tool 3.5 (continued)

Connections to *NGSS* Practices	• **QP** – Various perspectives and initial explanations that arise from students during discourse can better define and refine testable questions or engineering problems. Probing questions also help identify "the premises of an argument, request further elaboration, refine a research question or engineering problem, or challenge the interpretation of a data set (NRC 2012, p. 55)." • **E/S** – Explanations require scrutiny in a scientific community. Likewise students should develop their own explanations and they should be open to scrutiny, which is facilitated through discourse. Various explanations from students should be discussed and consensus explanations developed that best fit the data. • **AE** – Discourse is central to argumentation, whether it is to argue for the best possible explanation, experimental design, or interpretation of data as well as to select the most appropriate data analysis techniques. It would be difficult to find the best solution to an engineering problem without discourse in support of argumentation. • **OEC** – Though not all communication is done orally, oral communication is a fundamental practice in science. This communication must clarify thinking and justify arguments.
Classroom Implications	• Purposes of discussion during inquiry include eliciting ideas, planning investigations, and developing understanding. You can use surveys or questioning to assess students' pre-instructional ideas (Hartman and Glasgow 2002). Let your students talk in their own registers (not formally). Give them time and space to work productively without intruding (Keogh and Naylor 2007). • You can encourage deeper discussion and investigation with student-generated questions, open-ended questions, student choice of inquiry topics, and more time for student research and exploration (ASCD 2008). Use strategies like card sorts, concept cartoons, odd one out, and graphic organizers to effectively engage students and promote student self-motivation and self-sustaining conversation. Concept cartoons can provide a safe environment and focus the discussion during argumentation (Keogh and Naylor 1999, 2007). (See Instructional Tool 3.8, p. xx, for further information on concept cartoons.) • What is your role during small-group discussions? You should provide students with carefully selected materials to focus discussion and decision making, require them to operate at high cognitive levels, ensure that everyone has the opportunity to speak, and require student products that result from discussion (Hipkins et al. 2002). Avoid individual student worksheets because they often inhibit student talk; students tend to complete the worksheets instead of discussing (Keogh and Naylor 2007). Your questioning is essential when addressing a topic that requires thought and deduction. Use cognitive, speculative, affective, and management questions, each asked at various levels (Hartman and Glasgow 2002). Avoid questions that elicit factual recall because such questions bring discussion to a halt. It is more productive to use questions that probe thinking, build problem-solving steps, encourage participation, and sequence conversation (ASCD 2008; Hipkins et al. 2002). • Subsidiary questions and elaborative feedback are critical to build productive discourse, and evaluative or corrective feedback quickly shuts down discourse (Lee 2012). Further information about feedback is included in Chapter 4 of this book.

Instructional Tool 3.5 (continued)

<table>
<tr>
<td></td>
<td>

- When forming groups, keep in mind that role allocation in mixed-gender groups promotes more conceptual dialogue; role allocation is less necessary in same-gender group. (Hipkins et al. 2002). You can promote accountable talk by pushing students to (1) clarify and explain, (2) require justification for proposals, (3) recognize and challenge misconceptions, (4) demand evidence from peers for claims and arguments, and (5) interpret and use one another's statements (ASCD 2008).

- Some prompts you might teach students to use during accountable talk include: I agree/disagree with _____ because ..., One thing I noticed ...., What do you think?, I'm struggling to understand that idea, my idea about that is a little different, it says in this resource that ..., etc. Modeling these types of statements and encouraging student use of them should increase accountable talk.

- The informal cooperative learning structures should be taught so that students gain understandings and abilities in sharing and supporting their ideas. Remember that everyone needs to contribute yet individual accountability is important (Johnson and Johnson 1986, Kagan 1989).

</td>
</tr>
<tr>
<td>Application Example</td>
<td>On a rotating basis, all students in a class are assigned responsibilities for facilitating both large- and small-group discussions. The discussions should have a variety of goals and outcomes and occur at multiple points in the learning cycle.</td>
</tr>
<tr>
<td>Technology Applications</td>
<td>Voice annotations in PowerPoint are simple applications. More powerful and collaborative options are available with VoiceThread (http://voicethread.com).</td>
</tr>
<tr>
<td>Resources</td>
<td>

- *Quality Questioning: Research-Based Practice to Engage Every Learner* (Walsh and Sattes 2005)

- *Open-Ended Questioning: A Handbook for Educators* (Freedman 1994)

- *The Teacher's Guide to Leading Student-Centered Discussions: Talking About Texts in the Classroom* (Hale and City 2006)

- *Meaning Making in Secondary Science Classrooms* (Mortimer and Scott 2003)

- The "Tools for Ambitious Science Teaching" website (*http://tools4teachingscience.org*) provides a guide on managing talk in the classroom as well as a variety of discourse tools

</td>
</tr>
</table>

Instructional Tool 3.5 (continued)

Student Questioning Student questioning includes not only questioning during small- and large-group discussions but also questioning as an essential feature of inquiry.	
The Research	• Formulating questions is the key to metacognitive knowledge, including strategic knowledge, knowledge about cognitive tasks, and self-knowledge (Walsh and Sattes 2005). Student-generated questions, open-ended questions, student choice of inquiry topics, and more time for student research and exploration encourage deeper discussion and investigation (ASCD 2008). Question generation by students is linked to retention of content, higher conceptual achievement, and improvement of problem-solving abilities (Colbert, Olson, and Clough 2007). • "If teachers merely elicit and run with students questions without framing overarching curricular goals and essential questions to support them, then there can be no guaranteed and viable curriculum (McTighe and Wiggins 2013)." However, helping them become aware of what constitutes a good question makes them better learners and improves their questioning abilities (Hartman and Glasgow 2002). • Teachers can improve student questioning by encouraging students to formulate their own questions and by responding positively to students' spontaneous questions (Harlen 2001). They can also improve student questioning by providing examples of testable questions (Krajcek et al. 1998), materials that stimulate questions (Chin et al. 2002; Harlen 2001), opportunities to explore information related to their questions (Krajcek et al. 1998), and feedback and the opportunity to change factual questions into testable questions or to generate new questions (Harlen 2001; Krajcek et al. 1998). • Question posing (question generation and evaluation) is essential to scientific inquiry. If we want students to pose questions, we must let them interact with phenomena and create situations where question posing is required. To create an environment that encourages question posing, we need to establish a "question focus," not "the answer focus" typical of traditional instruction that concentrates on the products of inquiry (Milne 2008). • Web-based discussion boards have been effective at increasing student-generated questions and interactions about content (Colbert, Olson, and Clough 2007).
Connections to *NGSS* Practices	• **QP** – The definition and refinement of testable questions or problems is central to science and engineering. Probing questions also help identify "the premises of an argument, request further elaboration, refine a research question or engineering problem, or challenge the interpretation of a data set (NRC 2012, p. 55)." • **I** – Student questioning enhances all stages of investigation from formulating an initial question, determining appropriate data and tools, and identifying relevant variables and controls. Questioning during discourse leads to more thoughtful planning for investigation. • **E/S** – Questioning helps refine explanation or solution. • **AE** – Questioning helps identify the premises of an argument. • **OEC** – Questioning is core to evaluating information.

Instructional Tool 3.5 *(continued)*

Classroom Implications	• Provide your students with models of good questions and discuss effective questioning strategies. Also give them opportunities to practice questioning and provide them with feedback on their questions (Hartman and Glasgow 2002; Walsh and Sattes 2005). Consider using reciprocal teaching and paired-problem-solving strategies because they require students to ask and answer questions and thus promote questioning abilities and metacogntive knowledge (Walsh and Sattes 2005). • Posed questions are not all investigable. Some are too broad to investigate and some are too narrow and can be quickly answered by using reference materials (Milne 2008). "Why" questions are common but often not investigable; "how" questions, on the other hand, lend themselves to scientific inquiry (NRC 2000). • As you help students learn how to develop scientific questions, consider a skill or process to teach this. For example, for a question to be considered scientific it must address these four criteria: there should be a way to test it, it should be a question about which predictions can be made, there should be an objective way to measure whether or not the prediction was met, and knowing the answer should contribute important information to our understanding of the natural world (National Institutes of Health 1998). • A questioning tree that helps students generate engaging and investigable questions is found at the Southwest Center for Education and the Natural Environment (SCENE) website (*http://scene.asu.edu/habitat/inquiry.html#4*). Questions are sorted first by unanswerable questions (why) and answerable questions (how, what, when, who, and which), then by interesting and uninteresting, next by comparative and non-comparative, and finally by manipulative or observational (Southwest Center for Education and the Natural Environment 2004). • Question-generating activities engage students in phenomena and ask that they generate a list of possible questions that are derived from their prior knowledge and the experiences. Possible activities include laboratory explorations, demonstrations, reading scientific studies, and making observations in the community. Regardless of what the teacher uses as a starter, the goal is to generate a range of questions.
Application Example	Have students interact with materials related to the content you are teaching. Examples could range from the study of osmosis (various stations with specific activities to complete) to the observation of animals. Set up organized stations at which students can complete set interactions with materials designed to explore various aspects of the content. Have students observe and generate questions. Then, work with students to identify which questions are investigable, which are not, and which might be changed to make them investigable (e.g., they can change why questions into how questions, or broad questions can be narrowed and made easier to control).

Instructional Tool 3.5 (continued)

Technology Applications	• Inspiration software can be used to brainstorm/generate questions related to content investigations. This activity, coupled with analysis of each question using the questioning tree (see Resources below), can help students quickly generate questions and then reduce the questions to those that are engaging and investigable. In addition, those questions that are informative and perhaps foundational to the content, but not investigable, might be answered via a web search. This can be done individually, in small groups, or as a whole class. • E-mail/webconferencing/blogging with scientists and other students could be used to generate questions. • Microblogging, like Twitter (*http://twitter.com*), can be used to capture students' questions.
Resources	• The SCENE website (*http://scene.asu.edu/habitat/inquiry.html#4*) offers a nice overview of a questioning cycle in inquiry-based science and a questioning tree (Figure 3 on that site) that guides thinking about questions that are both engaging and investigable. • Resources that should help support quality questioning, for teacher use and to model for students, are *Essential Questions: Opening Doors to Student Understanding* (McTighe and Wiggins 2013), *Quality Questioning: Research-Based Practice to Engage Every Learner* (Walsh and Sattes 2005), and *Open-Ended Questioning: A Handbook for Educators* (Freedman 1994). Walsh and Sattes include a chapter with specific suggestions on how to teach students to generate questions. • A helpful reference that includes a process students can use to develop scientific questions for student designed investigations is *Students and Research: Practical Strategies for Science Classrooms and Competitions, 4th* Ed. (Cothron, Giese, and Rezba 2000).

Communication

Discourse and presentation in the science classroom require the ability to communicate ideas, clearly share accurate information and listen with understanding. While there are general communication skills related to both speaking and listening, there are also skills specific to science. These include among others an understanding of science ideas and vocabulary; correct interpretation of terms like "If…, then…," "causes," "correlated with," etc.; and interpretation and communication of tabular and graphic data (AAAS 1990). Communication is one of the five features of inquiry. As is the case with scientists, students need to be able to communicate scientific and specifically inquiry-based information orally and/or in writing. This can also include graphical representations, written explanations, and oral presentations. Communication is a core activity of scientists and is represented in the NGSS documents as part of the Science and Engineering Practices: Obtaining, Evaluating and Communicating Information.

Instructional Tool 3.5 (continued)

The Research	• Within the NSES from 1996, students engaged in inquiry at the high school level should be able to communicate and defend a scientific argument (NRC 1996). According to the document, "Results of scientific inquiry—new knowledge and methods—emerge from different types of investigations and public communication among scientists. In communicating and defending the results of scientific inquiry, arguments must be logical and demonstrate connections between natural phenomena, investigations, and the historical body of scientific knowledge. In addition, the methods and procedures that scientists used to obtain evidence must be clearly reported to enhance opportunities for further investigation (p. 176)." • In the Common Core ELA standards for speaking and listening Anchor 1 asks students to "prepare for and participate effectively in a range of conversations and collaborations with diverse partners, building on others' ideas and expressing their own clearly and persuasively." To meet this anchor, students must pose specific questions by making comments that contribute to the discussion; respond to questions that relate to the current discussion; and pose and respond to questions that probe reasoning and evidence. (NGSS Lead States 2013, Appendix p. 160). This standard speaks directly to the importance of discussion about science questions that lead to solutions and explanations. • In the Common Core ELA standards for speaking and listening Anchor 3 asks students to evaluate a speaker's point of view, reasoning, and use of evidence to evaluate the soundness of the reasoning and the sufficiency of the evidence and identifying when irrelevant evidence is introduced. This aligns with the SEPs as students learn how to analyze and critique the explanation provided from the perspective of the evidence provided and the reasoning that was communicated (NGSS Lead States 2013, Appendix p. 160). • In a study of science communications skills and their relationship to student performance, students were taught the following skills: information retrieval; scientific reading and writing; listening and observing; data representation; and knowledge presentation. Results indicated that student achievement increased in all cases, even when only one strategy was taught as compared to the control group. What was clear was that students did not learn these strategies spontaneously and systematic teaching of the skills does make a difference (Spektor-Levy, Eylon, and Scherz 2009).
Connections to NGSS Practices	• **AE** – Speaking and listening skills are central to both sharing ideas and hearing alternative ideas. Argumentation would be impossible without these skills. • **OEC** – Evaluating information is enhanced through quality discourse, dependent on both speaking and listening skills. However, the strong tie to OEC is oral communication and is certainly central to describing observations, clarifying thinking, and justifying arguments. Engineers must also communicate ideas and exchange information in developing solutions to problems.

Instructional Tool 3.5 (continued)

Classroom Implications	• When students share their explanations in the classroom, it provides opportunities for the other students to know the question being investigated, the procedures followed, the evidence that was gathered, and the proposed claims and explanations. Students can then review the information that was communicated and discuss alternate explanations. Without this communication, students would not have the chance to review the findings, analyze the evidence, and recognize faulty reasoning and where the explanations go beyond the evidence. Students, just as scientists do, can then generate revised explanations that are based on the evidence. With communication and discussion, students can resolve differences in explanations and develop scientifically aligned explanations. This is a critical part of a conceptual change process where students' initial ideas must be challenged.
	• Teaching procedures for scientific argumentation will help students incorporate this science practice. The implication here is that we should be more about students engaging in communication and less on our "teaching"—telling students—about science ideas. As a result our classrooms will be less about science teaching and more about science communication.
	• Students need to be prepared to participate effectively in collaborative conversations in a variety of settings. They need to be able to communicate persuasively in an environment that makes it safe for students to disagree with one another without reprisal. Sharing personal "theories," evidence, and explanations is a skill that is essential to develop understanding of the science concepts that are the focus of your lessons. In particular with the new vision of learning described in the *NGSS*, students need to regularly engage in this process of making observations based on investigable questions, analyze the findings and determine the useful evidence that can then lead to communication that is part of the process of making student thinking visible.
	• Teaching communication strategies should be part of instruction. Students need 20–24 opportunities to practice procedures before they can use them with automaticity (Dean et al. 2012). This means that teaching students communications strategies, modeling them, reinforcing them and discussing how they are working in the classroom with our students should be part of student learning experiences in multiple settings with a variety of contexts.
Application Example	When students analyze large data sets, they may provide statistical explanations to show their reasoning (Duschl 2000). In an example from the recent research study of novice teachers (Windschitl, Thompson, and Braaten 2008), a teacher asked students in her biology class to examine two data sets, (1) a map of Africa showing allele frequencies of sickle cell anemia and (2) a map of Africa showing epidemiology data about malaria. Students were asked to study and analyze the relationship between the two diseases. At the end of their study they were expected to develop an explanation of the genetic existence of a harmful gene in the population. The reasoning should include a discussion of probabilities that cite evidence from the data sets and use their understanding of inheritance patterns and evolutionary theory to support their claims. Students would communicate their explanations both orally and with graphic and mathematical representations. The other students would ask questions and critique the evidence based on explanations and propose alternative explanations if the evidence extended beyond a reasonable interpretation. As the teacher you would need to help students by modeling how they would use statistics to support their reasoning. In this case there may also be mediating factors in the environment that occur in natural systems so that explanations are not always as simple as data might suggest. With discussion and communication students can come to a broader understanding of complex science phenomena.

Instructional Tool 3.5 (continued)

Technology Applications	• E-mail/webconferencing/blogging with scientists and other students could be used to communicate outside of the classroom. • Microblogging, like Twitter (*http://twitter.com*), can be used to capture students' discussion threads. • Interactive science notebooks that use technology can be shared among students and used for communication of findings and explanations based on analysis and reasoning. • The sites in the resources can be used to generate graphical and pictorial representations that can be used as part of communication methods to be provided in addition to oral and written materials.
Resources	• Find graphic organizer generators at *www.teach-nology.com/web_tools/graphic_org* • A good resource on all types of graphic organizers is *A Field Guide to Using Visual Tools* (Hyerle 2000). • The "Tools for Ambitious Science Teaching" website (*http://tools4teachingscience.org*) provides a guide on managing talk in the classroom as well as a variety of discourse tools • Check out the discourse tools and discourse primer at *http://tools4teachingscience.org/tools/discourse_tools/* • Read the Arguing to Learn article from AAAS' *Science* magazine, *www.sciencemag.org/content/328/5977/463.abstract* • Read the article, Supporting and Promoting Argumentation Discourse in Science Education (Duschl and Osborne 2008) in Studies in Science Education *www.sciencemag.org/content/328/5977/463.abstract*

Note: The abbreviations for practices referenced in the tool are:
- Asking questions and defining problems (QP)
- Developing and using models (M)
- Planning and carrying out investigations (I)
- Analyzing and interpreting data (D)
- Using mathematics and computational thinking (MCT)
- Constructing explanations and designing solutions (ES)
- Engaging in argument from evidence (AE)
- Obtaining, evaluating, and communicating information (OEC)

Instructional Tool 3.6

Sense-Making Approaches: Nonlinguistic Representations—Six Kinds of Models

Mathematical Models	
Mathematical models can be used to model events, objects, and relationships. Using mathematical models includes learning to abstractly represent things, logically manipulate the abstract representations, interpret the results, and determine the model's appropriateness (AAAS 2001a). Mathematical models come in various forms that include formulas, equations, figures, graphs, and tabular data models (Gilbert 2011, Gilbert and Ireton 2003).	
The Research	• Students of all ages often interpret graphs as literal pictures rather than symbolic representations, but little is known about how graphic skills are learned or about how graph production is related to graph interpretation. It is known than MBLs (microcomputer-based laboratories) improve students' abilities to interpret graphs (AAAS 2001a). • Graphic representations can help students discover and communicate patterns not seen as easily in tables or equations (AAAS 2001a). • Students often miss the more global features of graphs and instead pay attention to details. Examples include cases where they confound slope—interpreting it as maximum and minimum rather than a measure of rate and interpreting the graph point by point rather than considering maximum and minimum values, intervals over which a function changes or levels off, or rates of change (AAAS 2001a).
***NGSS* Connections to Practices**	• **M** – Mathematical models serve as functional analogs that can be used to develop explanations, generate data, make predictions, and communicate ideas. • **MCT** – The effective use of mathematical models depends upon and promotes mathematical and computational thinking. • **E/S** – Explanations and solutions work hand-in-hand with models, since explanations are both built from models and refine existing models.
Classroom Implications	• Mathematical models have a structure and function through analogy, but all require that students apply rules of interpretation. For tabular data models you should stress with your students that each table element corresponds to some element of the target, and you should work with your students so they can map models to targets. When working with formulas and equations, you should also make certain that students understand the symbols and placements of the models rather than just plug in numbers, which leads to rote learning (Gilbert 2011). • You should regularly and frequently involve your students in creating graphs, and you should do so purposefully as models. It is just as important to have them explain their models as to make them so that you can see that they understand what their abstract model—a graph—represents in concrete terms (Gilbert and Ireton 2003). Gilbert (2011) suggests that for students to clearly understand the elements of the model they should: (1) precisely explain the model in words and ask questions about elements they do not understand, (2) mentally visualize the model and make concrete representations when possible, and (3) work concretely with the model. • Students should be adequately exposed to various types of relationships demonstrated in graphs that include maximums and minimums, direct proportions, and inverses, so that it is more habitual for them to consider how the quantities are related. In high school, these relationships should include cases not only to the amount of some other quantity but also to its rate of change (AAAS 2001a).

Note: See legend on bottom of page 126.

Instructional Tool 3.6 (continued)

Application Example	Your students are studying enzyme action, comparing rates of reaction at two different temperatures. Have them build a data table (a tabular data model) to summarize collected data. Have students explain the various elements of the table and relationships to the target system. Have them make predictions and build explanations based on the model, and discuss its limitations and precision to represent the studied system.
Technology Applications	Microcomputer-based laboratories (MBLs) let students conduct experiments and generate graphs representing results, spending their time on what the representation means rather than on the creation of the representation itself.
Resources	• *Math: Stop Faking It!* (Robertson 2006) • The National Library of Virtual Manipulatives (*http://nlvm.usu.edu/en/nav/vLibrary.html*)
Physical Models Physical models are concrete models that are either two-dimensional (such as diagrams found in textbooks) or three-dimensional (such as working models and scaled models). When students explain their mental models in public, those models become "expressed models."	
The Research	• Generating a concrete representation creates an "image" of the knowledge in a student's mind. Physical models, which are one form of concrete representations, enhance nonlinguistic representations and students' understanding of content (Dean, Hubbell, Pitler and Stone 2012). • If conceptual and metacognitive learning are linked, students' attitudes as they work with physical models improve (Hipkins et al. 2002). When students develop their own models, they are likely to see connections between ideas; student-developed models also help teachers recognize gaps in student understanding. They can be shared and critiqued, as well as changed, as students learn more (Windschitl 2008). • The incorrect use of models can lead to misconceptions. Textbook diagrams can be misleading because they depict in two dimensions something that is actually three-dimensional (Andersson 1990). Furthermore, students tend to see models as concrete mini-representations of larger concrete objects, not understanding the abstract concepts they represent (Grosslight et al. 1991).
NGSS Connections to Practices	• **M** – Physical models serve as structural analogs that can be used to develop explanations, generate data, make predictions, and communicate ideas. • **E/S** – Explanations and solutions work hand-in-hand with models, since explanations are both built from models and refine existing models.
Classroom Implications	• Student practice in construction, critique, and use of their own models enhances conceptual development (Hipkins et al. 2002). It is important to have students compare various models to the actual concept they represent, clarifying the similarities and differences between the model and the target concept (Gilbert and Ireton 2003). • Campell, Neilson and Oh 2013) suggest these tips for modeling instruction. Identify a modeling focus that (1) can lead to many different student investigations, (2) has the potential to be explained in multiple ways, and (3) elicits students' prior knowledge. They use an example of rockets, but these criteria are helpful for many concepts.

Instructional Tool 3.6 (continued)

Application Example	Bottle Biology TerrAqua Columns (Ingram 1993) are one of many physical models for ecosystems. Use these models in the classroom prior to and/or in conjunction with fieldwork. Be certain to discuss with students the strengths and deficiencies of the models as compared to actual ecosystems.
Technology Applications	3-D printing affords students the opportunity to construct their own physical models that represent concepts in science or serve as prototypes in engineering.
Resources	• *Understanding Models in Earth and Space Science* (Gilbert and Ireton 2003) • *Models-Based Science Teaching* (Gilbert 2011)
Verbal Models: Analogies Creating analogies as a process identifies relationships between pairs of concepts, identifying relationships between relationships (Dean, Hubbell, Pitler and Stone 2012).	
The Research	• Creating analogies is an effective way to identify similarities and differences. They help students see that things that seem to be dissimilar are also similar in some ways. This insight increases understanding of new information and improves student achievement (Dean, Hubbell, Pitler and Stone 2012). Using analogies may also be motivational because they tend to provoke student interest (Cawelti 1999). • Analogies help students construct more accurate conceptions of complex ideas (Hartman and Glasgow 2002), especially when students have alternative conceptions (Cawelti 1999). They help familiarize students with concepts that are outside their previous experiences (Cawelti 1999). • Using multiple analogies in a bridging sequence can help students make sense of initially counterintuitive ideas (Cawelti 1999).
NGSS Connections to Practices	**M** – Analogies serve as functional analogs that can be used to develop explanations and communicate ideas. **E/S** – Explanations and solutions work hand-in-hand with models, since explanations are both built from models and refine existing models.
Classroom Implications	• Analogies can be effective models but can also lead the learner astray so make certain that you include a critical analysis of the analogy in your lessons. • You can use analogy creation to help students identify similarities and differences, a prime strategy to improve achievement. Creating analogies is effective both when you explicitly guide analogy creation and when students independently create their own analogies. Direct instruction is more effective if there are specific similarities and differences on which you want your students to focus. On the other hand, if divergence in student thinking is your goal, have students work independently (Dean, Hubbell, Pitler and Stone 2012). • Finding good analogies to develop conceptual understanding is more difficult when relationships are abstract. You can make ideas more concrete by helping your students form concrete mental images based on personal life experiences and then having them create their own analogies (Hartman and Glasgow 2002).

Instructional Tool 3.6 (continued)

	• To be effective, analogies must be familiar to students so they can determine if the features and functions of the analogies are congruent with the targeted concept. Spend time with students discussing the similarities and differences between the analogy and the target. For your students to fully understand the effectiveness of an analogy, it might be helpful to compare multiple analogies to a single learning target (Cawelti 1999). • Another reason that the framework used for an analogy has to be familiar to students is that, otherwise, analogies can be biased socially, experientially, or culturally (e.g., the analogy of cell to city certainly makes more sense to students who are city dwellers) (Hartman and Glasgow 2002). • Discussion of analogies helps students build understanding (Cawelti 1999). You can provide students with an analogy to a concept and have them discuss its relevance and limitations (Hartman and Glasgow 2002).
Application Example	Dean, Hubbell, Pitler and Stone (2012) share a teacher-directed analogy that gives structure to students as they learn to use analogies: "thermometer is to temperature as odometer is to distance" (p. 26). They then suggest that the teacher ask the students to explain how the two relationships are similar. Try this in your classroom. After helping students with several additional analogies, have students use an analog graphic organizer to develop analogies related to the current topic of study in your classroom. Make sure to discuss the strengths and weaknesses of developed analogies. (*Note:* Analog organizers can be found online. One example is at TeacherVision [*www. teachervision.fen.com/graphic-organizers/printable/48386.html*]. For more, google "graphic organizer and analogies.")
Technology Applications	Elaborate analogies can be built using various combinations of text, audio, video, animation, interactivity, and hyperlinks.
Resources	• *Using Analogies in Middle and Secondary Science Classrooms: The FAR Guide–An Interesting Way to Teach with Analogies* (Harrison and Coll 2008) • *Models-Based Science Teaching* (Gilbert 2011) • *Teaching-with-Analogies Model* (*www.coe.uga.edu/twa/PDF/Glynn_2007_article. pdf*) • *Metaphors and Analogies: Power Tools for Teaching Any Subject* (Wormeli 2009); online at *http://pwoessner.com/2009/12/17/ metaphors-and-analogies-power-tools-for-teaching-any-subject*
Verbal Models: Metaphors Creating metaphors is a process that identifies a general or basic pattern in a topic and finds another topic that seems different but has the same general pattern. The relationship between the two items in the metaphor is abstract (Dean, Hubbell, Pitler and Stone 2012).	
The Research	Creating metaphors is an effective way to identify similarities and differences, a strategy shown to be effective at increasing student achievement (Dean, Hubbell, Pitler and Stone 2012).
***NGSS* Connections to Practices**	• **M** – Metaphors serve as functional analogs that can be used to develop explanations and communicate ideas. • **E/S** – Explanations and solutions work hand-in-hands with models, since explanations are both built from models and refine existing models.

Instructional Tool 3.6 (continued)

Classroom Implications	• Metaphors can be effective models but can also lead the learner astray so make certain that you include a critical analysis of the analogy in your lessons.
	• You can use metaphor creation to help students identify similarities and differences, a prime strategy to improve achievement. Metaphors are effective both when you explicitly guide metaphor creation and when students independently create them. Direct instruction is more effective if there are specific similarities and differences on which you want your students to focus. On the other hand, if divergence in student thinking is your goal, have students work independently (Dean, Hubbell, Pitler and Stone 2012).
	• When you direct the creation of the metaphor, provide the first element and the abstract relationship. That will scaffold the student's construction of the metaphor. Once students become familiar with these abstract relationships, you can give students the first portion of the metaphor and let them develop the second component as well as identify the relationship (Dean, Hubbell, Pitler and Stone 2012).
	• Because the relationship in a metaphor is abstract, it is important that the instructional strategies you use that involve metaphors address this abstract relationship (Dean, Hubbell, Pitler and Stone 2012).
Application Example	When studying how organisms get the energy they need to live, use various metaphors to develop understanding. Examples might be comparing an organism to a factory or to a car. Use multiple examples with your students and have them identify the strengths and weaknesses for each set of comparisons.
Technology Applications	If the metaphor is supported by images, standard graphics programs can be used.
Resources	• *Understanding Models in Earth and Space Science* (Gilbert and Ireton 2003) • *Models-Based Science Teaching* (Gilbert 2011) • *Metaphors and Analogies: Power Tools for Teaching Any Subject* (Wormeli 2009); online at *http://pwoessner.com/2009/12/17/metaphors-and-analogies-power-tools-for-teaching-any-subject*

Visual Models: Graphs, Pictures, and Diagrams

Visual models include images of actual objects (photographs) or graphics of objects, graphs, and or other two-dimensional representations of ideas or data. Different visual models are more or less effective for different purposes and yield different understandings (e.g., photographs to capture images and graphs to display relationships).

The Research	Graphic models can be used to determine students' preconceptions. Presenting a graph, table, or figure and asking students to describe/interpret the image allows you to determine what they know and/or misunderstand about the represented ideas. This can be done individually or in small groups (Wright and Bilica 2007). If you have your students share and critique their developed models, it will help them see connections between ideas and help you recognize gaps and changes in student understanding (Windschitl 2008).

Instructional Tool 3.6 *(continued)*

NGSS Connections to Practices	• **M** – Visual models serve as functional analogs that can be used to develop explanations, generate data, make predictions, and communicate ideas. • **E/S** – Explanations and solutions work hand-in-hand with models, since explanations are both built from models and refine existing models.
Classroom Implications	• Diagrammatic and pictorial models help students visualize whole systems and require that students interact with them as they interpret the model. Work with students to help them build these models, analyze them and create networks among them, as this helps them develop their mental models of the concepts (Gilbert 2011). • Give students multiple opportunities to develop figures, graphs, and charts during experimentation because they are part of, and help explicate, students' mental models. When students share their models in small groups, alternative explanations will be heard, allowing students to confront their preconceptions and those of their peers. • *Before-during-after* drawings can be completed at the beginning of a unit of study and then again later on. They are composed of three parts that demonstrate what happens before, during, and after an event or process. As study continues, the drawings can be modified or redrawn. In biology, you can ask students to draw what they would see if they had microscopic eyes. A whole-class consensus model can also be generated. (Windschitl and Thompson 2013). • Windschitl and Thomson (2013) also suggest using sticky notes and sentence frames when making whole-class consensus models, having students add comments in various categories like adding ideas, revising ideas, or posing questions. They suggest each category is represented on a different color sticky note. • Passmore et al (2013) suggest the following hints to anchor instruction and discourse: (1) place the cognitive load on your students, requiring them to always be engaged with a question, problem, investigation, model or explanation, (2) use models that have explanatory power, (3) keep a public, explicit record of the model in the classroom, (4) clearly distinguish between reasoning about and reasoning with a model.
Application Example	For almost any topic, you can display pre-instructional graphic models and ask your students to explain them in their science notebooks. As the lesson progresses, have them revisit the models and explain again. You can also show a picture (e.g., a cell membrane with various ions on either side) and ask students to explain what they think will happen and provide their reasoning for the explanation.
Technology Applications	• Google SketchUp (*http://sketchup.google.com*) and Dabbleboard (*www.dabbleboard.com*) can be used to develop visual models and lets students work collaboratively on these models. • Free graphing software is available, including *http://nces.ed.gov/nceskids/createagraph*. • The Computations Science Education Reference Desk (*www.shodor.org/refdesk*) is a resource for free tools for creating graphs, calculators that plot changes between dependent variables, lots of simulations and computational models, and software for creating computational models.

Instructional Tool 3.6 (continued)

Resources	*Models-Based Science Teaching* (Gilbert 2011) *The Modeling Toolkit* (Windschitl and Thompson 2013) is an article in the 2013 issue of *The Science Teacher* and describes various ways in which to work with pictures and diagrams as models, including those explained above and more. *Making Sense of Natural Selection: Developing and Using the Natural Selection Model as an Anchor for Practice and Content* is an article in the 2013 issue of *The Science Teacher* that nicely outlines ways to work with students on modeling.
Dynamic Models Dynamic models are visualization and analysis tools that help students detect patterns and understand data. Examples are simulations, computer-based models, geographical information systems, and animations.	
The Research	Computer simulations can enhance students' conceptual understandings, as well as improve achievement with complex concepts, more quickly than traditional instruction (Cawelti 1999). They are helpful when instruction involves scientific models that are difficult or impossible to observe and can simplify complex systems. Simulations are most effective when used by students individually or in small groups, resulting in better conceptual understanding. Use of simulations also appears to increase problem-solving and process skills (Cawelti 1999).
NGSS Connections to Practices	• **M** – Dynamic models serve as functional analogs that can be used to develop explanations, generate data, make predictions, and communicate ideas. • **I** – Investigations that are difficult or impossible to conduct in the laboratory can be completed using simulations. • **D** – Many dynamic models generate data that students must then interpret and apply to explanations. • **E/S** – Explanations and solutions work hand-in-hand with models, since explanations are both built from models and refine existing models.
Classroom Implications	• You can use simulations to probe students' pre-instructional ideas (Hand 2006). Simulations can be used in whole-class, small group, and individual instruction, though they are most effective in small groups or when used by individual students (Cawelti 1999). But be careful. Use of simulations can promote misconceptions unless you explicitly work with your students to identify limitations of the simulated model (Cawelti 1999). • If you lack expensive laboratory equipment, use virtual manipulatives (Hartman and Glasgow 2002).
Application Example	See Chapter 3 for a description of protein folding activities found in the *Molecular Workbench* (referenced below).
Technology Applications	• Model-It, developed at the University of Michigan, is designed specifically to make systems diagramming and modeling software accessible to precollege students. Information and research about modeling in general and Model-It in particular are available at *www.umich.edu/~hiceweb/modelit/index.html*. • Remote and Virtual Labs extend capacity to conduct experiments for which you do not have laboratory or equipment access.

Instructional Tool 3.6 (continued)

Resources	Concord Consortium's Molecular Workbench (*http://mw.concord.org/modeler/*) offers a rich set of interactive simulations, many of which apply to the biological sciences. LabShare (go.nmc.org/labs) provides a network of shared remotely accessible laboratories. NYU-Poly Virtual Lab (go.nmc.org/vlab) is a virtual laboratory (i.e., one that can be accessed via the Internet through a browser interface). Some virtual manipulative sites are Utah Education Network at *www.uen.org/7-12interactives/science.shtml* Global Classroom's collection at *www.globalclassroom.org/ecell00/javamath.html*

Note: The abbreviations for practices referenced in the tool are:

- Asking questions and defining problems (QP)
- Developing and using models (M)
- Planning and carrying out investigations (I)
- Analyzing and interpreting data (D)
- Using mathematics and computational thinking (MCT)
- Constructing explanations and designing solutions (ES)
- Engaging in argument from evidence (AE)
- Obtaining, evaluating, and communicating information (OEC)

Instructional Tool 3.7

Sense-Making Approaches: Nonlinguistic Representations—Visual Tools

Brainstorming Webs

Brainstorming webs are open systems that help students think "outside the box," creating webs related to their own personal thinking, yet also allowing them to move from idea generation to organization and transformation. Brainstorming is used at the beginning of a process but webs can be revised as thinking changes (Hyerle 2000). Webs are holistic and usually unstructured. They start with a central idea and support free association to create a graphic that reflects relations with other ideas (Young n.d.). Types of webs include clustering, mind mapping, and circle maps.

The Research	• Clustering as a pre-writing strategy builds strong links among associative thinking, drawing, creativity, and fluency of thinking (Rico 2000). Mind mapping connects brain hemispheres, drawing on creativity and logic in the development and support of memory and depth of understanding.
	• Mapping draws from students' prior knowledge, which is essential for student transfer of information to new contexts, and allows them to connect new information to their maps (Hyerle 2000).
	• Circle maps purposefully make no connections, leaving the brain free to brainstorm and later make connections (Hyerle 2000). Generating group circle maps enhances accountability and shared ownership (Lipton and Wellman 1998).
NGSS Connections to Practices	• **QP** – Webs and maps are good ways to begin a research project since they help graphically represent and organize ideas, show how they are related to one another, help translate ideas into a topic, and generate questions on which research can focus.
	• **M** – Webs can serve as visual models and functional analogs that can be used to develop explanations, generate data, make predictions, and communicate ideas.

Note: See legend on bottom of page 133.

Instructional Tool 3.7 (continued)

Classroom Implications	• *Clustering* uses ovals and words to generate ideas, images, and feelings around a stimulus word. Begin with an idea, write a phrase or word in the central oval, branch out to other ovals and add words, and extend these ovals by adding details or new ideas. Clustering requires no drawing abilities and is a good starting point to develop mental fluency. Teachers may use clustering as a whole-class, small-group, or individual activity. Clustering is not a structured organization of ideas but rather a network of associations. Initial clusters can serve as a foundation for the development of more-focused webs that lead to greater clarity of thinking and writing.
	• *Mind mapping*, a more specific technique than clustering, supports creativity and memory by using both words and images to represent relationships and conceptual knowledge (thus, it depends on both sides of the brain). Students start in the middle of the page with a word or drawing for a concept, write on arched lines to build connections between ideas, and draw connections among parts of the map. This strategy determines prior knowledge but can also be used as a lesson progresses. It is useful when studying content-area textbooks, as students can take notes that show both the big picture and the details. It also shows interrelationships among concepts over the course of a text (Hyerle 2000).
	• *Circle maps* help students focus on a topic and brainstorm related ideas while framing them in context, thereby helping students understand their own and other students' points of view. Circle maps are best used in small groups using chart paper (Lipton and Wellman 1998). They consist of a circle within a circle on a sheet of paper. Students place a word, symbol, or picture that represents the concept or idea being studied in the center circle. In the outer circle, students list words and phrases that relate to that concept or idea. Students write, outside the circle, information about their lives that provides the context for their ideas (Hyerle 2000). They list in the upper right-hand corner categories of ideas generated so far. In the lower left-hand corner they list frames of reference. To help students list frames of reference, teachers ask questions such as, "What types of things influence your point of view—for example, prior knowledge or personal and cultural influences?" (Lipton and Wellman 1998).
Application Example	Student groups begin with the word *survival* in the center of a circle map. They brainstorm, adding words, phrases, and/or pictures in the outer circle. They list categories in the upper right-hand corner for the words they listed in the outer circle. This gives you an idea of their prior knowledge and helps launch the lesson, based on their current understandings.
Technology Applications	Software is available for each of these strategies (see Resources below). Content Clips from NSDL is a website that allows import of photos and video and provides tools to sort and organize conceptually. Regardless of software used, interactive whiteboards are always an option for sharing and interacting.
Resources	• Rapid Fire in Inspiration (*www.inspiration.com*)
	• Inspiration Pro (*www.inspiration.com*)
	• Content Clips (*www.contentclips.com*)
	• Clustering software (*http://bonsai.hgc.jp/~mdehoon/software/cluster/software.htm*)
	• iMindMap (*www.imindmap.com*)
	• FreeMind (*http://freemind.sourceforge.net/wiki/index.php/Main_Page*)

Instructional Tool 3.7 (continued)

Task-Specific Graphic Organizers
Task-specific organizers help students see the big picture about the content you want them to learn. They help organize the mind and promote "thinking inside the box" (Hyerle 2000). They are applied in formal, rule-based ways and used for defined tasks or in a specific knowledge area. Emphasis is on organization as specified by the teacher rather than the learner's creative organization (Young n.d.). Organizers can be descriptive, sequential, process or cause and effect, categorical, comparison or relational, and problem solution.

The Research	• These highly structured tools help develop habits of mind such as persistence, self-control, and accuracy and precision of language and thinking. Their step-by-step nature results in concrete models, thereby providing scaffolds for students who might otherwise give up. They offer a global view of a process as well as an end-point and help students stay "inside the box" of learning that is the target. They also result in a written display of students' ideas, which students can reflect on and perhaps use to modify their thinking. Once students become familiar with these structured organizers, they are more able to create their own organizers and control their own thinking (Hyerle 2000).
	• Flowcharts, a type of sequential organizer, can represent simple one-way processes (e.g., a simple experiment), more complex scientific processes with loops and decision points, and cause and effect (Gore 2004).
	• Venn-Euler diagrams help students learn to compare and contrast because the abstract is made visible, supporting reasoning (Gore 2004).
NGSS Connections to Practices	• **I** – Task specific organizers can help identify questions for possible investigation, record observations and questions/predictions that arise from them, organize a plan for investigation, and more.
	• **E/S** – Graphic organizers often depict relationships among facts, terms, and ideas as well as use visual symbols to express ideas and concepts, so they are useful in both developing and communicating explanations and solutions.

Instructional Tool 3.7 (continued)

Classroom Implications	• *Descriptive organizers* describe persons, places, events, and things. They have a central, main idea with subcategories or properties radiating out from the center (Dean, Hubbell, Pitler and Stone 2012; Gregory and Hammerman 2008). They are simple visual representations of key concepts and related terms and ideas; they let students see relationships among ideas and how the ideas link together. They also help students represent abstract ideas, show relationships, and organize ideas for storage and recall. • *Sequential organizers* organize events in a sequence and include flowcharts, timelines, and cycle diagrams. Flowcharts are useful when teaching a process with several steps. They can represent simple one-way processes (e.g., a simple experiment), more complex scientific processes, or cause and effect (Gore 2004). • *Process or cause-and-effect organizers* show either cause-and-effect relationships or a sequence of causal events. They describe how events affect one another in a process. Students identify and analyze the cause(s) and effect(s) of an event or process. There are many different types of cause-and-effect organizers so it is important to select the one that best fits the content you are teaching. • *Categorical organizers* are used for classification and have a horizontal or vertical treelike organization, showing a system of things ranked one above another or left to right. They can be used at the beginning of a project to visually arrange interrelated and sequentially ordered sections within a whole. Projects, term papers, and study of systems all work well with hierarchical organizers. They are used to show causal information, hierarchies, or branching procedures. • *Comparison or relational organizers* identify similarities and differences or comparisons among objects or events. Overlapping Venn-Euler diagrams are used to compare and contrast (when they overlap). They help develop logic, deductive reasoning, and cognitive processes by having students look at various relationships among classes (i.e., a set, a collection, or a group of words or concepts) such as circles within circles and circles outside of circles. Matrixes help when two or more things are compared and contrasted (Gore 2004). • *Problem-solution organizers* identify a problem and possible solutions. They show the problem-solving process by defining the parts of the problem and possible solutions. They structure a process to identify a problem, identify a goal and ways to perceive the goal, identify constraints and effects on the problem context, and generate solutions and text alternatives. The organizer "gives a flow of possible solutions and pathways back when a solution is not immediately apparent" (Hyerle 2000, pp. 71, 74).
Application Example	Students use a cause-effect organizer to develop their understandings about the relationship of photosynthesis and respiration. The term *photosynthesis* is listed at the far left and *respiration* at the far right. Down the center are several entries that include the terms *carbon dioxide, oxygen, water, glucose,* and *energy*. Empty boxes with arrows are included between these terms and respiration, as well as between these terms and photosynthesis. Students must complete the boxes that explain the relationships, thus cause and effect.

Instructional Tool 3.7 (continued)

Technology Applications	Regardless of software and/or strategy used, Smart Boards are always an option for sharing and interacting.
Resources	Find organizers at • *http://edhelper.com/teachers/Sequencing_graphic_organizers.htm* • *www.educationoasis.com/curriculum/graphic_organizers.htm* • *www.eduplace.com/graphicorganizer/* • *http://udltechtoolkit.wikispaces.com/Graphic+organizers* • *http://cooltoolsforschools.wikispaces.com/Organiser+Tools* • Find graphic organizer generators at *www.teach-nology.com/web_tools/graphic_org.* • A good resource on all types of graphic organizers is *A Field Guide to Using Visual Tools* (Hyerle 2000).

Thinking-Process Maps

Thinking-process patterns grow out of and synthesize brainstorming webs and graphic organizers, and they support "thinking about the box." They help students define specific thinking processes as recurring patterns that can be transferred across disciplines, and they guide the building of simple and complex mental models. Students focus on evaluating their own and peers' mental models and reflect on their own meaning making (Hyerle 2000). The term *thinking-process map* refers to concept maps, systems diagrams, and thinking maps.

The Research	• Concept maps show relationships between concepts; students derive meaning from seeing these relationships (Novak 1996). Concept mapping helps people learn how to learn, differentiate misconceptions from accurate conceptions, decrease anxiety, and improve self-confidence (Fisher, Wandersee, and Moody 2000; Hartman and Glasgow 2002). It has positive effects on student attitudes and achievement (Horton et al. 1993; Cawelti 1999). It promotes metacognition, especially when used with interviews, as it provides a frame of reference for students to analyze their own thinking, identify their strengths and weaknesses, and set learning targets. Mapping also increases student motivation (Stow 1997).
	• Students in classes involved with group concept mapping outperform students in classes where maps are created individually or not at all (Brown 2003). There appears to be no difference in achievement when the maps are made by teachers or by students. However, greater achievement is demonstrated if students supply the key words in concept map construction (Cawelti 1999). Concept maps are especially effective when working with concept-rich units (Brown 2003). They are also useful assessments because they help determine changes in understandings of concepts and connections among them (Cawelti 1999).
	• For most people, it is easier to interpret and remember images than text, and drawing a diagram shows the linkages among concepts or variables better than text. Connecting new information to existing knowledge by using diagrams helps stimulate thinking about the situation. Diagramming can also overcome language barriers. Constructing diagrams as a group aids brainstorming, analysis, communication, and understanding (ICRA n.d.; Vasquez, Comer, and Troutman 2010).
	• Thinking maps focus on forms of concept development and reflection, so it is important for teachers to facilitate the development of four habits of mind: questioning, multisensory learning, metacognition, and empathic listening (Hyerle 2000).

Instructional Tool 3.7 (continued)

NGSS Connections to Practices	• **M** – Thinking-process maps help guide the development of simple and complex mental models. They allow students to evaluate their own and peers' mental models. • **D** – Scientists use a range of tools to identify patterns in data. These include visualization, which can be supported through the use of thinking-process maps. • **MCT** – Scientists and engineers can test ideas from the thinking-process maps through the use of computer simulations and mathematical approaches to test predictions of the behavior of systems (e.g., modeling global climate change). • **E/S** – Since thinking-process maps help students identify recurring patterns, they help support the development of explanations and solutions. • **OEC** – Thinking process maps help students brainstorm, analyze, communicate, and understand information. The maps themselves can be used to help communicate explanations to others.
Classroom Implications	• *Concept maps* are useful at any point during a unit of study and are effective evaluation tools because they require high levels of synthesis and evaluation (Novak 1996). You can make a concept map, have students make them individually, or ask small groups to develop them (Cawelti 1999). You should model concept mapping for students. Then use cooperative learning to let students model the techniques for others (Hartman and Glasgow 2002). Group construction of concept maps makes evident students' misunderstandings, allows them to correct one another's mistakes, and develops deeper understanding (Brown 2003). Have students work individually on maps and then work collectively to merge their maps into a more comprehensive group map. This process provokes dialogue and debate and keeps students on task during collaborative efforts (Novak 1996). Questioning enhances the effectiveness of concept maps. A good focus question leads to a richer map; questions you ask during map construction should probe student thinking and guide instruction (Cañas and Novak 2006). • *Systems diagrams* are used to represent ideas about complex situations and help us make sense of the world. They are used to describe either a structure or process, but not both. Systems diagrams have many forms and uses, but when studying a system they can be considered a "model" (Mind Tools n.d.). You can use them for brainstorming, but they are also helpful as students try to understand connectivity in a system. They can also be used to diagnose, plan and implement, and communicate (Mind Tools n.d.). Simple diagrams with 5 to 10 elements are best, though it is difficult to limit a diagram to these few elements. Include only essential elements and use single words and short phrases. A variety of systems diagrams are available with different purposes (see Resources on next page). • *Thinking maps* are often highly structured and thus resemble task-specific organizers. However, they are different in that they help students see the big picture because students must analytically organize material. Thinking maps are, in many ways, a synthesis of brainstorming webs and graphic organizers. The eight different kinds of thinking maps—circle maps (define context), tree maps (classify/group), bubble maps (describe with adjectives), double bubble maps (compare/contrast), flow maps (sequence and order), multi-flow maps (analyze cause and effect), brace maps (identify part/whole relationships), and bridge maps (draw analogies)—focus students' thinking on the map as well as on what influences the creation of the map. Implementation requires use of recurrent thinking patterns and reflective questioning (Hyerle 2000). The maps can be used individually or in concert.

Instructional Tool 3.7 (continued)

Application Example	Provide students with 10 to 20 concepts to map for a given topic of study. Ask them to determine which concepts are the most significant (superordinate concepts) and also identify the subordinate concepts and appropriate linking words to describe the concept relationships. Give them multiple opportunities to synthesize and evaluate their maps. Then ask them to add to their maps several more related concepts. These steps require them to recall, synthesize, and evaluate (Novak 1998).
Technology Applications	• Inspiration mapping software can be used in a variety of ways. Information can be found at the Inspiration Software, Inc. website: *www.inspiration.com*. • Content Clips is an NSDL project that provides for import of photos and video and then offers tools for sorting and conceptual organizing (*www.contentclips.com*). • Model It, developed at the University of Michigan, is designed specifically to make systems diagramming and modeling software accessible to precollege students. Information and research about modeling and Model It are available at *www.umich.edu/~hiceweb/modelit/index.html*. • There are free online versions of cognitive mapping tools such as *bubbl.us* (*http://bubbl.us*) and Mind Meister (*www.mindmeister.com*). • ThinkingMaps software is also available (see Resources below). • Regardless of software used, interactive whiteboards are always an option for sharing and interacting.
Resources	• See *Learning, Creating, and Using Knowledge: Concept Maps as Facilitative Tools in Schools and Corporations* (Novak 1998). • Go to *http://systems.open.ac.uk/materials/t552/index.htm* for an excellent tutorial that will help you learn about various systems diagrams and their uses and construction. • Open learning tutorials on systems diagramming are available at *www.open.edu/openlearn/science-maths-technology/computing-and-ict/systems-computer/systems-diagramming/content-section-0*. • *A Field Guide to Using Visual Tools* by David Hyerle (2000) includes a full chapter (Chapter 6) on thinking maps. • Go to *www.thinkingmaps.com/htthinkmap.php3* for information about the eight thinking maps developed by David Hyerle (2000): circle maps for defining context, tree maps for classifying and groups, bubble maps for describing with adjectives, double bubble maps for comparing and contrasting, flow maps for sequencing and ordering, multi-flow maps for analyzing causes and effects, brace maps for identifying part-to-whole relationships, and bridge maps for seeing analogies.

Note: The abbreviations for practices referenced in the tool are:
- Asking questions and defining problems (QP)
- Developing and using models (M)
- Planning and carrying out investigations (I)
- Analyzing and interpreting data (D)
- Using mathematics and computational thinking (MCT)
- Constructing explanations and designing solutions (ES)
- Engaging in argument from evidence (AE)
- Obtaining, evaluating, and communicating information (OEC)

Instructional Tool 3.8

Sense-Making Approaches:
Nonlinguistic Representations—Drawing Out Thinking

Drawings and Annotated Drawings The act of drawing uses the right brain to visualize and solve problems and allows thinking in a visual language. Teachers can use students' drawings to assess science concept knowledge, observational skills, and ability to reason. Drawings also allow students to explore their own understanding about a concept.	
The Research	• Our brains receive 80–90% of information visually. When drawing, students are free to express much of this information, enhancing the nonlinguistic representations in our students' minds as well as their content understandings (Hyerle 2000; Dean, Hubbell, Pitler and Stone 2012). • Drawings are considered fairer (less biased) than more structured approaches to expression because students can choose what they draw and because drawings are related to students' own experiences (McNair and Stein 2001). • Annotated drawings are an alternative form of expression that allows those of our students who may understand a concept, but find it difficult to express themselves in words, to convey what they know. They also allow students to reveal understandings that might not otherwise be revealed (Atkinson and Bannister 1998). • We can use drawings to uncover how our students perceive objects and the degree to which they perceive and represent details. In addition, the drawing process elicits student questions on points that their peers and the teacher can then clarify (McNair and Stein 2001). • Drawings are a good way to promote conceptual change because they provide information about specific misconceptions, help our students grapple with their own ideas and questions, and provide information that shows development of ideas over time (Edens and Potter 2003; Stein and McNair 2002). • Drawing activities, coupled with interviews, help teachers explore students' ideas about abstract concepts (Köse 2008). Drawing tasks and annotated illustrations support the processes of selection, organization, and integration—all cognitive processes necessary for meaningful learning. But keep in mind that the effectiveness of drawings may depend on the level of a student's prior understandings (Edens and Potter 2003).
NGSS Connections to Practices	• **M** – Drawings can serve as visual models. Visual models serve as functional analogs that can be used to develop explanations, generate data, make predictions, and communicate ideas. • **D** – Drawings and annotated drawings serve as diagrammatic and pictorial models. These visual models serve as functional analogs and can be used to generate data. • **E/S** – Drawings and annotated drawings serve as diagrammatic and pictorial models that can be used to develop explanations and solutions. • **OEC** – Drawings and annotated drawings serve as diagrammatic and pictorial models that can be used to make sense of obtained information and communicate ideas.

Note: See legend on bottom of page 136.

Instructional Tool 3.8 (continued)

Classroom Implications	• Teachers typically present mental models and scientific thinking to their students using verbal language, but drawings are a viable alternative to explore their thinking and should be used more often. A combination of visual and verbal strategies may be best used when teaching nonobservable science concepts such as energy. This combination lets us look for a match between our students' verbal and visual representations and lets our students elaborate on their understandings more fully than if only one strategy were being used (Vasquez, Comer, and Troutman 2010).
	• Students can discuss what they have drawn and may, thereby, reveal their misconceptions (Edens and Potter 2003). They can later revisit their drawings, reconstruct earlier concepts, and use drawings to rethink an idea (McNair and Stein 2001).
	• Student-created drawings provide us with information that helps determine activities that will best serve students' learning needs (Stein and McNair 2002). Because drawings are based on students' own experiences, they help teachers to be responsive to students' interests, background knowledge, and emerging skills (McNair and Stein 2001).
Application Examples	"Talking drawings" translate mental images into simple drawings (McConnell 1993). Students are asked to (1) create mental images of their understandings of a topic prior to instruction; (2) draw pictures representing their mental images and label them; (3) after content instruction, draw a second round of pictures and label them; and (4) write about how their drawings have changed. This strategy makes student construction of knowledge visible, allows students to check their own understandings, and lets them adjust their thinking and study habits (Scott and Weishaar 2008).
Technology Applications	Any graphics program or photo-editing program can be used. There are many free ones.
Resources	*Science Formative Assessment: 75 Practical Strategies for Linking Assessment, Instruction, and Learning* (Keeley 2008)

Concept Cartoons

Concept cartoons are cartoon-style drawings that present alternative conceptions in science, elicit students' ideas, and challenge their thinking to promote further development of their ideas.

The Research	• Concept cartoons work in a variety of teaching situations and across grade levels. They call on students to focus on constructing explanations for the different situations in the drawings. Students must choose between the different explanations in the cartoon, either individually or in small groups. This process makes it evident to them the need for investigation to answer the questions. They become responsible for choosing what is appropriate to investigate, lessening our need as teachers to respond to each student individually about their ideas (Keogh and Naylor 1999).
	• Concept cartoons make it possible to elicit our students' ideas either concurrently or consecutively with the restructuring of their thinking, making the learning process more continuous (Keogh and Naylor 1999). The cartoons give our students the opportunity to discuss the causes of their misconceptions, create an environment where they can all participate during class discussion, and activate them to support their ideas (thereby remedying their misconceptions) (Ekici, Ekici, and Aydin 2007).

Instructional Tool 3.8 (continued)

NGSS Connections to Practices	• **QP** – Concept cartoons offer alternative explanations, which provide an entry into inquiry. Students can choose an explanation and then generate investigable questions to test their explanations. • **E/S** – Concept cartoons offer alternative explanations, which provide an entry into inquiry.
Classroom Implications	Concept cartoons are highly motivating because they present cognitive conflict through the alternative explanations shown in the cartoon (Keogh and Naylor 1999). Such cognitive conflict challenges students' misconceptions (an important part of our framework). It also provides a wonderful entry into inquiry experiences because students have to consider what they might investigate to clarify their understandings and determine which explanation in the cartoon is most correct. The authors encourage you to study the various concept cartoons that align with the various standards you address when teaching and use them as entries into inquiry. You can also create your own concept cartoons quite easily once you are familiar with the common misconceptions about a topic.
Application Examples	**Concept Cartoon: *Where Does a Plant's Mass Come From?*** **Question:** This large tree started as a little seed. What provided most of the mass that made the tree grow so large? This concept cartoon is from *Hard-to-Teach Biology Concepts* (Koba and Tweed 2009, p. 131). It was used to determine students' preconceptions about photosynthesis and as an assessment tool during and after instruction to determine how students' ideas had changed.
Technology Applications	Consider using ComicLife (*http://plasq.com/comiclife-win*) or Comic Creator (*www.readwritethink.org/materials/comic*) to create your own concept cartoons.
Resources	• Visit the Concept Cartoons website at *www.conceptcartoons.com*. A rich set of concept cartoons on evolution are available at *www.biologylessons.sdsu.edu/cartoons/concepts.html*. • See also *Concept Cartoons in Science Education* (Naylor and Keogh 2000a), *Concept Cartoons in Science Education* (CD) (Naylor and Keogh 2000b), and *Science Formative Assessment: 75 Practical Strategies for Linking Assessment, Instruction, and Learning* (Keeley 2008).

Note: The abbreviations for practices referenced in the tool are:
- Asking questions and defining problems (QP)
- Developing and using models (M)
- Planning and carrying out investigations (I)
- Analyzing and interpreting data (D)
- Using mathematics and computational thinking (MCT)
- Constructing explanations and designing solutions (ES)
- Engaging in argument from evidence (AE)
- Obtaining, evaluating, and communicating information (OEC)

Instructional Tool 3.9

Sense-Making Approaches: Nonlinguistic Representations—Kinesthetic Strategies

Hands-on Experiments and Activities and Manipulatives	
This category involves physical movement during a science learning experience. Examples include classroom experimentation that makes use of equipment such as probeware and requires movement from one part of the lab to another, use of other manipulatives (e.g., physical models), excursions into the field, and projects that require construction of materials.	
The Research	Kinesthetic activity involves physical movement; movement associated with specific knowledge builds a mental image of that knowledge in the learner's mind (Dean, Hubbell, Pitler and Stone 2012). Simply moving activates the brain, and if you add the requirement of communication to this action much more of the brain is involved in the learning experience (Lazear 1991). Concrete experiences engage more of the senses and activate multiple pathways to store and recall information (Wolfe 2001). Hands-on activities engage students, who must interact with materials and peers (Wolfe 2001).
NGSS Connections to Practices	• **QP** – Manipulation of materials and careful observation of them can lead to questions that can be investigated within the laboratory, research facilities or field (e.g., outdoor environments). Asking questions or identifying problems is also part of the inquiry process when using virtual manipulatives. • **I –** Excursions into the field are sometimes required for students to make initial observations that help plan investigations. Planning investigations requires that students determine what tools are needed to gather data. Physical manipulation of these tools might be required. Planning and carrying out investigations also occurs with the use of virtual manipulatives. • **M** – The construction of drawings and physical models engage students in thinking about the phenomenon the model represents. Role-playing allows for the human modeling of concepts. • **OEC** – Communication of information often requires movement and, in the case of demonstrating processes, can involve manipulation of materials. Movement and gesture can enhance communication and, in engineering fields, demonstration of models may better communicate ideas. Creating physical representations and graphical representations are also a component of communication.
Classroom Implications	Almost any concept or idea can be transformed to involve physical movement. Consider each lesson and determine ways in which experimentation and other hands-on activities can replace lecture and demonstration approaches. When students present research results, have them move in front of the class. Their movements can include demonstrations of techniques used and of samples gathered.
Application Examples	Begin a class period by modeling the use of sampling materials (e.g., collection nets, sampling containers, dissolved oxygen probes) that will be used in fieldwork outside the classroom. Students practice using the materials and then go out into the field to carry out the sampling. On their return to the classroom, they sort and identify samples. They determine the best way to present their materials during a poster presentation, prepare the posters, and present the posters in a round-robin sharing session. Notice that each step in this series of activities involves students in movement and manipulation.

Note: See legend on bottom of page 139.

Instructional Tool 3.9 (continued)

Technology Applications	• Using probeware to collect data requires movement on the part of students. • PhET Simulations also require students to physically manipulate the computer program to study the effects that occur with changes in variable. Check a variety of other computer manipulations, such as the populations biology simulations at *http://darwin. eeb.uconn.edu/simulations/simulations.html* • Use of handheld data collection devices can be as simple as using a phone to photograph images in the field. • Using 3D printers can be used to help meet the science and engineering practices in particular with the designing and planning an inquiry and designing a model. • Using NetLogo, students can design and create their own simulations to model biological systems—for example, modeling ant behavior finding food, leaving a chemical trail, and so on.
Resources	• Lesson Plans, Inc. (*www.lessonplansinc.com*) provides a variety of activities that make sure to address kinesthetic learners. • Read the article from The Journal about using Tech Tools for NGSS. *http://thejournal. com/articles/2013/10/17/5-tech-tools-for-the-next-generation-science-standards.aspx*

Physical Movement

These kinesthetic activities include the use of gestures, hand signals, and arm motions, as well as acting and role-playing. These are not science-specific actions, but work across the curriculum to improve acquisition and retention of information.

The Research	Kinesthetic activity involves physical movement; movement associated with specific knowledge builds a mental image of that knowledge in the learner's mind (Dean, Hubbell, Pitler and Stone 2012). Simply moving activates the brain, and if you add the requirement of communication to moving, even more of the brain is involved in the learning experience (Lazear 1991). Physical simulations and role-playing call for physical activity and the arousal of emotions, helping in the acquisition and retention of knowledge (Wolfe 2001). Changing locations during a lesson enhances acquisition of information and memory (Jensen 1998). Gestures associated with learning new information help students retain information (Jensen 1998).
***NGSS* Connections to Practices**	• **M** – Physical simulations and role-playing help in the acquisition and retention of knowledge. • **OEC** – Communication coupled with movement better engages the brain in learning.
Classroom Implications	• Learning almost any concept or idea can be made more engaging by requiring physical movement. Simple learning approaches such as the jigsaw technique require students to move around the room. Learning stations are another way to ensure movement. Role-playing can be used to illustrate concepts, practice skills, stimulate interest, and make ideas more concrete for discussion (Hartman and Glasgow 2002). During any lesson, students can indicate their levels of understanding by gestures (e.g., thumbs-up/thumbs-down). • Kinesthetic Learners like to click the mouse, move things around. Flash Technology, with lots of drag and drop functions, works well for kinesthetic learners—it's how the physical translates to the online.

Instructional Tool 3.9 (continued)

Application Examples	Use four-corner synectics (Walsh and Sattes 2005) to determine students' preconceptions about almost any concept. This activity requires you to use four different metaphors for a concept (e.g., the heart is most like a/an … bucket, pump, house, or engine), placing a label for one of the metaphors in each corner of the classroom. Students take a moment to personally think about and record which metaphor best reflects their current understandings of the concept. They then go to that corner and talk with the other students at the corner about why they selected that metaphor. Students then share their thinking as a whole class. These multiple metaphors give the teacher a glimpse into the students' preconceptions and start group and whole-class discussions about the concept.
Technology Applications	• Interactive technology such as gaming, drag-and-drop technology, interactive flash animations, simulations with 3D graphics, or virtual reality environments are likely to appeal to kinesthetic learners. • Rubberized iPads and iPods can measure levels of exertion. • Use an app like Field Notes LT to take observations, adding date, time, GPS location and photographs; add videos and voice recordings to field notes.
Resources	Lesson Plans, Inc. (*www.lessonplansinc.com*) has various activities for kinesthetic learners.

Note: The abbreviations for practices referenced in the tool are:
- Asking questions and defining problems (QP)
- Developing and using models (M)
- Planning and carrying out investigations (I)
- Analyzing and interpreting data (D)
- Using mathematics and computational thinking (MCT)
- Constructing explanations and designing solutions (ES)
- Engaging in argument from evidence (AE)
- Obtaining, evaluating, and communicating information (OEC)

Chapter 4

Connections to the Instructional Planning Framework and Resulting Lessons

"The major promise of our present time is to make an education that provides excellence for all. Both parts of the promise—*excellence* and *all*—are vital for our citizens, our society, and our place in the world."

—*Joyce and Calhoun 2012, p. 1*

Overview

A classroom and school climate that supports learning for all students is essential if we want *all* of our students to meet the expectations outlined in the *NGSS*. Even though Chapter 3 modeled use of the design process and provided tools to plan instruction for enhanced student conceptual understandings, further connections are required if we want excellence for all. We consider the "proteins and genes" unit we demonstrated in the previous chapter a bare-bones level of implementation of the *NGSS*, focusing entirely on the Foundation Boxes but *not* the Connections Box. Nor does it build other connections that we suggest you consider to further enhance a unit of study, making it more accessible to all students. Chapter 4 considers these additional connections and briefly addresses not only the *Common Core* aspects of the Connections Box but also three additional topics: formative assessment, STEM, and supporting students with diverse needs. For each of the topics, a brief summary of the research is provided and then applied to the same "proteins and genes" unit.

Formative Assessment

The Research

Formative assessment helps students meet the goals of lifelong learning—it helps promote higher achievement, increases access for *all* students, and builds their abilities to learn how to learn (CERI 2008). This clearly aligns with the expectations of the *NGSS* and the goals of most teachers. We want all of our students to achieve in our classrooms but as importantly, leave our classrooms with the understandings and abilities that prepare them to learn as adults and positively contribute to their community. However, different teachers interpret formative assessment in different ways.

What is formative assessment? There are various definitions, but we refer you to the following definition in a Southeast Comprehensive Center Briefing Paper (SCC 2012) that resulted from an analysis of a wide array of research. Formative assessment

- is a systematic, continuous process used during instruction by teachers;
- evaluates learning while it is developing;

- is indivisible with instruction and integrated with teaching and learning;

- actively involves both teacher and student;

- provides a feedback loop to adjust ongoing instruction and close gaps in learning;

- involves self- and peer assessment; and

- informs and supports instruction while learning is taking place. (p. 2)

How do we begin to translate this definition into practice? How do we assess for learning? "For assessment to be used to help learning means that teachers incorporate formative assessment strategies as part of their pedagogy rather than adding a series of mini-summative assessment events" (Harlen 2013, p. 3). In other words, it is essential to think of formative assessment as both planned for in our lessons and responsive in the moment. It must be an integral part of our ongoing instruction that helps respond to our students' learning needs—and not just periodic implementation of formative assessment strategies. This requires that we understand various models of teaching and when best to use each, how to differentiate instruction based on these needs, how to support their metacognitive learning and how to engage them in self-assessment (Heritage 2007). Heritage (2010) reinforces Harlen's statement above saying that we need to think of formative assessment not as a "finer-grained test (or tool as it is sometimes referred to), but as a practice involving both teachers and students" (p. 4).

Heritage (2007) talks about three types of formative assessment: curriculum-embedded assessment, on-the-fly assessment, and planned-for interaction. Curriculum-embedded assessments are designed to use at key points in instruction and are part of the ongoing activities (e.g., the systems diagrams explained in Chapter 3). She describes on-the-fly assessments as those the instructor spontaneously makes during instruction, perhaps in response to an observation or something heard during student discussion. Instead of proceeding directly with the planned lesson, the teacher modifies the lesson to address the students' learning needs. Planned-for instruction results from decisions made ahead of time about how students' ideas will be elicited during the lesson and includes things like predetermined questions that we might ask students to probe their thinking.

Sadler's (1989) seminal work on formative assessment says that we must identify the intended learning; elicit evidence of student understanding; interpret the evidence (the gap between that understanding and the intended learning); and act on that evidence. Action includes providing feedback to further students' understanding and revising our own instruction. Feedback is essential for student success and must meet four criteria – it must be timely and specific, understandable to the student, and designed to allow self-adjustment by the student (Wiggins 1998). Additionally, Heritage (2008) emphasizes that quality feedback is clear, descriptive, criterion-based feedback and indicates to students where they are in a learning progression, how their response differed from that reflected in the desired learning goal, and provides suggestions for how they can improve and move forward (OECD 2005). The Instructional Planning Framework described in Chapter 1 (Figure 1.1, p. 7) clearly aligns with this thinking, and the design process describes how to identify the intended learning and select strategies to elicit and confront student preconceptions and help them in sense making. However, ways in which to interpret and act on the evidence are quite sophisticated and require further description in this chapter. Let's look further at both feedback and student involvement in the assessment process.

What is known about feedback? Effective feedback is one of the most powerful ways to influence learning, with an effect size of 0.79 standard deviations. This effect size is greater than that of socioeconomic background, class size, and prior cognitive ability (Hattie and Timperly 2007). There are two main aspects of feedback for students—its form (comments vs. grades) and its content (focused on the task or judgmental). Comments are much more powerful than grades in positively impacting achievement (Harlen 2013), as long as teachers use "descriptive feedback in the form of ideas, strategies, and tasks the students can use to close the 'gap' between his or her current learning level and the next level" (Heritage 2010). Harlen (2013) provides the following recommendations about feedback:

- It should be in the form of comments with no marks, grades, or scores.

- Whether oral or written, comments on students' work should identify what has been done well, what could be improved and how to set about the improvement.

- Comments should help students to become aware of what they have learned.

- Teachers should check that students understand their comments.

- Time should be planned for students to read and, if appropriate, respond to comments. (p. 66)

Why is student involvement in the assessment process so important? "Students who are actively building their understanding of new concepts (rather than merely absorbing information), who have developed a variety of strategies that enable them to place new ideas into a larger context, and who are learning to judge the quality of their own and their peer's work against well-defined learning goals and criteria, are also developing skills that are invaluable for learning throughout their lives (CERI 2008)." If we want our students to not only understand the concepts we teach but also become lifelong learners our focus must be on the learning process (not the grade), we must engage them in self- and peer-assessment and they must learn how to learn—that is, we must enhance their metacognitive skills (CERI 2008; Harlen 2013; Heritage 2010; SCC 2012).

The research of Black et al. (2003) led them to three key recommendations about peer- and self-assessment: (1) the criteria must be clear to students including the aims of the work and what is meant by successful completion; (2) students must be taught collaborative habits and skills in peer-assessment; and (3) students should keep in mind the intent of the work and assess their own progress toward that intent if they are to become independent learners. In addition, they suggest that peer assessment is a necessary complement to and possibly prior requirement for self-assessment.

Implications for the Proteins and Genes Unit of Study

Planned vs. Implemented Learning Sequences

Up to this point in the book we modeled the process of *planning* for instruction and so far our planned formative assessment includes only the selection of formative assessment strategies, which Heritage calls curriculum-embedded assessments. Consider Figure 4.1 and notice that the "planned learning sequence" is linear in nature and appears quite predictable. However, as teachers know, things are not always so clear-cut during implementation. What if our formative assessments indicate that students just didn't "get it" or understood part of the concept but not all? Instead of a linear

Figure 4.1

Planned vs. Implemented Learning Sequence

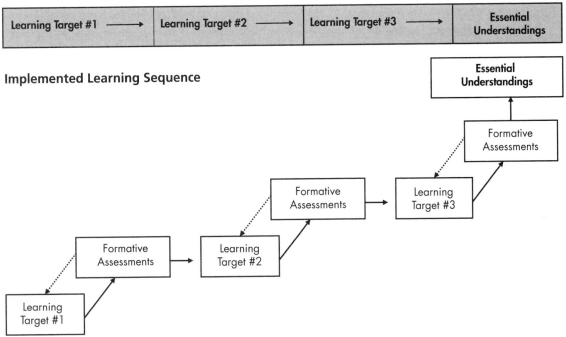

path through the lesson—charging ahead because we constructed the lesson in that way—we need to give students other experiences with the same content so that they have additional opportunities to learn. We must be as consistently aware of students' understandings throughout the lesson as possible.

Now, refer to the "implemented learning sequence" in Figure 4.1 (p. 145). Our planned formative assessments might indicate a need to provide additional instruction for some or all of the concepts for Learning Target #1. On the other hand, it is possible that we could proceed directly from the formative assessments to Learning Target #2 if *all* students demonstrate clear understanding of the first target. This is true for each of the learning targets and demonstrates the iterative nature of the responsive phase of our framework.

How does the earlier summarized research inform the actual implementation of the plan designed in Chapter 3? And how does it model the iterative nature of Phase 2, as just described? To this point we have mentioned the curriculum-embedded assessments, ones that we have embedded in the instructional plan. If we want to truly assess for learning, we need to further consider planned-for instruction that helps make on-the-fly assessments easier. This planning needs to consider specific ways in which to elicit student thinking as well as feedback that might best support student understanding. We also need to consider students' metacognitive engagement in the unit of study and ways in which they can self- and peer-assess.

Formative Assessment for Learning Target #1

This section of the chapter looks more deeply at formative assessment for Learning Target #1, feedback that might be appropriate, and ways in which to engage students in peer- and self-assessment. The focus of this learning target is to help students model the system that connects genotype to phenotype and develop explanations, based on their models, for how that occurs—not the details of DNA structure, transcription, translation, protein folding and so on. During Phase 2 of the planning process, we outlined the following types of formative assessment for the learning target:

- Teacher review of brainstorming webs (models)
- Teacher review of learning logs, including two-column notes, and systems diagrams (explanations and models)
- Analysis of systems diagram for inclusion of key ideas (models)
- Questions and probes during small-group work and class discussion (explanations)
- Final presentation critique using pre-established rubrics (models and explanations)

The criterion we established in Chapter 3 for Learning Target #1 was, "Develop and explain a model that illustrates the production of a protein and its action across levels of organization (cell, tissue, organ, organism), resulting in a particular phenotype." This requires that students demonstrate achievement of the targeted content as well as enhanced expertise in developing models and explanations.

As the lesson begins, it is essential that students are aware of the learning goal and criteria for evaluation. How might this be done? An essential question that drives investigation is one way to clarify the learning goal. This question should engage students in the lesson yet focus clearly on the targeted learning. A possible question for this lesson might be, "How does the production of proteins in my cells impact my appearance and how my body functions?"

To better engage your students in the practices, you might involve them in development of a base rubric for explanations that includes the components of claim, evidence and reasoning. Have them develop a description of each component and the criteria for a range of scores (0–2, 0–4, or whatever you choose). This involves students in the assessment process and makes explicit the learning goals and criteria for evaluation. You might also have them expand the base rubric into a rubric specific to the content of Learning Target #1, as well as develop a similar rubric for a graphic model.

The initial brainstorming web captures students' current explanations for the phenomenon of a phenotype that results from genes—a starting explanatory model. It is important for students to map their initial ideas about what proteins are and why they are important. But it is equally important for them to understand that this web is a graphic model that explains their best thinking about the question. If prior instruction in your class has not familiarized students with models and explanations, then you should provide specific instruction about each. In the Chapter 3 lesson we suggest that the brainstorming activity can be done as a whole class or in small groups. If your students are unfamiliar with models and with brainstorming webs in particular, it might be better to do this as a whole group. The resulting web and the discussion during its construction is an initial assessment of student thinking.

Next, showing the YouTube video "Protein Functions in the Body" provides a bit of background information. Small groups of students can then revise their model. To assess their understandings as they work, you should plan probing questions. Examples might include:

- What were the most important ideas in our original model? How do you know?

- Was there anything missing from our original model that you think should be added after watching the video?

- How did your model change and why did you change it?

- How well does your model explain why proteins are so important?

In addition, you should plan some probing questions about the use of models, for example:

- Why would I have you revise your model?

- How could you use your model to ask a question? What questions do you have based on your model?

Perhaps you find out during this activity that your students are confused about models. What input or feedback could you provide to scaffold their understandings? You might compare two simple models of a different phenomenon with which they are familiar and have students critique the model—which is better and why? How well does it explain the phenomenon? How do we represent the interactions? Have them revisit their model and revise appropriately.

Or perhaps you find out they still have misunderstandings about the content and need further support. They might have confusion about system functions or the concept of proteins, especially connecting the various levels of organization in living systems. One option might be to provide them further background information (text or video) to consider that describes a familiar situation to which they can easily connect.

Now we are ready to test their models and extend their understandings about the connections among genes, proteins, and phenotype. We ask students, "How could you use your model to predict what happens if there is a missing or ill-functioning protein?" They briefly discuss how their model supports a prediction and make their own prediction. We then share the YouTube video "Sickle Cell" and have students again revise their brainstorming webs. With each round of revision, students are referred to the rubrics developed earlier, facilitating self- and peer-assessment. Questions you might ask to facilitate this process include:

- How does your model explain the relationship of genes and phenotype? What evidence do you have to support that explanation?

- Is anything missing in your model that, if present, would help you make predictions?

- Is there further information you need before you can make a prediction using your model?

Assessment and feedback specific to explanations is also important. For example:

- Your explanation does a good job of _____ but what is your evidence?

- Perhaps you should look further into _____ to provide further evidence.

- You provide some evidence for your claim, but further information about _____ might improve your reasoning.

Your probing questions and feedback should be specific to the criteria in the developed rubrics to enhance content understanding as well as understanding about models and explanations.

Now students are ready to test their model and revise their explanations while focusing on a specific disorder. In the Chapter 3 lesson, we suggested that students use two-column notes (Cornell notes) in their learning logs as they research the disorder and develop systems diagrams to formalize their models. If your students have never used either strategy, it is important that you provide specific instruction on each. Periodic review of learning logs lets you focus on each student's understanding of the DCI and associated practices and crosscutting concepts.

As students research their disorder and prepare to present their findings, you have the opportunity to eavesdrop on conversations and determine where each group is struggling and which groups have the most sophisticated explanation of the phenomena. Continue to ask probing questions, but also consider a sequence in which student groups present their findings to best support development of a class explanation of the relationship among genes, proteins, and phenotype. During the presentations, you might ask questions such as:

- What aspects of Group A's model best support its explanation? What is missing?

- How does this model compare to the model of Group B? What aspects better explain the phenomena?

- What levels of organization are represented in each model? Are the system interactions among the levels accurately represented?

- What are the commonalities across models? What is important about these commonalities?

Now facilitate a class discussion and create a single systems map that generalizes what your students currently know about proteins and genes and have student groups evaluate the model based on the rubric developed at the beginning of the lesson. Finally, ask them what further information is needed to enhance their model and explanation. The response to this question should point toward Learning Targets #2, #3, and #4.

Connections to the *Common Core*

NGSS and the *Common Core*

In order to realize a vision of science education where students develop a deep understanding of science, all three dimensions—science and engineering practices, crosscutting concepts, and disciplinary core ideas—need to be integrated into your curriculum, instruction, and assessment. This may cause some significant shifts in your planning and we have shared examples of how a teacher might do this in the invited chapters that follow.

With a focus on integration, another aspect of your planning that can be part of this integrated effort is to include the practices found in the *Common Core State Standards (CCSS), Mathematics and English Language Arts* that are clearly aligned with and incorporated into the *Framework* and *NGSS*. The crosscutting concepts and science and engineering practices align with many of the practices in the *CCSS* and there is significant convergence of the three sets of documents. As a whole they require a systematic shift in how teachers support reading, writing, and math practices in addition to integrating the science and engineering practices and crosscutting concepts for all students. The new vision of learning in science classrooms means that teachers need to make changes in how they practice teaching and learning. Most teachers will need support to make these shifts so we continue to recommend working with other teachers when planning with this broader view in mind.

What we do know is that this integrated approach is beneficial for students as it increases their opportunities to learn and significantly supports diverse learners. Elementary teachers who are pressed for instructional time for science can integrate the subject areas and strengthen the science learning for all students. The convergence of core ideas, practices, and crosscutting concepts across subject areas offers multiple opportunities to build and deepen student understandings. Many students arriving in biology classrooms can benefit from the repeated and contrasting experiences, which can lead to mastery of crosscutting concepts.

The Research

A comparison of the three content standards—*CCSS ELA, CCSS Mathematics,* and *NGSS*—was created in 2012 (Cheuk 2012). The resulting Venn diagram can be viewed at *http://ell.stanford.edu/content/science.* An adapted version was prepared and provided in an article that discussed the opportunities and challenges inherent in all three sets of next generation standards (Stage et al. 2013). The expectations that content teachers— science and social studies—will address literacy goals is a challenge that many states have already wrestled with. Of particular concern to many high school teachers are struggling readers. Learning biology is often compared to learning a foreign language because of the rich and extensive vocabulary included in this course. Clearly the literacy and math standards include practices that are difficult to teach in our classrooms without the support of teachers from other subjects. What are the overlapping goals in the *CCSS ELA* and *CCSS Mathematics*?

- *CCSS ELA* connections with *NGSS*
  - S8. Obtain, evaluate, and communicate information
  - E3. Obtain, synthesize and report findings clearly and effectively in response to task and purpose
- *CCSS Mathematics* connections with *NGSS*
  - S2. Develop and use models
  - M4. Model with mathematics
  - S5. Use mathematics and computational thinking
- Connections with *CCSS Mathematics, ELA,* and *NGSS*
  - E2. Build a strong base of knowledge through content-rich texts
  - E5. Read, write, and speak grounded in evidence
  - M3 and E4. Construct viable arguments and critique reasoning of others
  - S7. Engage in argument from evidence (Stage et al. 2013)
- Expanded connections with *CCSS ELA, Mathematics,* and *NGSS* also include
  - M2. Reason abstractly and quantitatively
  - S6. Construct explanations and design solutions
  - S8. Obtain, evaluate and communicate information
  - E3. Obtain, synthesize and report findings clearly and effectively in response to task and purpose
  - M5. Use appropriate tools strategically
  - E6. Use technology and digital media strategically and capably (NGSS Lead States 2013)

You might ask yourself, why are science teachers responsible for teaching reading, writing, and speaking? The answer is simple. An emphasis on literacy skills and reading (e.g., informational text) has an exponential effect on learning in *every* discipline. For students to read, write, and speak successfully in science (which is how they communicate and provide evidence of learning), teachers need to be better prepared to help their students understand that the ways they read in biology are different from the ways they read in English or mathematics. To develop their literacy skills, students must develop discipline-specific skills and strategies along with the core ideas of that discipline.

"There are approximately 8.7 million 4th through 12th graders in America whose chances for academic success are dismal because they are unable to read and comprehend the material in their textbooks" (Kamil 2003). And with technology's explosive growth in the last couple of decades, do students define the term *reading* differently than their parents and teachers? According to the *2010 Kids and Family Reading Report* (Scholastic 2010) based on a representative sample of 9- to 17-year-olds and their parents, 66% of students in the study reported that they wouldn't abandon printed books, even though e-books are available. Since 2010 that number has decreased to 58% as the number of children who have read an e-book has doubled (*http://mediaroom.scholastic. com/kfrr*). In an effort to determine adolescents' reading behaviors, researchers concluded that adolescents were reading and writing for many hours each day in "multiple, flexible, and varied ways and formats" (Pitcher et al. 2007). A key point of this study is an idea that most biology teachers would agree with: By increasing students' motivations to read and write you increase the odds of improving their achievement outcomes. Think about the following questions:

- What specific skills do students need to read, write, and speak effectively in biology?

- Which reading strategies are most appropriate to help students become more effective readers? How do I help students who get stuck when they read informational text to get unstuck?

- What kind of learning environment do I need to provide to promote literacy skills?

Connections to *CCSS ELA* are included in the discipline and grade band information of the *NGSS* document and specific information is incorporated in Appendix M. The *NGSS* document reinforces the need for teachers to incorporate literacy skills into instruction and support the science and engineering practices. Students need to be able to "read" and interpret charts and data to understanding science ideas. They also need to create representations to convey their understanding. Likewise, writing and communicating information orally is a key skill when making and defending claims based on evidence in science. Of critical importance in biology is the ability of students to engage in scientific argumentation. A framework that teachers can use has been developed by Sampson and Schleigh (2013) to illustrate the components of a scientific argument. They suggest, "it is important for students to understand that some forms of evidence and some types of reasons are better than others in science." Their framework provides resources for teachers to assist students as they develop arguments using a claim, evidence, and justification approach and then evaluate scientific explanations and arguments. These practices are critical to understanding hard-to-teach biology concepts.

A study of biology concepts includes both qualitative and quantitative information. As biology teachers we must link to *CCSS Mathematics* and mathematical practices that are relevant to biology topics. Appendix L of the *NGSS* provides the key topics in mathematics and when, in *CCSS M*, they are first expected to be mastered (*www.nextgenscience.org/sites/ngss/files/Appendix-L_CCSS%20Math%20Connections%20 06_03_13.pdf*). There are three mathematical practices that connect most directly to our work in biology: MP2. Reason Abstractly and Quantitatively; MP4. Model With Mathematics; and MP5. Use Appropriate Tools Strategically. We have already discussed the importance of models in Chapter 3 and now we need to determine connections to mathematical modeling. Additional examples are provided here that link to grade-level learning and disciplinary ideas.

Implications for the Proteins and Genes Unit

So what is next with our planning for the proteins and genes unit that connects to the *CCSS* practices? Opportunities for you to link to the practices from each of the three standards documents are numerous so we will provide a few examples to illustrate options for reinforcing the bulleted practices identified previously.

- Connections with *CCSS Mathematics*, *ELA*, and *NGSS*

  o E2. Build a strong base of knowledge through content-rich texts. What you could try: Expand your "library" of short informational texts related to proteins and their structures and functions. Students can benefit from multiple, flexible, and varied content materials. Get students to watch a video *http://education-portal.com/academy/lesson/proteins-i-chemical-structure. html#lesson* (Proteins 1: Structure and Function) and then make sense of the informational text using a reading strategy like reciprocal teaching (an explanation of the strategy can be found at *www.ncrel.org/sdrs/areas/issues/ students/atrisk/at6lk38.htm*) or a group summarization strategy.

  o E5. Read, write, and speak grounded in evidence: For example, have students gather information about the roles of proteins and then create a children's book that includes a focus on the importance of proteins to living things. Make sure that examples and evidence are included.

  o M3 and E4. Construct viable arguments and critique reasoning of others: For example, use the biological argumentation framework (claim, evidence, and justification) to answer questions about the daily percentage of protein that is needed to maintain a healthy diet and healthy weight. Have students create a poster to convince their classmates that their explanation is the best one. Engage in scientific discourse that includes analysis and critiquing of student arguments.

- o S7. Engage in argument from evidence: For example, use the online simulation called Eating and Exercise to have students gather data about the experimental individuals, BMI and weight changes over the course of a year (*http://phet.colorado.edu;/en/simujlation/weating-and-exercise*). Students can adjust a variety of variables to inform their explanations about the importance of proteins to proper functioning in humans. Have students analyze and challenge the explanations of the other students.

Connections to STEM

Definitions of STEM vary tremendously. The acronym was initially an abbreviation for the four disciplines and now is used to describe initiatives, education groups, programs, and practices (Bybee 2013). In this book we use the definition used by Vasquez, Sneider, and Comer (2013): "An interdisciplinary approach to learning that removes the traditional barriers separating the four disciplines of science, technology, engineering, and mathematics and integrates them into real-world, rigorous, and relevant learning experiences for students" (p. 4). These authors also nicely summarize practices in each area of STEM and encourage you to become familiar with them so you can purposefully integrate in ways that reflect the essence of STEM. Finally, they describe and model three ways in which you can integrate but in various degrees—multidisciplinary, interdisciplinary, and transdisciplinary—in that order, reflecting increasing integration. This aligns with Bybee's description of STEM 1.0, 2.0, 3.0, and 4.0, with each level integrating more disciplines until ultimately all four are integrated, and his discussion of a transdisciplinary approach that involves all disciplines and focuses on problems or issues.

Choices about the degree to which you integrate are made for various reasons. This might be driven by district policy. Or you might have more freedom to make integration decisions at the classroom level. Perhaps integration is new to you so you begin to integrate only two of the disciplines *or* a math colleague and you work well together so decide to begin with science and math *or* the particular unit of study drives the degree of integration. For instance, some topics lend themselves more easily to engineering integration than others. Regardless, there is research that outlines the need for and supports the effectiveness of STEM discipline integration.

The Research

The National Research Council (2011a) report, *Successful K–12 STEM Integration*, succinctly summarized research that established not only the link between K–12 STEM education and the leadership and economic growth of the United States but also the inadequate preparation of students to meet the economic demands. However, the NRC

(2011b) workshop summary that followed indicates that students are not generally motivated by economic arguments, suggesting that though STEM careers are lucrative, instruction rich in core knowledge and practices essential to science is required to reach students.

Both reports focus heavily on the science and mathematics components of STEM and outline effective STEM instruction. They propose that effective instruction allows students opportunities to engage in science, mathematics, and engineering practices throughout their school experiences and to ideally do so in ways that provide students opportunities to engage in these disciplines as they address problems with real-word applications. This clearly aligns with the *NGSS*, which embeds science and engineering practices and outlines the connections to mathematics.

Mathematics was discussed in the previous section on *Common Core* connections, so no further discussion is included here. The following section focuses instead on engineering and technology integration and their potential to enhance science teaching and learning.

"Any education that focuses predominantly on the detailed products of scientific labor—the facts of science—without developing an understanding of how those facts were established or that ignores the many important applications of science in the world misrepresents science and marginalizes the importance of engineering (NRC 2012, p. 58)." This statement implies that students should learn how scientific explanations are developed and how science is applied—and how engineering, technology, and science applications are distinct yet related (NRC 2012). Think of science as the understanding of a particular concept, technology as the implementation of the idea into practical application, and engineering as the connection between the two.

Technology in its broadest sense includes any tool or group of tools that help us solve a problem, achieve a goal, and improve the efficiency of our work as we perform particular tasks. Technology and science are intricately intertwined. Science catalyzes new technologies and new technologies have the potential to advance science—often with unanticipated outcomes. As classroom teachers we must expose our students to these relationships among science, technology, and society and give them opportunities to analyze cost, risk, impacts, and benefits of various technologies on society as related to the content we teach (NRC 2012).

Society is supported by technologies and their related systems, and science knowledge and engineering practices have contributed to them. Likewise, engineering and technology are influenced by society. Agriculture and medicine are only two of many fields impacted by advances in science, technology, and engineering (NRC 2012). Students' appreciation for these relationships is critical and the *NGSS* practices can lead to that if true integration of the DCIs and practices occurs in the classroom.

Modern biological sciences include several instances, including nanotechnology and synthetic biology, where lines are blurred among science, engineering, and technology. The impact of nanoscience and nanotechnology on society is difficult to predict, but they have applications in all areas of science and potential future impact that exceeds that of the silicon chip. Potential impact in the biological sciences includes the areas of medical diagnostics, environmental restoration, cancer treatment, and more (Madden et al. 2011). Nanoscience and nanotechnology beautifully represent the integration of science, technology, and engineering and your students' awareness of them opens doors to potential careers.

There are also standard tools and technologies used in science that students should be able to use—and we must help our students become proficient in using them (e.g., microscopes, balances, electrophoreses, and so on). As classroom teachers we need to make certain to either use the most modern of technologies with our students or access virtual and remote labs that model the use of these technologies in pursuit of science understandings.

Technology can make difficult concepts more accessible and engage students in the practices of science and engineering. Probeware and sensors can extend students' senses, help them quickly acquire real-time data (allowing them to focus more on core ideas and less on the mechanics of laboratory setup), and perform multiple experiments in short periods of time. Models and simulations can help build students' understanding by providing them access to processes too slow or fast to observe or phenomena at extremely small and large scales (Concord Consortium 2013).

Technologies continue to change at an increasingly rapid rate, with some of the most rapid changes in Information Services and Technology (IST)—the application of computers and telecommunications to transmit and manipulate data, as well as information distribution technologies, including computer-based information systems. Sadly, the use of these technologies in schools is not keeping up with students' use of them in their daily lives. Unless we effectively begin to integrate these technologies in ways to disrupt the current approaches in education, we will likely continue to lose students to other educational delivery systems (Joyce and Calhoun 2012).

The increased use of emerging technologies could transform the classroom, specifically the science classroom. However, if technology is seen as additive instead of a way to meet the needs of learners in new ways, the questions we ask as educators and our potential of impact change are diminished (Richardson 2013). Scherer (2011) suggests that a technology-rich school exhibits three qualities: full engagement of students' brains (with the teacher, peers, or the content), a compelling relevant assignment that allows for a variety of depths, and evidence of personalized learning (i.e., choice, assignments at the proper level and building on prior knowledge). Student engagement in STEM-based problems provides the potential to meet these three criteria and IST can further enhance that potential if appropriately used.

There are many ways in which IST can support students in development of the various practices, some of which include:

- 3-D Printers create models that represent concepts in science or serve as prototypes in engineering.

- Open source information access provides greater opportunities for students to obtain, evaluate, and put new information to use.

- Virtual laboratory sites provide access to real lab devices that extend investigative opportunities.

- Authentic research can be conducted using cloud-based tools (e.g., *go.nmc. org/sci*).

- Mobile devices encourage exploration since they provide opportunities to immediately search when questions arise and someone is curious as well as foster more and better collaboration and communication, both essential to doing science. They also can be used to gather data—photographs, video, and audio recordings—providing huge capacity in the field (Johnson et al. 2013).

Implications for the Proteins and Genes Unit

How can you integrate engineering and technology into the unit we outlined in Chapter 3? A natural tie to engineering is genetic engineering, since the unit asked students to investigate genetic disorders. The technologies of gene therapy that allow genetic engineering demonstrate the connections among science, technology, and engineering. Areas students might research, some of which are outlined by Lockhart and Le Doux (2012), include:

- whether the disorder they researched is treatable through gene therapy;

- the technologies required for gene therapy;

- how a therapy would be designed for the disease they researched;

- how one would measure or assess the success of a therapy;

- the costs and benefits of the gene therapy; and

- the related social, political, and ethical issues.

What are the IST applications we might consider to enhance the Proteins and Genes unit in meaningful ways?

- 3-D printing could help students explore the importance of protein shape when considering genetic disorders.

- There is extensive open source information related to the topic including the DNA Learning Center at Cold Springs Harbor Laboratory (*www.ygyh.org*).

- Mobile devices can enhance research access and collaboration and communication among student groups.

These are simple first steps to begin integration of science, engineering, and technology related to the Proteins and Genes unit. They both enhance the curriculum and allow greater flexibility (choice, enrichment) that promotes *excellence for all*, a goal of all concerned teachers and of the *NGSS*.

Connections to Support Students With Diverse Needs

Diversity and *NGSS*

Both the *NGSS* and *Framework* (NGSS Lead States 2013; NRC 2012) include sections that speak to the needs of diverse learners. The *NGSS* Appendix D: "All Standards, All Students": Making the *Next Generation Science Standards* Accessible to All Students includes seven case studies of diverse student groups, and provides teachers with suggestions for how to ensure that the *NGSS* are accessible to all students (*www.nextgen-science.org/appendix-d-case-studies*). With the increased rigor and connections included in the *NGSS* and the *Framework*, the goal of this appendix is to incorporate practical and useful strategies into the learning opportunities and challenges represented by the *NGSS* in classrooms, schools, homes, and communities. Each case study highlights one identified group (including economically disadvantaged students, students from major racial and ethnic groups, students with disabilities, and students with limited English language proficiency). In addition, student diversity also incorporated into the case studies includes three additional groups (e.g., girls, students in alternative education programs, and gifted and talented students). The members of the *NGSS* Diversity and Equity Team wrote the case studies, which represent examples across science disciplines and grade levels.

Review of the Research

The research for each student group tends to exist independently from each other so the case studies in *NGSS* Appendix D are intended to provide context in terms of demographics, science achievement, and education policies. For a recent review of effective strategies for non-dominant groups in science classrooms, see the Special Issue in *Theory into Practice*, 2013 and the work of Lee and Buxton, 2010.

In Chapter 11 of *A Framework for K–12 Science Education* (NRC 2012), an elaboration is provided to highlight equity issues related to student learning of science and engineering practices. To realize the vision of the *NGSS*, teaching and learning needs to be equitable and provide multiple opportunities for all students. A wide range of issues is discussed and can be summarized by the following categories. Reflect on the questions provided from your perspective and as they relate to your school and community of learners.

- Considering sources of inequity: Are differences in achievement related to differences in opportunities to learn that exist in your school, district, and community? Is instruction in your classroom inclusive and motivating for diverse learners?

- Capacity to learn science: Can all students aspire to the science and engineering learning goals outlined in the framework and *NGSS*? Do you have common expectations for all students (should they be expected to learn the core ideas and practices)?

- Equalizing opportunities to learn: Do I have lower expectations for minority and low-income student groups? Are there resource gaps or gaps in opportunities to learn amongst some or all of my students? Is my classroom redesigned to be "fair" for all students and serve the needs of academically at-risk students?

- Inclusive science instruction: Do I teach science in ways that actively involve students in the practices of science? Do I incorporate science practices that include scientific discourse, use scientific representations and tools, and focus on sense making?

- Approaching science learning as a cultural accomplishment: Students bring their cultural worldviews with them. Do I link to these everyday contexts and situations with my classroom instruction?

- Relating youth discourses to scientific discourses: Scientific sense making through discourse is a process that does not resemble discourse at home. Do I teach students effective ways to engage in scientific discourse, which includes argumentation and scientific inquiry?

- Building on prior interest and identity: Do I engage students' personal interest in learning science concepts? Do I help build student confidence and abilities to continue learning about issues, scientific and otherwise, that affect them and their communities?

- Leveraging students' cultural funds of knowledge: Students from different cultural groups acquire diverse knowledge and skills from their informal and formal science learning experiences. Do I consider these ideas and stories during instruction?

- Making diversity visible: Do you make an effort to include significant contributions of women and people from diverse cultures and ethnicities into your lessons? Do I build on the stories these examples provide?

• Value multiple modes of expression: Do the tests that you provide accurately show what students have learned? How do I support students whose first language is not English? When sharing performance expectations do I let students know what good work should look like?

Universal Design for Learning (UDL)

Universal Design for Learning (UDL) is a process that seeks to minimize barriers in all learning environments to improve outcomes and opportunities for all individuals, in our case, your science students. The primary difference between UDL and meeting the needs of students with disabilities is the focus on flexible learning experiences and environments. Meeting the needs of students with disabilities requires accommodations to the learning materials and the environment. The identification of limitations or "deficits" with corresponding adjustments allows the student to participate to a greater or lesser degree in the "normal" environment (Gill 1987; Hahn 1988; Jones 1996; Swain and Lawrence 1994). Accommodations that most teachers include involve extra time, various printed materials and notes that are provided, assistive technology and staff support, or accessibility to the workspace in a classroom. The accommodations are usually determined not by the teacher but by counselors or other pupil services staff. The process itself is responsive but not necessarily proactive.

Universal Design for Learning (UDL) processes engage the teacher proactively during the lesson planning process to reduce barriers and maximize accessibility of learning for all students. This ensures access for groups of diverse students and seeks to make it possible for everyone to participate in an inclusive learning environment where no one is singled out. Once you have set clear goals and completed the identifying essential content (Phase 1)—determine what students should know, do, and care about within a unit of instruction—then UDL principles should be considered with the Phase 2 work of planning for responsive action. To plan for flexible lessons, consider what barriers might exist for your students and how your procedures, policies, and learning experiences may be developed to support students. There are three primary principles of the UDL process to consider when planning instruction aligned to the learning targets that you have identified in your content storyline:

1. Content representations: Provide multiple ways to represent the content concepts. Think about the different ways you can provide representations, vary the supports for students, and determine how to make the vocabulary accessible. From what we understand about brain research, students connect differently with the content representations so consider several approaches.

2. Action and expression: Provide multiple means for students to engage with the learning, allow students to express their understanding in different ways, provide different models, and incorporate regular feedback processes. Refer

to the formative assessment information about the need for ongoing feedback between teachers and students and among students.

3. Engagement: Provide multiple means of intellectual engagement with the content, use different engagement strategies since students find different experiences engaging, foster risk-behaviors that support student metacognition and autonomous learning behaviors, and provide opportunities for reflection. Making connections between the learning and student's lives is an effective way to engage students. Letting them ask questions about what interests them is one strategy that supports relevant learning experiences that engage your students (*www.udlcenter.org/aboutudl/udlguidelines*).

An elaboration of the principles and guidelines for UDL can be reviewed at the websites previously identified. Teacher-friendly examples and resources that illustrate each of the UDL checkpoints are readily available and links to these resources are included in the resources section of the website. When you explore these examples and resources they will help clarify what "planning instruction that meets the needs of all students" means. In addition to students with identified disabilities, other beneficiaries of UDL processes include students with various learning styles, those whose native language is not English, and students who are ill or injured or are of atypical size or shape. The examples provided here are intended to provide ideas for you to think about as you incorporate UDL in your classrooms. The UDL checklist and guidelines components and principles can be summarized into seven larger categories. As you plan instruction, think about your individual lessons and what students need to have us do to support their learning. Let's think for a few minutes about what students say makes a difference to them in each of the seven categories.

- Class climate: Adopt practices that respect both diversity and inclusiveness. Welcome everyone; avoid stereotyping; be approachable and available; motivate all students; and address individual needs in an inclusive manner.

- Interaction: Encourage positive, effective opportunities for teachers and students to work together and communicate in ways that are accessible to all. Supplement in-person interactions with online communication; use straightforward language, avoid jargon and unnecessary complexity and use student names when communicating in-person, in writing and electronically; make interactions and discussions accessible to all students and be flexible when determining strategies for interactions; and teach and support procedures for cooperative learning to ensure active participation of everyone but expecting individual accountability for the learning.

- Physical environments and products: Plan the classroom space so activities, materials, and equipment are physically accessible by all students. Arrange the classroom, lab, and fieldwork site in ways to support a wide range of physical abilities; arrange seating to allow for comfort, inclusion, and cooperative learning; minimize distractions for students with a range of attention abilities; plan for access to assistive technology, laboratory technology, equipment and materials; and ensure safety.

- Delivery methods: Use a variety of instructional methods to support all learners. Carefully select text-based materials; incorporate technology that can be used flexibly by students and provide prompting, feedback opportunities for practice, background information, vocabulary, and other supports; provide Web-based unit plans; make content relevant; provide cognitive supports (e.g., outlines, class notes, study guides, and so on); provide multiple instructional methods and make each accessible to all students (e.g., lectures, collaborative learning, small group discussions, hands-on activities, internet-based communications, educational software, fieldwork, and so forth); and use large visual and tactile aids or manipulatives to demonstrate content.

- Information resources and technology: Ensure that resources are engaging, flexible, and accessible for all students. Select and prepare materials early; use multiple, redundant presentations of content that use multiple senses; make sure all materials are in accessible formats; accommodate and support a variety of reading levels using reading strategies; and check out availability of assistive technology for those students whose learning is supported in this way.

- Feedback: Provide specific and regular feedback aligned to the learning goals and success criteria. Determine a process for providing descriptive feedback to students at one-to-two week intervals; provide time for students to incorporate the feedback; teach strategies for peer- and self-assessment that includes feedback.

- Assessment: Using formative assessment processes, regularly assess student progress and adjust instruction accordingly. Provide multiple ways to demonstrate knowledge; interpret evidence of student learning and adjust instruction; minimize time constraints and provide due dates in advance; and provide scaffolded learning as needed by students and gradually withdraw the supports.

- Accommodation: If needed, plan for accommodations for students whose needs are not met by the instructional design. With pupil services support staff, determine needed accommodations and arrange for them to be implemented.

You may already incorporate some of these ideas when you plan your lessons. Continue to do what you know works with students and add strategies that you may not have previously considered but would help all of your students learn.

Implications for the Proteins and Genes Unit

When we consider the unit explored in Chapter 3, it is time to make some specific recommendations for lesson planning that incorporates UDL processes. If we return to the three primary principles, let's consider how we can meet student needs in each of the categories. The CAST (Center for Applied Special Technology) Universal Design for Learning (UDL) Lesson Builder (*http://lessonbuilder.cast.org*) provides a free tool for teachers to review sample lessons, create their own lessons, and add to and use a database of existing lessons. Refer to Table 4.1 (p. 164) to become familiar with the checklist and guidelines developed by CAST for UDL (CAST 2011).

Remember that the goal of UDL is to choose or create flexible materials and media to support teaching and learning by all students. You may find that your school or district is already focusing on UDL principles but if this is new to you, spending some time with the examples provided by the National Center for Universal Design will help you create a mental picture of what it would look like in a classroom. Remember the Learning Target #1: Proteins carry out the major work of cells and are responsible for both the structure and function of organisms. These proteins are made based on the code found in the organism's DNA (genotype) and result in the organism's traits (phenotype). How well the cellular system functions and interacts among cells and at the various levels of organization impacts the entire organism.

The first principle in the educator checklist is to provide multiple means of representation. When helping students have opportunities to learn, you have many strategies to try. For example you could include one or more of the following ideas:

- Create voice-over PowerPoint presentations so students could replay the slides on the roles that proteins plan both structurally and functionally at different levels of organization in an organism.

- Develop student comprehension using a variety of reading strategy supports such as Concept Definition Maps, Concept Circles, Frayer Models, Graphic Organizers, or other tools (Urquhart and Frazee 2012) to support student understanding of specific vocabulary around genotype and phenotypes.

- Provide more than one strategy to activate students' prior knowledge. We suggested using a brainstorming web but you could also let students choose another option to reveal their thinking. Possibilities include providing a nonlinguistic representation or responding to a probing question.

- To guide content processing, select a variety of content representations that help students build their own explanations. To help with this, use content representations such as analogies, metaphors, diagrams, charts, graphs, concept maps, models, and role-playing to make the science ideas more concrete and real for students.

Table 4.1

Universal Design Teacher Planning Template

Category I: Provide multiple means of representation.	Your ideas
A. Provide alternatives for visual, auditory, or customized displays of information	
B. Illustrate through multiple media, clarify vocabulary, and help to decode text and symbols	
C. Highlight patterns, make connections, activate background knowledge, and support visualization	
Category II: Provide multiple means for action and expression.	
D. Provide options for physical action including access to tools and options for navigation of information	
E. Provide options for communication including multimedia and build fluency with practice and performance	
F. Guide goal setting, provide structures for managing information and resources and support planning and progress monitoring	
Category III: Provide multiple means of engagement with the learning.	
G. Provide options for individual choice, encourage relevance, and minimize threats and distractions	
H. Provide options that support effort and persistence including collaboration, include opportunities to increase mastery and vary the demands and resources	
I. Promote metacognition and self-regulation, include options that motivate learning, facilitate coping skills, and provide opportunities for self-assessment and reflection.	

Note: The information provided in the table is adapted from the *UDL Guidelines—Educator Checklist Version 2,* National Center on Universal Design for Learning: *www.udlcenter.org/aboutudl/udlguidelines.*

With the second principle think about flexible ways that you can provide multiple means for action and expression. How can you provide materials with which all students can interact?

- Provide links to media software that allow students to share their ideas about the role that proteins play at each level of organization. They can also represent their understanding of genotypes and phenotypes. A few options include Sam Animation (*http://icreatetoeducate.com*) software for students to create and tell their stories, Animoto (*http://animoto.com*) to create a video of their ideas and Glogster (*http://edu.glogster.com*) for students to create interactive posters

- Use physical manipulatives (such as 3-D models) to represent student understanding of the how proteins are made based on the genotypic code.

- Use interactive simulations to develop an understanding of relationship between the genotypic code and the corresponding phenotypic appearance.

In the third principle, we focus on providing positive classroom environments where students are engaged intellectually and motivated to learn. What motivates one student may not "hook" another so incorporating a variety of strategies is still very important. You might try some of the following ideas or plan for others that you know work for your students.

- Encourage students to have choice when selecting a topic to investigate. In our example this would occur when students select a genetic disease to investigate to determine the errors in DNA coding that resulted in a change in the protein, which caused the phenotypic disease characteristics and the changes in function at some level within an organism.

- Make the learning experiences more relevant by connecting students to real scientists who study the Human Genome and those who specialize in genetic disorders.

- Take note of common classroom distractions and diversions and plan your lessons to support a safe and orderly environment where students know what is expected of them and have the resources that they need to meet the learning target. When students have a sense of comfort and order in the learning environment they are more intrinsically motivated to do the work needed to learn.

Resources

Technology Applications and Websites

- LabShare (*go.nmc.org/labs*) provides a network of shared remotely accessible laboratories.

- iLabCentral (*go.nmc.org/ilabs*) is a site to share and access real lab devices over the internet.

- NYU-Poly Virtual Lab (*go.nmc.org/vlab*) is a virtual laboratory (i.e., one that can be accessed via the internet through a browser interface).

- A great resource to help flip science with remote labs is found at *go.nmc.org/flipsci*

- FieldNotesLT app (*https://itunes.apple.com/us/app/fieldnoteslt/id443876537?mt=8*) is a note-taking tool for iPhone/iPad that can be taken on the road and into the field that makes easy geo-referenced data sharing and collection.

- Mendeley app (*www.mendeley.com*) helps organize research and collaborate with others online.

- Drosophila Virtual Lab (*go.nmc.org/flies*) provides a virtual lab bench for investigation about genetic inheritance, including an online notebook and class code for classroom use.

Build Your Library

- A great resource for formative assessments, which we used to construct the Instructional Planning Framework, is *Science Formative Assessment: 75 Practical Strategies for Linking Assessment, Instruction, and Learning* (Keeley 2008). You might also consider some of the *Uncovering Student Ideas in Science* volumes available through NSTA Press (*www.nsta.org/publications/press/uncovering.aspx*).

- *Assessment for Learning: Putting It Into practice* (Black et al. 2003) is a classic text that provides insights into assessment and includes teachers' examples of how they turned the ideas into classroom practice.

- A variety of assessment strategies and frameworks developed during National Science Foundation-funded research projects are included in the edited volume, *Assessing Science Learning: Perspectives from Research and Practice* (Coffey, Douglas, and Stearns 2008).

- Black and Harrison (2004) provide examples and suggestions for descriptive feedback for students in *Science Inside the Black Box: Assessment for Learning in the Science Classroom*.

- *STEM: Student Research Handbook* (Harland 2011) is a resource for high school teachers and students that outlines stages of large-scale research projects that supports science, technology, engineering, and mathematics student-researchers through the process.

- Though written for grades 3–8, *STEM Lesson Essentials: Integrating Science, Technology, Engineering, and Mathematics* (Vasquez, Sneider, and Comer 2013) provides tools and strategies for STEM implementation that are also applicable to high school.

- Though written for grades 6–8, *Everyday Engineering: Putting the E in STEM Teaching and Learning* (Moyer and Everett 2012) provides examples that help envision ways to integrate science and engineering.

- *E + S: Integrating Engineering + Science in Your Classroom* (Brunsell 2012) is a collection of 30 articles from the various NSTA journals that reinforce science content while illustrating a range of STEM skills, including six articles representing the life sciences.

- *Science by Design: Construct a Boat, Catapult, Glove, and Greenhouse* includes a project on building a greenhouse that integrates science, engineering, and technology with a conceptual focus on heat energy transfer, photosynthesis, plant metabolism, thermal regulation, and feedback control.

- *Welcome to Nanoscience: Interdisicplinary Environmental Explorations* (Madden et al. 2011) uses groundwater pollution as a theme to introduce an interdisciplinary approach to understanding the issue and introduces the fields of nanoscience and nanotechnology.

- *Realizing the Promise of 21st-Century Education: An Owner's Manual* (Joyce and Calhoun 2012) outlines a vision for advancing school reform with a focus on technology infusion across the curriculum.

- *The NMC Horizon Report: 2013 K-12 Edition* (Johnson et al. 2013) outlines near-term, mid-term, and far-term technologies and for each provides: (1) an overview, (2) relevance for teaching, learning, or creative inquiry including many resources, and (3) examples in practice at different educational institutions.

- Two issues of *Educational Leadership* are great resources: the February 2011 issue, "Teaching Screenagers" and the March 2013 issue, "Technology-Rich Learning."

PART II
Toolbox Implementation: The Framework and Strategies in Practice

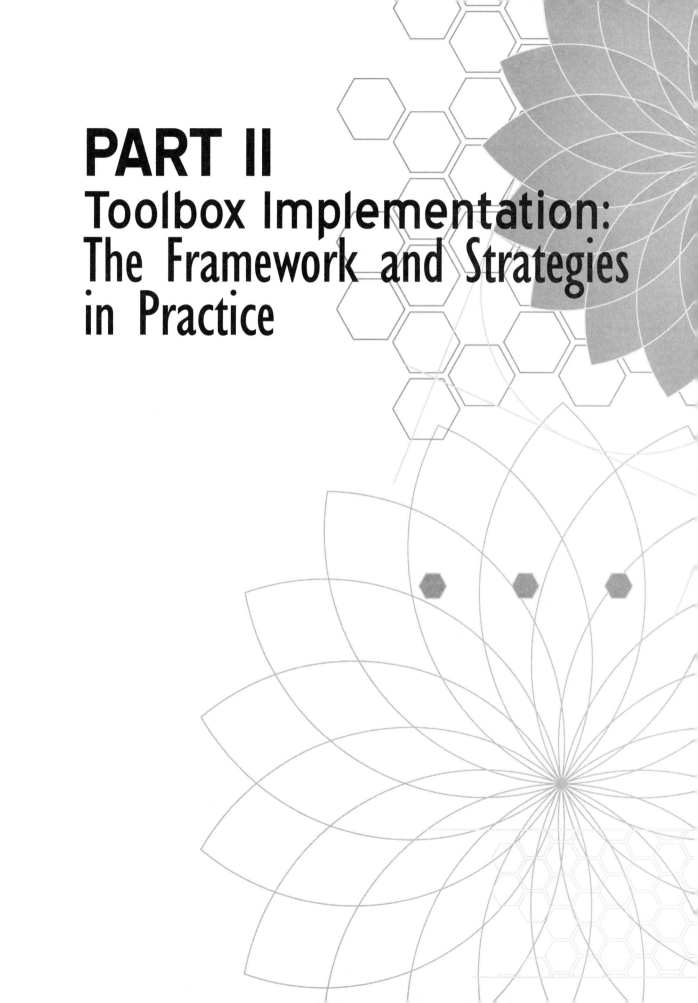

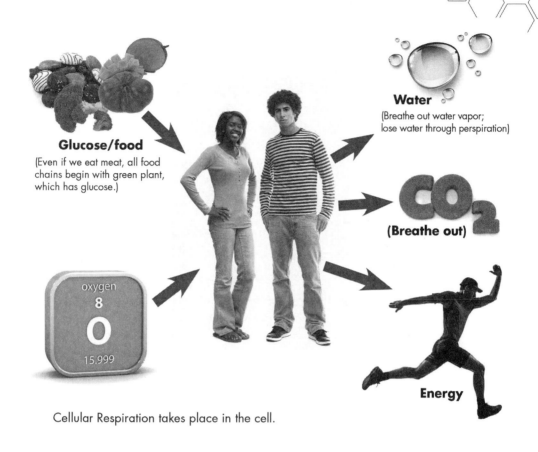

Glucose/food
(Even if we eat meat, all food chains begin with green plant, which has glucose.)

oxygen
8
O
15.999

Water
(Breathe out water vapor; lose water through perspiration)

CO₂
(Breathe out)

Energy

Cellular Respiration takes place in the cell.

Chapter 5

Matter and Energy in Organisms and Ecosystems

By Cynthia Long

As a high school science teacher, research scientist, curriculum and professional developer, and current director of education, one topic that consistently rises to the top of hard-to-teach concepts is cellular respiration, and in the bigger context, matter and energy.

Students need to make the connection between the abstract or unseen and concrete events that are obvious to them. For example, they know they somehow get energy (abstract) and they know they need to eat (concrete). Somehow this "magic" happens in their bodies, but since they can't see it, this leap from eating to energy becomes irrelevant. Helping students understand the relevancy and importance of matter and energy as it relates to human consumption and flow of matter through body systems into cells and then converted into usable energy for their everyday lives is at the heart of this chapter. With this topic, I want students to know why they are studying something called cellular respiration by connecting it to tasks they do every single day. Working with scientists, teachers, and curriculum developers throughout my career has afforded me access to different perspectives, expertise, and backgrounds. Over several years, I tried activities, refined ideas, altered the order, and so on. It was definitely iterative and not everything always worked the first time. Discussing ideas with colleagues and even the students influenced this development. All of this helped me as a teacher put together a unit of study for students that facilitated their learning and guided the development of their understanding in an authentic way. And, it ultimately helped me put together this chapter.

Phase 1: Identifying Essential Content

The study of cellular respiration in organisms is one of the most important but often difficult subjects to teach at the high school level. Concepts in biology are expansive and they all link together, so to identify the important content for this particular area of study I realized that I needed to identify the big idea that students should know. Then, I could break it down into chunks that students could grasp. These "chunks" needed to be in an order that made sense. Instead of giving students a thousand Lego pieces jumbled in a box and say, "build something," I would give them a few intentional pieces at a time. They could use these pieces to build a solid foundation for a structure. Additional pieces would fill in gaps and expand the structure. In this way, students would follow a storyline that makes sense, use what they already know, and add to their conceptual understanding of matter and energy.

The big idea for an area of study that challenges students year after year but yet is essential to life processes is the relationship between matter and energy—how we get and use energy for life. If we put a phrase to this, it is all about cellular respiration. But to say this phrase without meaning is just another intimidating biology term. This is how I set out to design a sequence of lessons for my biology students. This area of study comes directly after students have studied photosynthesis. After the lessons described in this chapter, students make connections between photosynthesis and cellular respiration, completing their understanding of matter and energy.

I generated a graphic to help me see connections among concepts, important questions to ask, and ideas to elucidate the big picture for energy and matter (see Figure 5.1).

Figure 5.1

The Big Picture

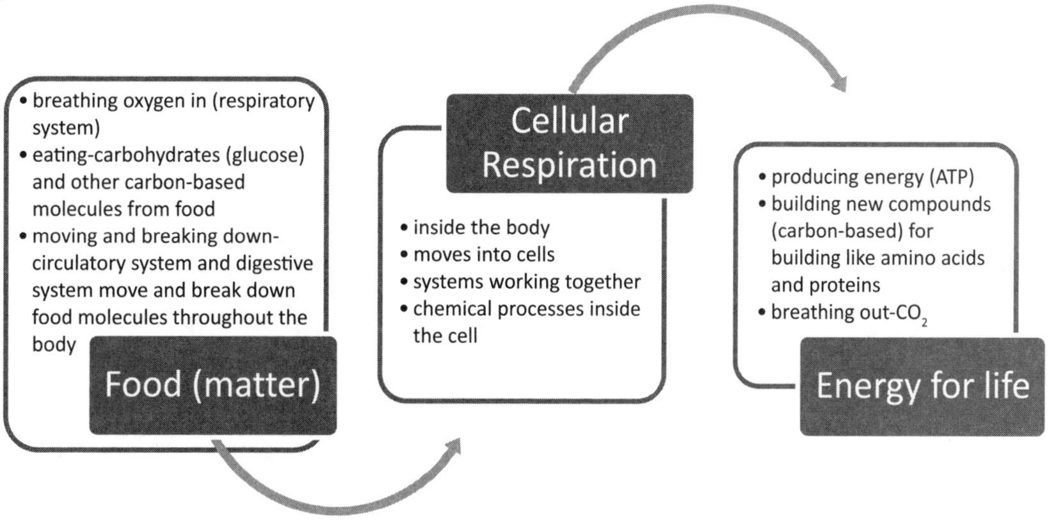

- **Food (matter)**
  - breathing oxygen in (respiratory system)
  - eating-carbohydrates (glucose) and other carbon-based molecules from food
  - moving and breaking down-circulatory system and digestive system move and break down food molecules throughout the body

- **Cellular Respiration**
  - inside the body
  - moves into cells
  - systems working together
  - chemical processes inside the cell

- **Energy for life**
  - producing energy (ATP)
  - building new compounds (carbon-based) for building like amino acids and proteins
  - breathing out-CO_2

- Why do we need to eat?
- How can we breath, move, and play?
- What does our body need to build, heal, and grow?
- How does everything work together in our whole body system (circulatory, respiratory, digestive, and so on) to provide us with the energy we need?
- Where does the energy come from? Where does it go?

Stage I: Identify Disciplinary Core Ideas, Practices, and Crosscutting Concepts

I turned to *A Framework for K–12 Science Education* (NRC 2012) and *Next Generation Science Standards* (NGSS Lead States 2013) to provide insight into the big ideas of cellular respiration. These documents helped me identify the performance expectation (PE), or cluster of PEs, that would be the focus for this area of instruction and would give me ideas on what students who demonstrate understanding would be able to do. I also pulled in related disciplinary core ideas (DCI), science and engineering practices (SEP), and crosscutting concepts (CCC). The disciplinary core ideas that drove instruction include Organization for Matter and Energy Flow in Organisms (LS1.C) and Cycles of Matter and Energy Transfer in Ecosystems (LS2.B). The practices that encourage and support the core ideas are Developing and Using Models and Constructing Explanations. The crosscutting concepts that provide an integrating theme that also supports student understanding are System and System Models and Energy and Matter.

These dimensions from *NGSS* helped me get my head around the idea that students need not memorize isolated facts about cellular respiration, but instead should develop a conceptual understanding through modeling, constructing evidence-based explanations, and understanding the relationship between matter and energy in a system. Table 5.1 shows the cluster of related PEs, DCIs, SEPs, and CCCs that I chose for the area of instruction focused on matter and energy: cellular respiration.

The following paragraphs are from *A Framework for K–12 Science Education* (NRC 2012). They helped me grasp main ideas and helped me steer away from detailed facts that may confuse students and inhibit their learning and understanding the big picture. Students can get lost in the minute detail if not connected to the question "Why do I need to know this?" I wanted to provide relevancy and purpose for their learning.

> *By the end of grade 12.* The process of photosynthesis converts light energy to stored chemical energy by converting carbon dioxide plus water into sugars plus released oxygen. The sugar molecules thus formed contain carbon, hydrogen, and oxygen; their hydrocarbon backbones are used to make amino acids and other carbon-based molecules that can be assembled into larger molecules (such as proteins or DNA), used for example to form new cells. As matter and energy flow through different organizational levels of living systems, chemical elements are recombined in different ways to form different products. As a result of these chemical reactions, energy is transferred from one system of interacting molecules to another. For example, aerobic (in the presence of oxygen) cellular respiration is a chemical process in which the bonds of food molecules and oxygen molecules are broken and new compounds are formed that can transport energy to muscles.

Table 5.1

PEs, DCIs, SEPs, and CCCs for Respiration

Performance Expectations
Students who demonstrate understanding can: HS-LS1-6. Construct and revise an explanation based on evidence for how carbon, hydrogen, and oxygen from sugar molecules may combine with other elements to form amino acids and/or other large carbon-based molecules. [Clarification Statement: Emphasis is on using evidence from models and simulations to support explanations.] *[Assessment Boundary: Assessment does not include the details of the specific chemical reactions or identification of macromolecules.]* HS-LS1-7. Use a model to illustrate that cellular respiration is a chemical process whereby the bonds of food molecules and oxygen molecules are broken and the bonds in new compounds are formed resulting in a net transfer of energy. [Clarification Statement: Emphasis is on the conceptual understanding of the inputs and outputs of the process of cellular respiration.] *[Assessment Boundary: Assessment should not include identification of the steps or specific processes involved in cellular respiration.]* HS-LS2-3. Construct and revise an explanation based on evidence for the cycling of matter and flow of energy in aerobic and anaerobic conditions. [Clarification Statement: Emphasis is on conceptual understanding of the role of aerobic and anaerobic respiration in different environments.] *[Assessment Boundary: Assessment does not include the specific chemical processes of either aerobic or anaerobic respiration.]*

Disciplinary Core Ideas	Science and Engineering Practices	Crosscutting Concepts
LS1.C: Organization for Matter and Energy Flow in Organisms • The sugar molecules thus formed contain carbon, hydrogen, and oxygen: their hydrocarbon backbones are used to make amino acids and other carbon-based molecules that can be assembled into larger molecules (such as proteins or DNA), used for example to form new cells. (HS-LS1-6) • As matter and energy flow through different organizational levels of living systems, chemical elements are recombined in different ways to form different products. (HS-LS1-6),(HS-LS1-7) • As a result of these chemical reactions, energy is transferred from one system of interacting molecules to another. Cellular respiration is a chemical process in which the bonds of food molecules and oxygen molecules are broken and new compounds are formed that can transport energy to muscles. Cellular respiration also releases the energy needed to maintain body temperature despite ongoing energy transfer to the surrounding environment. (HS-LS1-7) *LS2.B: Cycles of Matter and Energy Transfer in Ecosystems* • Photosynthesis and cellular respiration (including anaerobic processes) provide most of the energy for life processes. (HS-LS2-3)	• Developing and Using Models • Constructing Explanations	• System and System Models • Energy and Matter

Anaerobic (without oxygen) cellular respiration follows a different and less efficient chemical pathway to provide energy in cells. Cellular respiration also releases the energy needed to maintain body temperature despite ongoing energy loss to the surrounding environment. Matter and energy are conserved in each change. This is true of all biological systems, from individual cells to ecosystems. (NRC 2012, p. 148)

By the end of grade 12. Photosynthesis and cellular respiration (including anaerobic processes) provide most of the energy for life processes. Plants or algae form the lowest level of the food web. At each link upward in a food web, only a small fraction of the matter consumed at the lower level is transferred upward, to produce growth and release energy in cellular respiration at the higher level. Given this inefficiency, there are generally fewer organisms at higher levels of a food web, and there is a limit to the number of organisms that an ecosystem can sustain.

The chemical elements that make up the molecules of organisms pass through food webs and into and out of the atmosphere and soil and are combined and recombined in different ways. At each link in an ecosystem, matter and energy are conserved; some matter reacts to release energy for life functions, some matter is stored in newly made structures, and much is discarded. Competition among species is ultimately competition for the matter and energy needed for life.

Photosynthesis and cellular respiration are important components of the carbon cycle, in which carbon is exchanged between the biosphere, atmosphere, oceans, and geosphere through chemical, physical, geological, and biological processes. (NRC 2012, p. 154)

The performance expectation and the dimensions from *NGSS* informed the development of a learning goal that helped me focus this area of study: Living organisms obtain and break down matter to form new compounds and release energy that can be used or stored for life processes. So, what do students need to know to reach an understanding of this learning goal? They need to know that organisms "obtain" matter and what this matter is. Most students have this knowledge—they eat every day and know by high school that all living organisms must eat, either by ingesting or absorbing food (matter). Students need to know what it means to break down matter. A study of the digestive system and enzymes help students grasp that there are proteins (enzymes) in the body that break down food into smaller molecules that can be delivered to cells in the body via the blood stream. Once this "broken down matter" gets to cells, then what?

I think students have a tough time going from the idea that they eat to get energy to the process of cellular respiration. They often understand that the digestive system plays a role but may not realize that many of their body systems work together in this process. But what happens to the "food" to give them energy? How and why does it break down? What is this "energy"? The jump from food being digested to memorizing the steps in cellular respiration leaves a gap in understanding. Students don't understand the relevancy of cellular respiration. And memorizing steps doesn't necessarily get them there.

Stage II: Deconstruct DCIs, Create a Storyline, and Align Practices and Crosscutting Concepts

I grappled with stage II for a while, trying to decide how best to describe learning targets that capture the key ideas of this area of study without overwhelming students with details that don't necessarily contribute to their overall understanding. I used *NGSS* and my experience in teaching applied biology, general biology, and AP Biology (three very different levels) to arrive at the four learning targets below.

1. Matter and energy flow through different organizational levels of living systems. Living organisms consume food, which provides energy. Proteins, carbohydrates, and fats (lipids), which are large molecules (macromolecules), are food that can be converted to usable energy. These large molecules contain the elements carbon, hydrogen, and oxygen.

2. These large molecules are broken down by the digestive system into smaller molecules by mechanical processes and by enzymes. Examples of these smaller molecules are glucose (broken down from carbohydrates), amino acids (broken down from proteins), and lipids (into fatty acids and glycerol).

3. These small molecules are absorbed into the bloodstream (circulatory system) through the intestines and delivered to cells throughout the organism. These small molecules transport into cells and undergo chemical reactions. They can be building blocks for new large carbon-based molecules used for cell structure and processes or used in the process of cellular respiration.

4. Cellular respiration occurs inside cells. In cellular respiration, glucose is a common example of a molecule that goes through cellular respiration. Glucose is broken down further in a series of chemical reactions that result in the release and storage of energy. This energy is captured in a molecule called ATP. This process is used by living organisms to transfer energy released from molecules like glucose into a form of energy that is usable by living organisms. For example, the energy stored in ATP is released and used by muscles to contract.

For high-level biology courses, the following ideas will also be addressed through readings, activities, investigations, and explanations, but they are not part of the content storyline for a general life science or biology class. Resist the temptation to add facts and details into the storyline, which might divert students from the four learning targets.

- In a cell, the series of chemical reactions in cellular respiration include glycolysis (occurs in the cytoplasm), Krebs cycle (occurs in the mitochondria), and electron transport chain (occurs in the mitochondria).

- Aerobic respiration occurs in the presence of oxygen and generates a lot of ATP. Glucose + Oxygen → water + energy (ATP)

- Anaerobic respiration occurs in the absence of oxygen and follows a different, less efficient pathway. This generates less ATP.

  *Glucose → lactic acid + energy (small amount of ATP) <u>or</u>

  *Glucose → ethanol + carbon dioxide + energy (small amount of ATP)

Essential Understanding: The flow of energy and matter through different organizational levels of living systems recombine elements in different ways to form different products. As a result of these chemical reactions, energy is transferred from one system of interacting molecules to another. Photosynthesis and cellular respiration provide most of the energy for life processes. Energy and matter are both conserved.

Stage III: Determine Performance Expectations and Identify Criteria to Determine Student Understanding

See Table 5.2 (p. 179) for a summary of the work I did to complete Stage III of the design process.

Stage IV: Determine Nature of Science (NOS) Connections

Students know that they need to eat and that eating gives them energy. In thinking about where students are coming from in their understanding of matter and energy, most students have the empirical evidence that supports their understanding that food gives them energy. I thought about other evidence that would support student understanding of this occurring at the cellular level. Students can collect their own evidence, make sense of it, and connect it to science concepts through an investigation in the classroom. *NGSS* connects the Nature of Science to this standard through the following:

Scientific Knowledge Is Open to Revision in Light of New Evidence
• Most scientific knowledge is quite durable, but is, in principle, subject to change based on new evidence and/or reinterpretation of existing evidence. (HS-LS2-3)

I referred to *NGSS* Appendix H to glean a little more information about the Nature of Science. I focused on the matrix presented with the eight themes describing learning outcomes. For cellular respiration, a complex and abstract concept, I felt that students need to make observations and collect data that could provide evidence helping to

Table 5.2

Success Criteria and Aligned NGSS Performance Expectations

Learning Target	Success Criteria (from Science and Engineering Practices and Crosscutting Concepts)	Performance Expectations
#1: Matter and energy flow through different organizational levels of living systems. Food is consumed by living organisms. It contains proteins, carbohydrates, and fats (lipids), which are large molecules (macromolecules). These large molecules contain the elements carbon, hydrogen, and oxygen.	Construct a diagram (draw and write) that addresses the question, "How do you get energy to live, move, and grow?" Collect evidence to support the scientific idea about the composition and flow of matter.	HS-LS1-6. Construct and revise an explanation based on evidence for how carbon, hydrogen, and oxygen from sugar molecules may combine with other elements to form amino acids and/or other large carbon-based molecules.
#2: These large molecules are broken down by the digestive system into smaller molecules by mechanical processes and by enzymes. Examples of these smaller molecules are glucose (broken down from carbohydrates) and amino acids (broken down from proteins).	Create a diagram that describes the interactions of systems and how these systems process energy-containing large molecules (contained in food).	HS-LS1-7. Use a model to illustrate that cellular respiration is a chemical process whereby the bonds of food molecules and oxygen molecules are broken and the bonds in new compounds are formed resulting in a net transfer of energy.
#3: These small molecules are absorbed into the bloodstream through the intestines and delivered to cells throughout the organism. These small molecules transport into cells and undergo chemical reactions. They can be building blocks for new large carbon-based molecules used for cell structure and processes or used in the process of cellular respiration.	Develop and use a model to explain the carbon-based molecules protein, carbohydrates, and lipids and what components make up each.	
#4: Cellular respiration occurs inside cells. In cellular respiration, glucose is a common example of a molecule that goes through cellular respiration. Glucose is broken down further in a series of chemical reactions that result in the release and storage of energy. This energy is captured in a molecule called ATP. This process is used by living organisms to transfer energy released from molecules like glucose into a form of energy that is usable by living organisms. For example, ATP is used by muscles to contract.	Plan and carry out an investigation that examines yeast in the presence of sugar (sucrose) and oxygen, analyze and interpret the data, and construct an explanation about cellular respiration in this organism.	HS-LS2-3. Construct and revise an explanation based on evidence for the cycling of matter and flow of energy in aerobic and anaerobic conditions.

make this concept a little more concrete at the cellular level. At the high school level, students should understand that "Scientific argumentation is a mode of logical discourse used to clarify the strength of relationships between ideas and evidence that may result in revision of an explanation." I really wanted to focus on students looking at evidence and using evidence to support scientific ideas.

Stage V: Identify Metacognitive Goals and Strategies

Students need to identify what they already know, what gaps they have in their understanding, and what they can do to fill those gaps. That is easier to say than to actually teach. My goal is for students to recognize what they don't know and be able (and confident enough) to ask the questions that will help guide them. Then, I'd like students to be able to identify and use the resources necessary to fill the gaps in their understanding. In other words, I'd like students to develop self-regulated learning (MERC 2011). This connects well with what they already know about energy and matter at the macro level to developing an understanding and the cellular level. Students will develop a graphic organizer (concept map, mind web, etc.) at the beginning of a lesson and then revisit during and at the end of the lesson (Refer to Instructional Tool 3.7, pp. 127–133). In this way, they will identify what they know and then add to their thinking throughout the lesson using resources and activities that fill in the gaps.

The practices in this area of study include constructing and revising an explanation based on evidence. Teaching students to be metacognitive helps them connect what they already know to evidence and science concepts and ask questions and identify resources that will help them solidify or extend their understanding.

Phase 2: Planning for Responsive Action

Stage VI: Research Student Misconceptions Common to This Topic That Are Documented in the Research Literature

It is important for me to know what students are thinking as we began to study these concepts. In order to elicit student ideas, I felt that it was important to begin by asking questions and have students discuss, in small groups, what they understand about matter and energy, why they eat, and how they get energy from food. The misconceptions that my students typically had are reflected below. These align with what current research shows about this topic. Student misconceptions around this topic include:

- Cellular respiration means the same thing as breathing.

- Cellular respiration does not occur in plants (only in animals).

- Cellular respiration can only occur in the presence of oxygen.

- Food, calories, and energy are all equal. Students don't understand the relationship among these ideas.

I believe the first three misconceptions can be addressed during class discussion, activities, and student discourse. Breathing in oxygen, a molecule that is involved in the reactions of cellular respiration, and breathing out carbon dioxide, a waste product from cellular respiration, can be examined. Plants use energy for growth and other processes. Therefore they also need the glucose they make to be transferred to a usable form of chemical energy. This concept is important to discuss within this unit of study. However, the fourth misconception is overarching and identifies what I think is the most difficult part of this area of study. I really want students to understand the relationship between what they eat and how the organic molecules in their food contain chemical energy that is transferred to an energy storage molecule (ATP). This link asks students to look at what they know about eating and connect it to the abstract, that which they cannot "see."

Stage VII: Determine Strategies to Identify Students' Preconceptions

For this unit of study, I find it essential to check in and know what students are thinking by eliciting their preconceptions. I identified the Drawing Out Thinking approach as one that would work for my students at the start of Lesson 1, especially in making connections between the food they eat and the energy they get. The strategy I chose to use for this is Drawing and Annotated Drawings (see Instructional Tool 3.8, pp. 134–135). In Lesson 2, students use a diagram to draw what happens to the food they eat. Prior to students working on their diagram, I will provide an anticipation guide that aligns with the resources and expectations for the lesson. This will not only allow me to know what they already understand, but will help them look for important information that will help them develop understanding that they can use for their diagram. Lessons 3, 4, and 5 all begin with probing questions, whole-class and small- group brainstorms, and discourse, asking students to make connections from previous lessons. This approach worked in my classroom because students felt safe to share their ideas and they gained understanding by listening to others. Questions get students thinking about the topic in a way that helps them make connections between big ideas. Listening to student discussions lets me know where they are in their learning, what to focus on, and making explicit connections that may have been missed (see table 5.3, p. 182).

Stage VIII: Determine Strategies to Elicit and Confront Students' Preconceptions

Through a sequence of lessons, I wanted to create a coherent content storyline that would help students build their understanding about cellular respiration, energy, and matter.

Table 5.3

Strategies and Activities for Each Learning Target

Storyline				
Learning Target #1 →	**Learning Target #2 →**	**Learning Target #3 →**	**Learning Target #4 →**	**Essential Understanding**
Matter and energy flow through different organizational levels of living systems. Food is consumed by living organisms. It contains proteins, carbohydrates, and fats (lipids), which are large molecules (macromolecules). These large molecules contain the elements carbon, hydrogen, and oxygen.	These large molecules are broken down by the digestive system into smaller molecules by mechanical processes and by enzymes. Examples of these smaller molecules are glucose (broken down from carbohydrates) and amino acids (broken down from proteins).	These small molecules are absorbed into the bloodstream through the intestines and delivered to cells throughout the organism. These small molecules transport into cells and undergo chemical reactions. They can be building blocks for new large carbon-based molecules used for cell structure and processes or used in the process of cellular respiration.	Cellular respiration occurs inside cells. In cellular respiration, glucose is a common example of a molecule that goes through cellular respiration. Glucose is broken down further in a series of chemical reactions that result in the release and storage of energy. This energy is captured in a molecule called ATP. This process is used by living organisms to transfer energy released from molecules like glucose into a form of energy that is usable by living organisms. For example, ATP is used by muscles to contract.	The flow of energy and matter through different organizational levels of living systems recombine elements in different ways to form different products. As a result of these chemical reactions, energy is transferred from one system of interacting molecules to another. Photosynthesis and cellular respiration provide most of the energy for life processes. Energy and matter are both conserved.
In groups, answer "How do you get energy to move, live, and grow?" on chart paper with pictures, arrows, words. Generate whole-class diagram (class discussion). Students record food and exercise in a log and identify "categories" of food and exercise. Show pictures of carbohydrates, proteins, and lipids and students identify similarities.	Students create a diagram that shows the pathway of food to their cells, identifying breakdown and delivery of nutrients to cells. a. Students answer questions on an "exit slip." (What large molecules from food are broken down in your digestive system? b. How do the smaller molecules get to all the cells in your body?)	Opening poll question, "Do all of the cells of your body need energy? Why or why not?" Linking back to Target #2, students develop and use models to explain lipids, proteins, and carbohydrates. Jigsaw expert groups.	Show animation of an overview of cellular respiration. Inquiry investigation of cellular respiration using yeast, glucose, and water. Students identify what they've learned or understand (on green sticky notes), one thing they wonder or have a question about (on yellow sticky notes), and one thing they are still confused about (on pink sticky notes).	Revisit initial chart paper with pictures, arrows and diagrams. Add information that represents the flow of matter and energy through a system (from food to a molecule that stores chemical energy). What is the input? Purpose (why is it there?) How did it get there? What happens to it? What is the output?

I thought about activities that would help me elicit their ideas and preconceptions. Activities that I chose are outlined in Figure 5.2. The activities that I chose to help students confront possible misconceptions and help clarify and deepen their understanding draw on bringing relevance to their own lives and making them the focus of their own learning. They are active participants. I use probing questions; whole-class, small-group, and partner discourse; concept mapping; and drawing/diagramming with explanations, brainstorming, and an investigation. All of these strategies are aligned to the learning goal, or target, for the lesson and link to previous and subsequent ideas and concepts. They also focus on the student—what he/she eats, human digestive (and other related) system, cells in the larger system, and yeast as an example organism. Students will continually confront their ideas as we link together the concepts through discussion and action. These ideas are discussed in more detail in the last section, Instructions to Teachers.

Figure 5.2

Lesson Progression

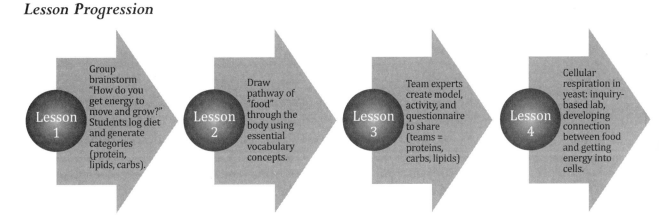

Lesson 1: Group brainstorm "How do you get energy to move and grow?" Students log diet and generate categories (protein, lipids, carbs).

Lesson 2: Draw pathway of "food" through the body using essential vocabulary concepts.

Lesson 3: Team experts create model, activity, and questionnaire to share (teams = proteins, carbs, lipids)

Lesson 4: Cellular respiration in yeast: inquiry-based lab, developing connection between food and getting energy into cells.

Stage IX: Determine Sense-Making Strategies

Sense-making strategies will be used in each lesson. This is critical for my students to really develop their understanding. Throughout all of the lessons, students will be using their science notebooks. They will answer questions, sketch ideas and diagrams (lessons 1 and 2), plan their lab investigation, make claims, collect evidence, and write explanations (lesson 4). They will also plan their model before building it (lesson 3) and generate notes for discussion from resources including an online animation. I believe science notebooks are key to students capturing their learning and have seen students gain confidence and knowledge through their notebooks. I will directly teach some vocabulary related to cellular respiration (e.g., enzymes, ATP, glucose, oxygen, carbon dioxide, cells, proteins, carbohydrates, lipids, and so on) and ask students to apply these terms to their understanding of the breakdown of their food into molecules that are distributed throughout the body and into cells to be broken down further. Students will communicate their ideas through pictures, words, and discussion throughout the four lessons, helping them assimilate and make sense of the concepts and connections presented. Using multiple strategies and allowing the time for students to make sense of what they are learning are critical in understanding these difficult and abstract concepts.

Stage X: Determine Responsive Actions Based on Formative Assessment Evidence

To get the most out of the formative assessment strategies I used in this area of study, I determined possible evidence that I could use to inform my instruction. What would the evidence that I collect tell me about my students? And what would I do with that knowledge? I identified ways to follow up with responsive actions that would help fill student gaps in their learning, affirm or extend their understanding. These ideas are shown in Table 5.4.

Table 5.4

Formative Assessments and Responsive Action for Each Learning Target

	Lesson Target #1	Lesson Target #2	Lesson Target #3	Lesson Target #4
Formative Assessment Strategy	Whole-group brainstorming web diagram	Exit slip with two questions: 1. What large molecules from food are broken down in your digestive system? 2. How do the smaller molecules get to all the cells in your body?	Peer-assessed questionnaire on models (proteins, carbohydrates, lipids)	Sticky note news: students write an idea they've learned and understand (green note), one thing they wonder or one question they have (yellow note), and one concept that they still don't understand (pink note).
Responsive Action	Make note of gaps and misconceptions based on student input; provide resources for next lesson that address these.	Categorize answers; address gaps in glass discussion at the beginning of the following class.	Review; Return to students for revision and corrections.	Review notes; Give students questions (yellow notes) to research (provide resources); Revisit pink notes.

In the next section, I've described each lesson as if I was giving instructions to a teacher (myself).

Instructions to Teachers

Lesson 1

Matter and energy flow through different organizational levels of living systems. Living organisms consume food, which provides energy. Proteins, carbohydrates, and fats (lipids), which are large molecules (macromolecules), are food that can be converted to usable energy. These large molecules contain the elements carbon, hydrogen, and oxygen.

In groups of three, ask students to gather around chart paper and draw pictures and write words in answer to the question, "How do you get energy to move, live, and grow?" Suggest using arrows and specific examples. Provide three different colored markers to each group and set the expectation that all students must contribute to this brainstorm.

After all groups have filled their chart papers, begin a diagram on the board (web, flow chart, visual representation). Ask each group to send one person to contribute to the diagram using ideas they generated in their smaller groups. Through this interactive, whole-class activity, formatively assess student preconceptions. Lead a discussion, eliciting their ideas. Students generate ideas such as eating food and the digestive system breaks down the food. Push students through questions such as, "then what happens?" and "how do you actually get the energy from the food?" "how is energy stored in food?" Accept all ideas and generate a curiosity and "need to know" as to how the energy from food is transferred into usable energy for living organisms. Identify misconceptions from student answers.

From the small-group brainstorm, diagrams, and class discussion, students should learn or remember that matter and energy flow through different levels of living systems (for example, from the Sun, to plants, to animals, to us (or from plants to us), and that the original source of energy is from the Sun). Students will have studied photosynthesis before this lesson. Verify that they know what is meant by the term *food* and the kinds of molecules that are considered food. And the light energy provided by the sun helps plants convert carbon dioxide and water into glucose (sugar).

Ask students to record everything that they eat and any type of exercise they do for two days in a log. Students compare their logs with a partner and identify categories of food and categories of exercises. They typically identify these categories: protein, carbohydrates, and fats (lipids) for foods. If not, help guide them to these categories and have them place the types of food they eat into these categories. For exercise, they typically come up with sports, walking or general movement. Using probing questions, ask them to think about how they get the energy to exercise. Some may identify breathing as an action that accompanies exercise (and living in general!) Ask students to remember breathing as part of the flow of energy and matter in cellular respiration (oxygen needed for aerobic respiration and carbon dioxide as a waste product).

Show pictures of the structure of a carbohydrate, a protein, and a lipid. Ask students to identify what all of these molecules (that make up foods) have in common. Students will identify carbon, hydrogen, and oxygen as elements that make up these molecules.

Lesson 2

These large molecules are broken down by the digestive system into smaller molecules by mechanical processes and by enzymes. Examples of these smaller molecules are glucose (broken down from carbohydrates), amino acids (broken down from proteins), and lipids (into fatty acids and glycerol).

Students are each given a large (11 × 17) outline of a person. They are given informational resources and work with a partner to draw the pathway of food from ingestion through digestion and excretion. Students are asked to identify where the food is broken down mechanically, where it is broken down chemically (by enzymes and acids), where it is absorbed into the bloodstream, and how it travels throughout the body. They are given a word list of essential vocabulary (pruned down to only terms that will help them understand the big ideas) to help them identify the key components of their diagram. The teacher checks in with groups as they are working and asks them to explain their diagrams. As an "exit slip," ask students two questions:

1. What large molecules from food are broken down in your digestive system?
2. How do the smaller molecules get to all the cells in your body?

Lesson 3

These small molecules are absorbed into the bloodstream through the intestines and delivered to cells throughout the organism. These small molecules transport into cells and undergo chemical reactions. They can be building blocks for new large carbon-based molecules used for cell structure and processes or used in the process of cellular respiration.

Open up this lesson by asking the question, "Do all of the cells of your body need energy? Why or why not?" Students should already know that cells make up tissues, tissues make up organs, organs make up systems, and systems make an organism. Take this opportunity to link to this past learning.

Transition by revisiting the food logs students created. Ask again what organic molecules make up food. Then, lead into a modeling activity by asking: What are proteins? What are lipids? What are carbohydrates?

Developing and using models for understanding and explanations is a science and engineering practice in *NGSS*. For this activity, group students in teams of three. Each team is assigned "carbohydrate," "protein," and "lipid." Provide materials such as clay, assorted craft supplies, and informational resources. Ask each team to provide a model of their assigned large molecule, identifying key components. Each team is responsible for showing what the large molecule is broken down into (during digestion) and the building of a new large molecule (and what the monomers are that make up the larger polymer—i.e., amino acids to proteins, glucose to carbohydrates). After becoming experts on their molecule, the team develops an activity (game, modeling, and so on) and creates a three to five-part "questionnaire."

Experts from each team join with an expert from another team until each new group has one person representing each large molecule (carbohydrate, protein, lipid). Students share their knowledge by explaining their model, facilitating the activity, and answering questions. At the end of the discussion, each person in the group takes all three developed "questionnaires." Students give their answers to the "expert" to grade. After reviewing student questionnaires, return them to students asking them to revise their answers (if incorrect) and self-assess their understanding.

Lesson 4

Cellular respiration occurs inside cells. In cellular respiration, glucose is a common example of a molecule that goes through cellular respiration. Glucose is broken down further in a series of chemical reactions that result in the release and storage of energy. This energy is captured in a molecule called ATP. This process is used by living organisms to transfer energy released from molecules like glucose into a form of energy that is usable by living organisms. For example, the energy stored in ATP is released and used by muscles to contract.

Show the "big picture" at the beginning of the lesson (*www.sumanasinc.com/webcontent/ animations/content/cellularrespiration.html*). Ask students if they know what yeast is. Lead a class discussion focused on if all living organisms (including yeast) require energy. Ask students how they could test this in yeast?

Give an overview of the yeast lab. Show students the equation $C_6H_{12}O_6 + O_2 \rightarrow 6\ CO_2 + H_2O$ and ask what is meant by each component. Ask them what they think is necessary for yeast to undergo cellular respiration (fermentation can also be introduced here). Facilitate an inquiry-based approach to this lab, supporting student questions regarding the necessary components and conditions for cellular respiration. Proceed with the lab. Students discuss the lab and what they discovered in a small group. They summarize the findings of the lab, how it relates to cellular respiration, and to energy required by living organisms. Students support their explanations by engaging in argumentation using results from their lab and connections to scientific knowledge. With their findings, students present a visual representation of this process of energy transfer.

Each student is given three sticky notes (green, yellow, and pink). They write down one idea they've learned and understand on green, one thing they wonder or one question they have on yellow, and one concept that they still don't understand on pink. They attach their notes to the stoplight chart on the way out of the room. Review the notes and have small-group discussions using the yellow and pink sticky notes to guide discussions. Bring any lingering questions from small groups to the whole class, clarifying understanding.

Essential Understanding: The flow of energy and matter through different organizational levels of living systems recombine elements in different ways to form different products. As a result of these chemical reactions, energy is transferred from one system of interacting molecules to another. Photosynthesis and cellular respiration provide most of the energy for life processes. Energy and matter are both conserved.

For a final assessment, begin by asking students probing questions to connect photosynthesis; cellular respiration; systems; and matter, energy, and life. Ask students to revisit their initial diagram and add to it pictures, words, and arrows to convey their understanding of the flow of energy and matter through a living organism. Their final diagram should include food, large molecules (protein, carbohydrates, lipids) and their components, digestion, cellular respiration, and how "food" is converted to energy in a living organism. By engaging in the four lessons and completing the final assessments, students meet the performance expectations for these standards and gain an understanding of why they need to eat and the processes necessary to convert food (matter) into usable energy for life.

Recommended Resources

Websites

- Cellular Respiration: *www.sumanasinc.com/webcontent/animations/content/cellularrespiration.html*

- Cellular Respiration and Photosynthesis: Important Concepts, Common Misconceptions, and Learning Activities: *http://serendip.brynmawr.edu/exchange/files/CellularRespPhotoOverview.pdf*

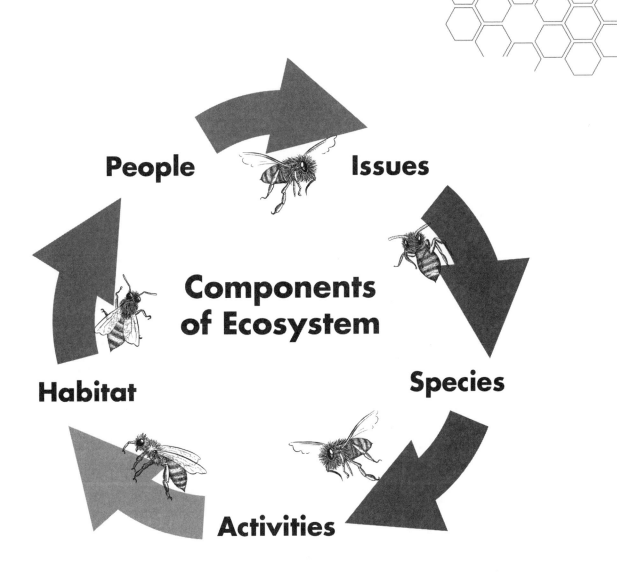

People

Issues

Components
of Ecosystem

Species

Habitat

Activities

Chapter 6

Ecosystems: Interactions, Energy, and Dynamics— An Issue-Based Approach

By Nancy Kellogg

"Man did not weave the web of life—he is merely a strand in it. Whatever he does to the web, he does to himself."

—*Chief Seattle, 1854*

"The race is now on between the technoscientific and scientific forces that are destroying the living environment and those that can be harnessed to save it …. If the race is won, humanity can emerge in far better condition than when it entered, and with most of the diversity of life still intact."

—*Edward O. Wilson, The Future of Life*

Overview

The honeybee Colony Collapse Disorder (CCD) mystery provides a great opportunity for high school teachers to introduce ecosystem stability and resilience that includes the importance of biodiversity in ecosystems. This issue can easily encompass all three dimensions of *NGSS*: scientific and engineering practices (SEP), crosscutting concepts (CCC), and disciplinary core ideas (DCI). The storyline for this chapter builds around the honeybee CCD issue and incorporates the three *NGSS* dimensions.

I have found that addressing a current ecosystem issue better engages students as they learn the disciplinary core ideas on ecosystem stability and resilience than the standard traditional approach to teaching these ideas. Other critical issues that impact ecosystem stability and resilience include invasive species, overharvesting fish, clear-cutting forests, drought conditions resulting in wildfires, and many other local issues. You can choose any issue that excites students to learn the important content, practices, and crosscutting concepts using the model provided in this chapter. You can also link students with a citizen science project. Citizen science projects that engage the public in environmental research are springing up all over the world (Dickinson and Bonney 2012).

So why did I choose honeybees and CCD for my focus? Every day, news commentators and writers discuss issues related to ecosystems. For example, a recent *Time* magazine cover boldly stated, "A World Without Bees: The price we'll pay if we don't figure out what's killing the honeybee." The feature article is titled "The plight of the honeybee, Mass deaths in bee colonies may mean disaster for farmers—and your favorite foods" (Walsh 2013, p. 24). Commercial beekeeping is a critical industry in the United States and throughout the world.

Honeybees, *Apis mellifera,* are currently dying on a large scale and hives are abandoned. Colony collapse disorder (CCD) is a fairly recent phenomenon for which scientists are looking for evidence about why honeybees are dying on a large scale. The CCD mystery has several possible explanations and may involve more than one reason. Some possible explanations include neonicotinoids (chemical pesticides), American foulbrood (bacterial disease), the Varroa microscopic mite, and a fungal infection. As a result, the agricultural industry is impacted. For example, almond trees are 100% pollinated by bees. According to the *Time* article about one-third of the food we consume relies somewhat on pollination by honeybees (Walsh 2013).

I completed further research on the history of the honeybee in the United States that illuminated some historical aspects of importance to biodiversity. European immigrants introduced the honeybee to North America along with European agriculture crops in the early 1600s. Native bees, other insects, and some vertebrates (e.g., birds) pollinated native plants prior to the introduction of the honeybee. With the advent of modern agriculture during the 1950s, farmers began planting more monoculture crops with little or no native vegetation. This decreased the wild bee populations due to lack of good nesting sites. Most native bees build nests on the ground in thick grass, soil, and dead wood (Horn 2005, 2008; Kellar, n.d.). Figure 6.1 is an example of how teachers might start a student citizen science project around an issue such as the population decline in bees.

Figure 6.1

Student Citizen Scientist

Wanted

Student Citizen Scientist

to study bee biodiversity in a local ecosystem

The native (wild) bee story is not well known. To learn more about wild native bees I talked with two scientists at the University of Colorado Museum of Natural History. Over 900 species of native bees including wasps are found in Colorado. Think about the numbers for North America! These many species pollinate different plants than honeybees (e.g., tomatoes, peppers, squash, eggplant, blueberries, and raspberries). In some cases native bees pollinate only one plant species. Native bees are also on the decline. A citizen science project "The Bees Needs" is in the pilot year and is sponsored by the CU Museum (see Figure 6.1). In addition to project research on the diversity of native bees that nest in hollow wood, project leaders educate the volunteers about the importance of native bees related to biodiversity and how to improve management of habitats to increase species of all organisms for more stable ecosystems (Interviews with Alexandra Rose and Virginia Scott 2013).

A compelling reason to study an ecosystem issue such as the honeybee population decline is that this issue offers an opportunity for authentic science research. Ecosystem issues are messy and generally do not have simple solutions. Roth (1995) gives five qualities of authentic science, which allows students to work in the way scientists work. This applies to an ecosystem issue-based approach.

1. Learn in contexts constituted in part by ill-defined problems.
2. Experience uncertainties, ambiguities, and the social nature of scientific work and knowledge.
3. Engage in learning (curriculum), which is predicated on, and driven by their current knowledge state.
4. Experience themselves as part of communities of inquiry, in which knowledge, practices, resources, and discourse are shared.
5. Participate in classroom communities, in which they can draw on the expertise of more knowledgeable others, whether those others are peers or advisors.

A framework for socioscientific issues also supports the model unit of study shared in this chapter. Learner experiences, one part from the Socio-Scientific Issues (SSI) Framework, is highlighted here. The authors of the framework give a vision for scientific literacy that includes "real-life" situations that are scientific in nature but are influenced by other factors (e.g., political, ethical, social). This aligns with NGSS, especially the scientific practices. The authors state "Socio-scientific issues (SSI), open-ended social problems with substantive connections to science (e.g., climate change, gene therapy, and nuclear power) represent the kinds of situations in which many individuals will be challenged to exercise their scientific literacy" (Presley et al. 2013, p. 28). *NGSS* science and engineering practices align with learner experiences and opportunities from the SSI as shown in Figure 6.2.

Personal Experience and Student Tips

My own teaching experience demonstrates the power of experiential education using community-based issues that engage high school students. For example, my students designed and constructed a Braille nature trail in a local park that had a relatively stable ecosystem. To accomplish this task, the students learned the major concepts related to the ecosystem, met with park employees, collaborated with visually impaired students, and engaged in engineering practices for trail design and construction. Throughout the project, students were highly motivated to learn the science content and engineering design skills. The project culminated with a trail opening ceremony with local media coverage.

Figure 6.2

SSI Framework Learner Experiences Align With NGSS Science and Engineering Practices

The second core aspect in the SSI framework describes necessary learner experiences and opportunities, which include:

1. Engaging in higher-order practices (e.g., reasoning, argumentation, decision making and/or position taking).

2. Confronting scientific ideas and theories related to the issue being considered.

3. Collecting and/or analyzing scientific data related to the issue being considered.

4. Negotiating social (e.g., political and economic) dimensions of the issue being considered.

Source: Presley et al. p. 28

As a professional development provider for many K–12 science teachers, I facilitated sessions where teachers learned content and skills to work with their students on local issue-based projects. Examples include monitoring river quality using specific protocols; planning for and constructing a school garden; restoring a natural wetland near their school; and evaluating energy use in their school with follow-up strategies for energy conservation. To achieve their project goals, students had to engage in authentic research. These projects might take several weeks or be an ongoing endeavor.

Based upon my experience, I have some tips for planning and implementation.

- Ask for input from your students about what issue(s) they would like to study (at least one month prior to studying the issue).

- Narrow down to one issue, preferably one that has active inquiry-based investigation and different potential solutions or outcomes.

- Determine local community resources to support work on the issue.

- Plan for students to interview experts and local community people on the issue. This means students will need to plan for an interview (in person, phone, Skype). Students generally need guidance on what kinds of questions to ask to gather the information they want (who, what, when, where, why, how) and ways to record their information.

- Students will need to practice their interview questions, strategies for collecting and compiling their information, and reporting what they learned.

- Assess other student skills that might be needed and give them practice opportunities if there are deficits (e.g., reading, constructing, and interpreting graphs and mathematical representations).

- Students begin research on their issue. Stages and phases are outlined later in this chapter to guide the process.

- Students need to check in frequently with you on the progress they are making. Some groups may need more structure than other groups.

- At the end of the unit it is important to honor the student research with a celebration. Community resource people that have worked with the students should be invited.

Why This Topic?

The ecosystems topic I chose is ecosystem resilience and stability emphasizing the importance of biodiversity. The impact of human activity on ecosystem stability is incorporated into this unit. Prerequisite student knowledge of this unit includes genetics related to speciation and natural selection, hierarchal structure of ecosystems, cycles of matter, energy transfer in ecosystems, basic population dynamics, and carrying capacity.

There are so many topics that could serve well for an issue-based project, so why this one? The *National Science Education Standards* state "Human beings live within the world's ecosystems. Increasingly, humans modify ecosystems as a result of population growth, technology, and consumption. Human destruction of habitats through harvesting, pollution, atmospheric changes, and other factors is threatening current global stability, and if not addressed, ecosystems will be irreversibly affected" (NRC 1996, p. 186).

Hazen and Trefil (1991) give one rule that emerges from studies of ecosystems, "You can't change just one thing" or more grandiloquently, it is: "The Law of Unintended Consequences." They further state that the rule comes down to this: "In a complex system it is not always possible to predict what the consequences of any change will be, at least with the present state of knowledge. This means that seemingly small changes in ecosystems can cause large effects, while huge changes might leave the system pretty much as it was" (p. 262).

E. O. Wilson, an evolutionary biologist, introduced the term *biodiversity* in 1988 (Wilson 1988). There are numerous publications that address the importance of biodiversity on the stability and resilience of ecosystems. Scientists are still determining the number of species on Earth. A group of leading biodiversity experts issued a statement in 2006 warning that the world is "on the verge of a major biodiversity crisis" and that governments and private actors needed to take the issue more seriously. They stressed the scientific complexity of biodiversity. "By definition, biodiversity is diverse: it spans several levels of biological organization (genes, species, ecosystems); it cannot be measured by simple universal indicators such as temperature and atmospheric CO_2 concentration; and its distribution and management are more local in nature" (Loreau et. al. 2006, pp. 245–246). The NSDL-AAAS Science Literacy Map (The Living Environment, Diversity of Life) states, "A great diversity of species increases the chance that at least some living things will survive in the face of large changes in the environment" (NSDL n.d.).

Scientists track extinction rates, population decline, and habitat loss trends to estimate the change in biodiversity levels. Habitat loss is generally viewed as the largest single cause of biodiversity loss worldwide. When habitats are developed into roads and other human uses, they become fragmented and the undeveloped parts may become too small or isolated to support viable populations of species. Biodiversity is also threatened by invasive species because native species are not adapted to compete with them. Other human activities that threaten biodiversity are overharvesting plants and animals, pollution, and global climate change (Annenberg n.d.).

There are three statements in the NSDL-AAAS Science Literacy Map (The Living Environment, Interdependence of Life) that are important for this chapter:

- "Ecosystems can be reasonably stable over hundreds of thousands of years. As any population grows, its size is limited by one or more environmental factors: availability of food, availability of nesting sites, or number of predators."

- "The human species has a major impact on other species in many ways reducing the amount of the Earth's surface available to those other species, interfering with their food sources, changing the temperature and chemical composition of their habitats."

- "If a disturbance such as flood, fire, or the addition or loss of species occurs, the effected ecosystem may return to a system similar to the original one, or it may take a new direction, leading to a very different type of ecosystem." (NSDL n.d.)

Students and the population in general have misconceptions about ecosystem resilience and biodiversity. Social, economic, and political issues come into play in the decisions humans make about their environment. Human decisions greatly impact their environment and need to be based upon scientific evidence not pseudoscience, beliefs, or opinions.

Some students have misconceptions about ecological relationships. "Students' understanding of ecological relationships depends on their concepts of 'plant' and 'animal', and also on their knowledge of habitats and use of physical principles" (Driver et al. 1994, p. 61). Few students from ages 5 to 16 used the idea of interdependence to explain their selection of organisms for a balanced community (Driver et al. 1994, p. 62). Figure 6.3 (p. 196) addresses some specific misconceptions about biodiversity.

Figure 6.3

Misconceptions About Biodiversity

Misconception:	Not all species are important, so some have to be sacrificed.
Clarification:	It is difficult to fully appreciate the complete impact a species has on an ecosystem. The complex interactions between species are often unknown. Example: loss of a single pollinator could have profound effects seen all the way up through the food chain.
Misconception:	If there is a lot of one species, then the ecosystem must be healthy.
Clarification:	To have a healthy ecosystem diversity is necessary with populations that approach the carrying capacity of the ecosystem.
Misconception:	It is common for people to think that large empty spaces that are unmanaged are wastelands (e.g., desert, abandoned city lot).
Clarification:	These spaces are very necessary to provide habitat for many species.
Misconception:	Biodiversity is not important to humans. Humans are not part of the ecosystem.
Clarification:	Humans are part of the ecosystem and are an important part in the web of life. Therefore biodiversity is important for humans.

Adapted: Annenberg Biodiversity Decline Professional Development Guide. *www.learner.org*

Given that students hold these misconceptions, there is a compelling reason to illustrate a model unit of study for ecosystem resilience and stability emphasizing biodiversity and human impact in this chapter.

Application of Phase 1 to Ecosystem Resilience and Stability, Biodiversity, and Human Impact Unit

Stage 1: Identify Disciplinary Core Ideas, Practices, and Crosscutting Concepts

The DCIs for my model unit come from more than one standard, namely HS-LS2.C (Ecosystem Dynamics, Functioning, and Resilience), part of HS-LS4.D (Biodiversity and Humans), and part of HS-ETS1.B (Developing Possible Solutions). I selected these DCIs as a bundled group since ecosystem resilience and stability are heavily impacted by human activities that often decrease biodiversity. The engineering DCI focuses on solutions to enhance biodiversity and ecosystem stability, limiting the focus to a selected ecosystem that students want to study.

As I began the design process, I started with HS-LS2-C, Ecosystem Dynamics, Functioning, and Resilience. Refer to the center column of the foundation boxes that describe the DCIs (Figure 6.4, p. 198) and look at both bullets under HS-LS2.C for major content for this chapter. Bullet 1 addresses ecosystem stability and resilience while bullet 2 focuses on human activity in the environment that can disrupt an ecosystem. The next DCI I included is HS-LS4.D (Biodiversity and Humans). Bullet 2 focuses on how humans depend on the living world for resources with emphasis on the importance of sustaining biodiversity.

The Engineering Design DCI, HS-ETS1.B (Figure 6.5, p. 199), is a nice fit that is bundled with the above DCIs in this model unit of study. I selected this engineering design DCI to engage students in evaluating solutions with an emphasis on constraints, benefits, and environmental impacts.

I next selected the best crosscutting concepts (See the right-hand column of Figure 6.4). System and System Models is an obvious choice. From the *Framework*, "The parts of a system are interdependent, and each one depends on or supports the functioning of the system's other parts. Yet the properties and behavior of the whole system can be very different from those of any of its parts, and large systems may have emergent properties, such as the shape of a tree, that cannot be predicted in detail from knowledge about the components and their interactions" (NRC 2012, p. 92).

A second crosscutting concept that fits well is Stability and Change. In the *Framework*, "Any system has a range of conditions under which it can operate in a stable fashion, as well as conditions under which it cannot function. For example, a particular living organism can survive only within a certain range of temperatures, and outside that span it will die. Thus elucidating what range of conditions can lead to a system's stable operation and what changes would destabilize it (and in what ways) is an important goal" (NRC 2012, p. 99). The *Framework* also discusses that stability is always a balance of competing effects. "An understanding of dynamic equilibrium is crucial to understanding the major issues in any complex system—for example, population dynamics in an ecosystem or the relationship between the level of atmospheric carbon dioxide and Earth's average temperature" (NRC 2012, p. 100).

I completed Stage I by determining appropriate science and engineering practices (See the first column in Figure 6.4). There are three practices that fit well for this model unit: "using mathematical and computational thinking" (bullet 2), "constructing explanations and designing solutions" (bullet 2), and "engaging in argument from evidence" (bullet 1). "Mathematics (including statistics) and computational tools are essential for data analysis, especially for large data sets. The abilities to view data from different perspectives and with different graphical representations, to test relationships between variables, and to explore the interplay of diverse external conditions all require mathematical skills that are enhanced and extended with computational skills" (NRC 2012, p. 65). For example, as students view ecosystem biodiversity, they will view computer data sets, make comparisons, and look for mathematical trends in the data.

Figure 6.4

HS-LS2: Ecosystems: Interactions, Energy, and Dynamics Foundation Boxes

Science and Engineering Practices

Developing and Using Models
Modeling in 9–12 builds on K–8 experiences and progresses to using, synthesizing, and developing models to predict and show how relationships among variables between systems and their components in the natural and designed worlds.

- Develop a model based on evidence to illustrate the relationships between systems or components of a system. (HS-LS2-5)

Using Mathematics and Computational Thinking
Mathematical and computational thinking in 9-12 builds on K-8 experiences and progresses to using algebraic thinking and analysis, a range of linear and nonlinear functions including trigonometric functions, exponentials and logarithms, and computational tools for statistical analysis to analyze, represent, and model data. Simple computational simulations are created and used based on mathematical models of basic assumptions.

- Use mathematical and/or computational representations of phenomena or design solutions to support explanations. (HS-LS2-1)
- Use mathematical representations of phenomena or design solutions to support and revise explanations. (HS-LS2-2)
- Use mathematical representations of phenomena or design solutions to support claims. (HS-LS2-4)

Constructing Explanations and Designing Solutions
Constructing explanations and designing solutions in 9–12 builds on K–8 experiences and progresses to explanations and designs that are supported by multiple and independent student-generated sources of evidence consistent with scientific ideas, principles, and theories.

- Construct and revise an explanation based on valid and reliable evidence obtained from a variety of sources (including students' own investigations, models, theories, simulations, peer review) and the assumption that theories and laws that describe the natural world operate today as they did in the past and will continue to do so in the future. (HS-LS2-3)
- Design, evaluate, and refine a solution to a complex real-world problem, based on scientific knowledge, student-generated sources of evidence, prioritized criteria, and tradeoff considerations. (HS-LS2-7)

Engaging in Argument from Evidence
Engaging in argument from evidence in 9–12 builds on K–8 experiences and progresses to using appropriate and sufficient evidence and scientific reasoning to defend and critique claims and explanations about the natural and designed world(s). Arguments may also come from current scientific or historical episodes in science.

- Evaluate the claims, evidence, and reasoning behind currently accepted explanations or solutions to determine the merits of arguments. (HS-LS2-6)
- Evaluate the evidence behind currently accepted explanations to determine the merits of arguments. (HS-LS2-8)

- -
Connections to Nature of Science

Scientific Knowledge is Open to Revision in Light of New Evidence

- Most scientific knowledge is quite durable, but is, in principle, subject to change based on new evidence and/or reinterpretation of existing evidence. (HS-LS2-2),(HS-LS2-3)
- Scientific argumentation is a mode of logical discourse used to clarify the strength of relationships between ideas and evidence that may result in revision of an explanation. (HS-LS2-6),(HS-LS2-8)

Disciplinary Core Ideas

LS2.A: Interdependent Relationships in Ecosystems
- Ecosystems have carrying capacities, which are limits to the numbers of organisms and populations they can support. These limits result from such factors as the availability of living and nonliving resources and from such challenges such as predation, competition, and disease. Organisms would have the capacity to produce populations of great size were it not for the fact that environments and resources are finite. This fundamental tension affects the abundance (number of individuals) of species in any given ecosystem. (HS-LS2-1),(HS-LS2-2)

LS2.B: Cycles of Matter and Energy Transfer in Ecosystems
- Photosynthesis and cellular respiration (including anaerobic processes) provide most of the energy for life processes. (HS-LS2-3)
- Plants or algae form the lowest level of the food web. At each link upward in a food web, only a small fraction of the matter consumed at the lower level is transferred upward, to produce growth and release energy in cellular respiration at the higher level. Given this inefficiency, there are generally fewer organisms at higher levels of a food web. Some matter reacts to release energy for life functions, some matter is stored in newly made structures, and much is discarded. The chemical elements that make up the molecules of organisms pass through food webs and into and out of the atmosphere and soil, and they are combined and recombined in different ways. At each link in an ecosystem, matter and energy are conserved. (HS-LS2-4)
- Photosynthesis and cellular respiration are important components of the carbon cycle, in which carbon is exchanged among the biosphere, atmosphere, oceans, and geosphere through chemical, physical, geological, and biological processes. (HS-LS2-5)

LS2.C: Ecosystem Dynamics, Functioning, and Resilience
- A complex set of interactions within an ecosystem can keep its numbers and types of organisms relatively constant over long periods of time under stable conditions. If a modest biological or physical disturbance to an ecosystem occurs, it may return to its more or less original status (i.e., the ecosystem is resilient), as opposed to becoming a very different ecosystem. Extreme fluctuations in conditions or the size of any population, however, can challenge the functioning of ecosystems in terms of resources and habitat availability. (HS-LS2-2),(HS-LS2-6)
- Moreover, anthropogenic changes (induced by human activity) in the environment—including habitat destruction, pollution, introduction of invasive species, overexploitation, and climate change—can disrupt an ecosystem and threaten the survival of some species. (HS-LS2-7)

LS2.D: Social Interactions and Group Behavior
- Group behavior has evolved because membership can increase the chances of survival for individuals and their genetic relatives. (HS-LS2-8)

LS4.D: Biodiversity and Humans
- Biodiversity is increased by the formation of new species (speciation) and decreased by the loss of species (extinction). *(secondary to HS-LS2-7)*
- Humans depend on the living world for the resources and other benefits provided by biodiversity. But human activity is also having adverse impacts on biodiversity through overpopulation, overexploitation, habitat destruction, pollution, introduction of invasive species, and climate change. Thus sustaining biodiversity so that ecosystem functioning and productivity are maintained is essential to supporting and enhancing life on Earth. Sustaining biodiversity also aids humanity by preserving landscapes of recreational or inspirational value. *(secondary to HS-LS2-7) (Note: This Disciplinary Core Idea is also addressed by HS-LS4-6.)*

PS3.D: Energy in Chemical Processes
- The main way that solar energy is captured and stored on Earth is through the complex chemical process known as photosynthesis. *(secondary to HS-LS2-5)*

ETS1.B: Developing Possible Solutions
- When evaluating solutions it is important to take into account a range of constraints including cost, safety, reliability and aesthetics and to consider social, cultural and environmental impacts. *(secondary to HS-LS2-7)*

Crosscutting Concepts

Cause and Effect
- Empirical evidence is required to differentiate between cause and correlation and make claims about specific causes and effects. (HS-LS2-8)

Scale, Proportion, and Quantity
- The significance of a phenomenon is dependent on the scale, proportion, and quantity at which it occurs. (HS-LS2-1)
- Using the concept of orders of magnitude allows one to understand how a model at one scale relates to a model at another scale. (HS-LS2-2)

Systems and System Models
- Models (e.g., physical, mathematical, computer models) can be used to simulate systems and interactions—including energy, matter, and information flows—within and between systems at different scales. (HS-LS2-5)

Energy and Matter
- Energy cannot be created or destroyed—it only moves between one place and another place, between objects and/or fields, or between systems. (HS-LS2-4)
- Energy drives the cycling of matter within and between systems. (HS-LS2-3)

Stability and Change
- Much of science deals with constructing explanations of how things change and how they remain stable. (HS-LS2-6),(HS-LS2-7)

Figure 6.5

HS-ETS1: Engineering Design Foundation Boxes

Science and Engineering Practices	Disciplinary Core Ideas	Crosscutting Concepts
Asking Questions and Defining Problems Asking questions and defining problems in 9–12 builds on K–8 experiences and progresses to formulating, refining, and evaluating empirically testable questions and design problems using models and simulations. • Analyze complex real-world problems by specifying criteria and constraints for successful solutions. (HS-ETS1-1) **Using Mathematics and Computational Thinking** Mathematical and computational thinking in 9-12 builds on K-8 experiences and progresses to using algebraic thinking and analysis, a range of linear and nonlinear functions including trigonometric functions, exponentials and logarithms, and computational tools for statistical analysis to analyze, represent, and model data. Simple computational simulations are created and used based on mathematical models of basic assumptions. • Use mathematical models and/or computer simulations to predict the effects of a design solution on systems and/or the interactions between systems. (HS-ETS1-4) **Constructing Explanations and Designing Solutions** Constructing explanations and designing solutions in 9–12 builds on K–8 experiences and progresses to explanations and designs that are supported by multiple and independent student-generated sources of evidence consistent with scientific ideas, principles and theories. • Design a solution to a complex real-world problem, based on scientific knowledge, student-generated sources of evidence, prioritized criteria, and tradeoff considerations. (HS-ETS1-2) • Evaluate a solution to a complex real-world problem, based on scientific knowledge, student-generated sources of evidence, prioritized criteria, and tradeoff considerations. (HS-ETS1-3)	**ETS1.A: Defining and Delimiting Engineering Problems** • Criteria and constraints also include satisfying any requirements set by society, such as taking issues of risk mitigation into account, and they should be quantified to the extent possible and stated in such a way that one can tell if a given design meets them. (HS-ETS1-1) • Humanity faces major global challenges today, such as the need for supplies of clean water and food or for energy sources that minimize pollution, which can be addressed through engineering. These global challenges also may have manifestations in local communities. (HS-ETS1-1) **ETS1.B: Developing Possible Solutions** • When evaluating solutions, it is important to take into account a range of constraints, including cost, safety, reliability, and aesthetics, and to consider social, cultural, and environmental impacts. (HS-ETS1-3) • Both physical models and computers can be used in various ways to aid in the engineering design process. Computers are useful for a variety of purposes, such as running simulations to test different ways of solving a problem or to see which one is most efficient or economical; and in making a persuasive presentation to a client about how a given design will meet his or her needs. (HS-ETS1-4) **ETS1.C: Optimizing the Design Solution** • Criteria may need to be broken down into simpler ones that can be approached systematically, and decisions about the priority of certain criteria over others (trade-offs) may be needed. (HS-ETS1-2)	**Systems and System Models** • Models (e.g., physical, mathematical, computer models) can be used to simulate systems and interactions—including energy, matter, and information flows— within and between systems at different scales. (HS-ETS1-4) *Connections to Engineering, Technology, and Applications of Science* **Influence of Science, Engineering, and Technology on Society and the Natural World** • New technologies can have deep impacts on society and the environment, including some that were not anticipated. Analysis of costs and benefits is a critical aspect of decisions about technology. (HS-ETS1-1) (HS-ETS1-3)

Reviewing current competing explanations for a phenomenon such as the honeybee colony collapse disorder will give students experience evaluating the strengths and weaknesses for each explanation and how well each explanation fits with the available data. "Because scientists achieve their own understanding by building theories and theory-based explanations with the aid of models and representations and by drawing on data and evidence, students should also develop some facility in constructing model- or evidence-based explanations" (NRC 2012, p. 68). This also gives students the opportunity to develop an engineering design solution to enhance biodiversity in ecosystems changed by humans.

Reasoning and argumentation are critical parts of science and engineering and fit well with the honeybee issue. "Becoming a critical consumer of science is fostered by opportunities to use critique and evaluation to judge the merits of any scientifically based argument. In engineering, reasoning and argument are essential to finding the best possible solution to a problem" (NRC 2012, pp. 71–72). Cost-benefit analysis, and analysis of risk are important parts of the process.

Constructing Explanations and Designing Solutions is an important component for the honeybee issue. I selected the Engineering Design Foundation Box under Constructing Explanations and Designing Solutions (Figure 6.5) with bullets 1 and 2. This Science and Engineering Practices foundation box breaks out "design a solution" in bullet 1 and "evaluate a solution" in bullet 2. This fits better with the honeybee unit I developed rather than using Constructing Explanations and Designing Solutions bullet 2 for the HS-LS2: Ecosystems: Interactions, Energy, and Dynamics Foundation Boxes (Figure 6.4). In HS-LS2 "design, evaluate, and refine a solution" are all coupled in one bullet.

Figure 6.6 gives a summary of all selected disciplinary core ideas, crosscutting concepts, and science and engineering practices.

Figure 6.6

Summary of Selected Disciplinary Core Ideas, Crosscutting Concepts, and Science and Engineering Practices

Disciplinary Core Ideas

LS2.C: Ecosystem Dynamics, Functioning, and Resilience
Bullet 1: A complex set of interactions within an ecosystem can keep its numbers and types of organisms relatively constant over long periods of time under stable conditions. If a modest biological or physical disturbance to an ecosystem occurs, it may return to its more or less original status (i.e., the ecosystem is resilient) as opposed to becoming a very different ecosystem. Extreme fluctuations in conditions or the size of any population, however, can challenge the functioning of ecosystems in terms of resources and habitat availability.

LS2.C: Ecosystem Dynamics, Functioning, and Resilience
Bullet 2: Moreover, anthropogenic changes (induced by human activity) in the environment—including habitat destruction, pollution, introduction of invasive species, overexploitation, and climate change—can disrupt an ecosystem and threaten the survival of some species.

LS4.D: Biodiversity and Humans
Bullet 2:Humans depend on the living world for resources and other benefits provided by biodiversity. But human activity is also having adverse impacts on biodiversity through overpopulation, overexploitation, habitat destruction, pollution, introduction of invasive species, and climate change. Thus sustaining biodiversity so that ecosystem functioning and productivity are maintained is essential to supporting and enhancing life on Earth. Sustaining biodiversity also aids humanity by preserving landscapes of recreational or inspirational value.

ETS1.B: Developing Possible Solutions
Bullet 1: When evaluating solutions, it is important to take into account a range of constraints including cost, safety, reliability, aesthetics, and to consider social, cultural, and environmental impacts.

Figure 6.6 (continued)

Crosscutting Concepts

Systems and System Models:
Bullet 1: Models (e.g., physical, mathematical, computer models) can be used to simulate systems and interactions—including energy, matter, and information flows—within and between systems at different scales.

Stability and Change
Bullet 1: Much of science deals with constructing explanations of how things change and how they remain stable.

Science and Engineering Practices

Using Mathematical and Computational Thinking
Mathematical and computational thinking in 9–12 builds on K–8 experiences and progresses to using algebraic thinking and analysis, a range of linear and nonlinear functions including trigonometric functions, exponentials and logarithms, and computational tools for statistical analysis to analyze, represent, and model data. Simple computational simulations are created and used based on mathematical models of basic assumptions. (HS-LS2)

Bullet 2. Use mathematical representations of phenomena or design solutions to support and revise explanations.

Constructing Explanations and Designing Solutions
Constructing explanations and designing solutions in 9–12 builds on K–8 experiences and progresses to explanations and designs that are supported by multiple and independent student-generated sources of evidence consistent with scientific ideas, principles, and theories. (HS-ETS1)

Bullet 1. Design a solution to a complex real-world problem, based on scientific knowledge, student-generated sources of evidence, prioritized criteria, and tradeoff considerations.

Bullet 2. Evaluate a solution to a complex real-world problem, based on scientific knowledge, student-generated sources of evidence, prioritized criteria, and tradeoff considerations.

Engaging in Argument From Evidence
Engaging in argument from evidence in 9–12 builds on K–8 experiences and progresses to using appropriate and sufficient evidence and scientific reasoning to defend and critique claims and explanations about the natural and designed world(s). (HS-LS2)

Bullet 1. Evaluate the claims, evidence, and reasoning behind currently accepted explanations or solutions to determine the merits of arguments.

Source: NGSS Lead States 2013.

Stage II: Deconstruct DCIs, Create a Storyline, and Align Practices and Crosscutting Concepts

I first reviewed *A Framework for K–12 Science Education: Practices, Crosscutting Concepts, and Core Ideas* (NRC 2012, pp. 150–157) to determine the intent of the *Framework* authors and the progression of learning. I then reviewed the DCI foundation box for the entire

HS-LS2 standard, Ecosystems: Interactions, Energy, and Dynamics and determined what needs to precede this model unit for student understanding. I finally studied the selected DCI bullets and crosscutting ideas and determined the essential understanding for the unit of study. Refer to the first two columns of Table 6.1.

I then broke down the essential understanding into smaller content "chunks"—the *learning targets*—and put them in a sequence to create a storyline (Column 3 of Table 6.1). These targets build on previous learning in earlier grades and in biology (prerequisite knowledge/skills). Review how the first three columns (bullets selected from LS2.C, essential understanding, and learning targets) in Table 6.1 are aligned. Finally, I decided the most important concepts and vocabulary and "prune" parts (Column 4 of Table 6.1) since they are not essential for student understanding. Notice how the engineering DCI ETS1.B fits into the essential understanding and learning target progression for the unit. The two crosscutting concepts given in Figure 6.4 (p.198) are embedded in the essential understanding and learning targets.

The storyline is further developed in Table 6.2 (pp. 205–206) and brings in science and engineering practices (see column 3). The honeybee issue is shown in the storyline under column 4. You may choose any environmental issue that is compelling for students and important in your community.

Stage III: Determine Performance Expectations and Identify Criteria to Determine Student Understanding

I next determined criteria, one for each target, to measure student understanding. These criteria are used to determine student success and need to reflect understanding of the content of the learning target and the intersection of the target with the science and engineering practices and crosscutting concepts. Table 6.3 (p. 207) gives the set of criteria for each learning target.

As shown in Table 6.4 (p. 208) there is close alignment between the criteria for this unit and the selected high school performance expectations. For example, criteria that have students conduct research on honeybee population change and use mathematical representations such as statistics and graphs to show evidence for ecosystem stability or change aligns with PE HS-LS2. This PE states "use mathematical representations to support and revise explanations based on evidence about factors affecting biodiversity and populations in ecosystems of different scales."

Stage IV: Determine Nature of Science (NOS) Connections

The *NGSS* Nature of Science connection with HS-LS2 has the following two statements:

> Most scientific knowledge is quite durable, but is, in principle, subject to change based on new evidence and/or reinterpretation of existing evidence.

Table 6.1

Deconstructing DCIs Into Learning Targets

Bullets Selected From HS-LS2.C	Essential Understanding	Learning Targets	Pruning
LS2.C A complex set of interactions within an ecosystem can keep its numbers and types of organisms relatively constant over long periods of time under stable conditions. If a modest biological or physical disturbance to an ecosystem occurs, it may return to its more or less original status (i.e., the ecosystem is resilient), as opposed to becoming a very different ecosystem. Extreme fluctuations in conditions or the size of any population, however, can challenge the functioning of ecosystems in terms of resources and habitat availability.	Even though ecosystems naturally change, greater resilience and stability depend upon biodiversity. A greater number of species leads to more ecosystem resilience. Extreme biological and/or physical changes can result in a different ecosystem. Human activities can disrupt an ecosystem, threaten the survival of some species, and decrease biodiversity. Possible solutions to ecosystem changes by human activity need to take into account the impact on the environment.	1. Ecosystems are complex, interactive systems that include biological and physical components that change over time. Populations fluctuate based upon interaction with other organisms and physical factors in the environment. The number of species in an ecosystem impacts the overall health and resilience of an ecosystem. The more species that live together, the more stable and productive the ecosystem.	An understanding of the hierarchal structure of ecosystems (species, populations, communities, ecosystem, biosphere), cycles of matter, energy transfer, and carrying capacity should precede these learning targets. Variation of traits, genetics related to speciation, and natural selection are pruned (should have preceded this learning target).
Moreover, anthropogenic changes (induced by human activity) in the environment—including habitat destruction, pollution, introduction of invasive species, overexploitation, and climate change—can disrupt an ecosystem and threaten the survival of some species.		2. Human activities impact ecosystems that threaten the survival of some species and decrease biodiversity. Overharvesting of animals, plants, and other organisms and environmental destruction destabilizes the ecosystem.	Limit examples of extinction to local or regional ecosystems.

Table 6.1 (continued)

Bullets Selected From HS-LS4.D and HS-ETS1.B	Essential Understanding	Learning Targets	Pruning
LS4.D. Humans depend on the living world for the resources and other benefits provided by biodiversity. But human activity is also having adverse impacts on biodiversity through overpopulation, overexploitation, habitat destruction, pollution, introduction of invasive species, and climate change. Thus sustaining biodiversity so that ecosystem functioning and productivity are maintained is essential to supporting and enhancing life on Earth. Sustaining biodiversity also aids humanity by preserving landscapes of recreational or inspirational value.		3. Some human activities enhance biodiversity while other activities decrease biodiversity. Humans depend on ecosystems for resources and other benefits provided through ecosystem biodiversity.	Select human activities to illustrate human impact on specific ecosystems and the importance of maintaining a stable ecosystem by enhancing biodiversity (e.g., agriculture practices, habitat loss to urban development, impact of invasive species, human overpopulation, impact of climate change, pollution)
ETS1.B When evaluating solutions, it is important to take into account a range of constraints, including cost, safety, reliability, and aesthetics, and to consider social, cultural, and environmental impacts.		4. There are various, potential solutions to enhance biodiversity and ecosystem stability in a selected ecosystem that include constraints, benefits and environmental impacts. These must be evaluated to determine the best possible solution.	Limit the potential solutions to one selected example with evidence to support the best options.

Table 6.2

Developing a Storyline for Ecosystem Stability and Resilience

Learning Targets	Crosscutting Concept Addressed in Learning Target	Science and Engineering Practices to Include	Possible Aligned Representations or Activities to Include
1. Ecosystems are complex, interactive systems that include biological and physical components that change over time. Populations fluctuate based upon interaction with other organisms and physical factors in the environment. The number of species in an ecosystem impacts the overall health and resilience of an ecosystem. The more species that live together, the more stable and productive the ecosystem.	Models (e.g., physical, mathematical, computer models) can be used to simulate systems and interactions—including energy, matter, and information flows—with and between systems at different scales.	Use mathematical representations of phenomena or design solutions to support and revise explanations.	Research an issue related to ecosystem stability such as the decline of the honeybee populations and their impact on plant pollination. Use mathematical representations (e.g., statistics, graphs) to illustrate the decline of the honeybee population and their impact on plant pollination.
2. Human activities impact ecosystems that threaten the survival of some species and decrease biodiversity. Overharvesting of animals, plants, and other organisms and environmental destruction destabilizes the ecosystem.	Much of science deals with constructing explanations of how things change and how they remain stable. Models (e.g., physical, mathematical, computer models) can be used to simulate systems and interactions—including energy, matter, and information flows—with and between systems at different scales.	Evaluate the claims, evidence, and reasoning behind currently accepted explanations or solutions to determine the merits of arguments. Use mathematical representations of phenomena or design solutions to support and revise explanations.	Evaluate the claims, evidence, and reasoning for several accepted explanations related to honeybee population decline and colony collapse disorder. Analyze arguments for the different explanations including the use of statistics and other mathematical representations.

Table 6.2 (continued)

Learning Targets	Crosscutting Concept Addressed in Learning Target	Science and Engineering Practices to Include	Possible Aligned Representations or Activities to Include
3. Some human activities enhance biodiversity while other activities decrease biodiversity. Humans depend on ecosystems for resources and other benefits provided through ecosystem biodiversity.	Much of science deals with constructing explanations of how things change and how they remain stable. Models (e.g., physical, mathematical, computer models) can be used to simulate systems and interactions—including energy, matter, and information flows—with and between systems at different scales.	*Design a solution to a complex real-world problem, based on scientific knowledge, student-generated sources of evidence, prioritized criteria, and tradeoff considerations.	Design a solution to increase biodiversity and reduce negative human impacts (e.g., agriculture practices) on the honeybee population. Student groups collaborate on different solutions. Solutions should include criteria and tradeoff considerations.
4. There are various, potential solutions to enhance biodiversity and ecosystem stability in a selected ecosystem that include constraints, benefits and environmental impacts. These must be evaluated to determine the best possible solution.	Much of science deals with constructing explanations of how things change and how they remain stable.	*Design a solution to a complex real-world problem, based on scientific knowledge, student-generated sources of evidence, prioritized criteria, and tradeoff considerations. *Evaluate a solution to a complex real-world problem, based on scientific knowledge, student-generated sources of evidence, prioritized criteria, and tradeoff considerations.	Groups share and evaluate different group solutions. Groups refine their solutions to increase biodiversity, enhance ecosystem stability, and increase the honeybee population. Groups share and evaluate their refined solutions.

With an asterisk (*): These Science and Engineering Practices are from HS-ETS1: Engineering Design Foundation boxes (bullets 1 and 2).

Without an asterisk (*): The other Science and Engineering Practices are from HS-LS2: Ecosystems: Interactions, Energy, and Dynamics Foundation Boxes.

Table 6.3

Criteria to Determine Student Understanding of Ecosystem Stability, Resilience, Biodiversity, and Human Impact on the Environment

Learning Target	Criterion
1. Ecosystems are complex, interactive systems that include biological and physical components that change over time. Populations fluctuate based upon interaction with other organisms and physical factors in the environment. The number of species in an ecosystem impacts the overall health and resilience of an ecosystem. The more species that live together, the more stable and productive the ecosystem.	Conduct research on an issue related to ecosystem stability (e.g., honeybee population change) and show evidence using mathematical representations such as statistics and graphs that the ecosystem is changing or stable. Include evidence on the biodiversity of the ecosystem and if biodiversity is increasing, stable, or decreasing.
2. Human activities impact ecosystems that threaten the survival of some species and decrease biodiversity. Overharvesting of animals, plants, and other organisms and environmental destruction destabilizes the ecosystem.	From review of research evaluate explanations including mathematical representations for the selected environmental issue based upon claims, evidence, and reasoning. After analysis of arguments for different explanations, determine which ones are the strongest.
3. Some human activities enhance biodiversity while other activities decrease biodiversity. Humans depend on ecosystems for resources and other benefits provided through ecosystem biodiversity.	Develop a design with other students for a solution to a selected real-world problem (issue) to increase biodiversity and reduce negative human impacts. The design must include criteria and tradeoff considerations (constraints, benefits, and environmental impacts).
4. There are various, potential solutions to enhance biodiversity and ecosystem stability in a selected ecosystem that include constraints, benefits and environmental impacts. These must be evaluated to determine the best possible solution.	Evaluate small-group solution designs for the selected issue to increase biodiversity and reduce negative human impacts. Refine and evaluate solution designs for the selected issue to increase biodiversity and reduce negative human impacts.

Table 6.4

Alignment of Assessment Criteria and Selected NGSS Performance Expectations

Developed Criteria for Unit of Study	Aligned HS-LS2 Performance Expectations
Conduct research on an issue related to ecosystem stability (e.g., honeybee population change) and show evidence using mathematical representations such as statistics and graphs that the ecosystem is changing or stable. Include evidence on the biodiversity of the ecosystem and if biodiversity is increasing, stable, or decreasing.	HS-LS2-2 Use mathematical representations to support and revise explanations based on evidence about factors affecting biodiversity and populations in ecosystems of different scales.
From review of research evaluate explanations including mathematical representations for the selected environmental issue based upon claims, evidence, and reasoning. After analysis of arguments for different explanations, determine which ones are the strongest.	HS-LS2-6 Evaluate the claims, evidence, and reasoning that the complex interactions in ecosystem maintain relatively consistent numbers and types of organisms in stable conditions, but changing conditions may result in a new ecosystem.
Develop a design with other students for a solution to a selected real-world problem (issue) to increase biodiversity and reduce negative human impacts. The design must include criteria and tradeoff considerations (constraints, benefits, and environmental impacts).	HS-LS2-7 Design, evaluate, and refine a solution for reducing the impacts of human activities on the environment and biodiversity.
Evaluate small group solution designs for the selected issue to increase biodiversity and reduce negative impacts. Refine and evaluate the solution design for the selected issue to increase biodiversity and reduce negative impacts.	

Scientific argumentation is a mode of logical discourse used to clarify the strength of relationships between ideas and evidence that may result in revision of an explanation.

Both these NOS statements align well with the selected performance expectations for my model unit of study. The first NOS statement aligns with the PE HS-LS2.2, "Use mathematical representations to support and revise explanations based on evidence about factors affecting biodiversity and populations in ecosystems of different scales."

The second NOS statement aligns with the PE HS-LS2.6, "Evaluate the claims, evidence, and reasoning that the complex interactions in ecosystems maintain relatively consistent numbers and types of organisms in stable conditions, but changing conditions may result in a new ecosystem."

Stage V: Identify Metacognitive Goals and Strategies

Step 1

I chose "Creative Thinking and Learning" as a good fit since it includes working on complex tasks without obvious solutions and generating ideas about solving a problem. In my model unit students research current thinking and explanations on the honeybee population decline and biodiversity in the selected ecosystem. Based upon their previous research in Learning Targets #1 and #2, Learning Target #3 students are engaged in designing a solution to the problem that requires looking at trade-offs.

Step 2

I selected number 3 from Instructional Tool 3.1 (pp. 57–62) to support student metacognition. Since students are involved in an issue involving an environmental impact decision on increasing the honeybee population and overall biodiversity in their selected ecosystem, they can use the "Options Diamond" routine to develop their solution to the issue. This routine focuses the student groups to look at trade-offs and solutions for each trade-off. Reflection is an important component for this routine addressing what the group has learned through the process.

Step 3

The criterion I developed to determine understanding of the decision-making process is to have each group develop a written solution for their best option to increase both the honeybee population and biodiversity in an ecosystem by evaluating trade-offs for their options. Groups share and write critiques for other group's solutions. Individuals write about what they learned about this process. Figure 6.7 (p. 210) summarizes the three steps.

Figure 6.7

Metacognition Approach

Creative Thinking and Learning

Step 1. Description
"Creative thinking and learning includes engaging intensely in tasks even when answers or solutions are not immediately apparent, pushing the limits of your knowledge and abilities, generating, trusting, and maintaining your own standards of evaluation, and generating new ways of viewing a situation outside the boundaries of standard conventions." (Marzano 1992, p.134)

Step 2. Routine That Supports Creative Thinking and Learning
Options Diamond Description: Groups draw large diamond and write the decision to be made in the center (e.g., how to increase both the honey bee population and biodiversity in an ecosystem). Write one or two main trade-offs in the left and right corners related to a particular decision.

Groups brainstorm:
1. solutions for each trade-off;
2. compromises between trade-offs (write at bottom point of diamond)
3. clever solutions that combine what seems to be opposites from right and left corners (write at top of diamond).

At the end of the process groups and individuals reflect on the process. (*www.pz.harvard.edu/vt*)

Step 3. Determine Criterion for Student Understanding
Criterion for decision-making process: Each group develops a written solution to increase both the honeybee population and biodiversity in an ecosystem by evaluating trade-offs for at least two options. Groups share and critique solutions.

Application of Phase 2 to "Ecosystem Resilience and Stability, Biodiversity, and Human Impact"

Stage VI: Research Student Misconceptions Common to This Topic That Are Documented in the Research Literature

As I began this stage I referred back to "Why This Topic" (pp. 194–196) on some misconceptions about ecosystem resilience and biodiversity. Figure 6.3 (p. 196) gives specific misconceptions about biodiversity. Since ecosystem stability and resistance, biodiversity and human impact on ecosystems are my areas of focus, those are the misconceptions that will be addressed for the chapter. This includes student understanding of interrelationships, populations, systems, and changes to the physical environment.

According to *Benchmarks for Science Literacy* (AAAS 1993, p. 355), "some research has found that student misconceptions about certain subjects can arise from their difficulty in recognizing natural phenomena as groups or systems of interacting objects." Some students have difficulty recognizing the complexity of interrelationships in ecosystems.

Some misconceptions related to this unit from the National Science Education Standards (NRC 1996) are the following:

- "Other misconceptions center on a lack of understanding of how a population changes as a result of differential reproduction (some individuals producing more offspring), as opposed to all individuals in a population changing." (p. 184)

- "Most high school students have a concept of populations of organisms but they have a poorly developed understanding of the relationships among populations within a community and connections between populations and other ideas such as competition for resources." (p. 193)

- "Few students understand and apply the idea of interdependence when considering interactions among populations, environments and resources." (p. 193)

As reported by Driver et al., M. J. Brody interviewed students between ages 9 and 16 and found that some important misconceptions about pollution were held by at least half of the large sample. "They included: anything natural is not pollution; biodegradable materials are not pollutants; the oceans are a limitless resource; solid waste in dumps is safe; the human race is indestructible as a species" (Driver et al. 1994, p. 68).

Other student misconceptions from the Environmental Literacy Council (2007) about biodiversity are the following:

- Biodiversity refers simply to all species living in a certain area and does not take into account species abundance and richness.

- The scope of plant biodiversity may be minimized.

- Biodiversity of other species is not relevant to humans.

- Biodiversity loss on the Earth is inevitable.

- A lot of biodiversity means the ecosystem is healthy.

- Not all species are important, so some have to be sacrificed.

- Scientists know exactly how many species exist at any given time.

- Extinction is not a "natural" phenomenon.

A study conducted by S. M. Smith (2003) looked at students' understanding of interdependency concepts that included food chains, pollination, and seed dispersal. No published research prior to this study pertained directly to students' understanding of pollination interdependency. The results showed students across grade levels 3, 7, 10 and college varied in their everyday and scientific understanding of the three interdependency concepts. Students at grades 3, 7, and 10 were in transition from everyday to scientific understanding on pollination and food chains. Students had more difficulty with scientific understanding of seed dispersal. This study sheds light on the honeybee issue in this chapter and the importance of interdependence.

Table 6.5 summarizes the misconceptions I found for ecosystems and includes some instructional ideas.

Instructional Strategic Selection Tool: Identifying Strategies for Phase 2 Planning

Stage VII: Determine Appropriate Strategies to Identify Your Students' Preconceptions

My next step was to determine instructional strategies that help identify, elicit, and confront students' preconceptions and help them make sense of their experiences. I referred to the following instructional tools and figures to guide me through the process selected for this chapter.

- Instructional Tool 3.2, Instructional Strategy Selection Tool (pp. 67–68)

- Instructional Tool 3.7, Sense-Making Approaches: Nonlinguistic Representations—Visual Tools (pp. 127–133). This tool gives more details to support Tool 3.2.

- Appendix 2, Hints and Resources for the Design Process (p. 311)

- Figure 6.4, Ecosystems: Interactions, Energy, and Dynamics Foundation Boxes (p. 198)

- Table 6.5, Misconceptions and Instructional Ideas for Ecosystem Stability and Resilience, Biodiversity, and Human Impact (p. 213)

From Instructional Tool 3.2 I selected the Visual Tools approach, brainstorming web strategy to determine student preconceptions. I decided to start with "ecosystem components and interrelationships" where students use clustering as a strategy to generate their ideas. You should select a local ecosystem so students have a better opportunity to draw on their own knowledge for an ecosystem that is familiar. Ecosystem components and interrelationships are an important place to begin in order

Table 6.5

Misconceptions and Instructional Ideas for Ecosystem Stability and Resilience, Biodiversity, and Human Impact

Learning Target #1. Ecosystems are complex, interactive systems that include biological and physical components that change over time. Populations fluctuate based upon interaction with other organisms and physical factors in their environment. The number of species in an ecosystem impacts the overall health and resilience of an ecosystem. The more species that live together, the more stable and productive the ecosystem.

Misconceptions	Instructional Ideas
• Students of all ages think that some populations of organisms are numerous in order to fulfill a demand for food by another population (Leach et al. 1992). • From an early age, children can draw linear 'who eats what' food chains. However, few pupils understand the integration of chains into webs and cycles even at the end of secondary school, nor that food chains and webs are intended as models of all feeding relationships in a habitat (Leach et. al. 1992). • Most students over 13 know that animals cannot survive without plants, but lack an understanding of why plants are crucial. Few can relate the importance of plants in the ecosystem to the Sun as the source of life (Leach et al. 1992). • Students age 13 and older have a concept of populations of organisms in the wild, but their 'explanations' of relationships are merely descriptions of nature (birds live in trees, foxes eat rabbits). It is not until much later that students think in terms of populations of organisms in the wild competing for scarce resources (Leach et al. 1992). • Although secondary students are able to draw and manipulate food chain and food web diagrams, they often think of the components as individual organisms. They focus on linear food chains in predicting the effects of a change in one component and do not recognize the far reaching effects on the whole food web and whole ecosystem (Leach et al. 1992). • Students' meanings of ecological terms were related to everyday usage rather than to scientific definitions such as 'community' to mean a group of people living together with similar ideas (Adeniyi 1985). • Generally, pupils are unaware of the role that microorganisms play in nature, especially as decomposers and as recyclers of carbon, nitrogen, water and minerals. They conceptualize decomposition as the total or partial disappearance of matter (Leach et al. 1992).	• Pupils need to bring quantitative ideas about populations into their thinking about food chains. Preoccupation with the conventions of drawing diagrams (food chains horizontally, pyramids vertically) may interfere with grasping the concepts. Pupils need to realize that a pyramid of numbers assumes a snapshot view of an ecosystem. They need to imagine a closed ecosystem within which the numbers of organisms could theoretically be counted. They need to apply the model to a variety of ecosystems (Leach et al. 1992). • The challenge of fitting 'decay' into the cycling of matter, depends on realizing that material does not 'disappear' when bodies or excretory products decay. Pupils need to recognize that the chemical interactions involved in decay are the respiration processes of microbes, and that decay is one route for the conversion of the matter of living things to environmental matter. The authors advocate not teaching the concepts of plant nutrition, animal nutrition, respiration and decay as isolated topics but should be taught as an integrated approach (Leach et al. 1992). • The idea of interdependence needs to be linked with the idea of a system in which change in any part affects the whole, and brings it to a new equilibrium. The model of interdependence needs to be used for understanding cycling of specific elements, population dynamics and flow of energy. Ecosystems are 'open' in that in all ecosystems energy and gas exchange is global (Leach et al. 1992).

Table 6.5 (continued)

Learning Target #2. Human activities impact ecosystems that threaten the survival of some species and decrease biodiversity. Over harvesting of animals, plants, and other organisms and environmental destruction destabilizes the ecosystem.

Misconceptions	Instructional Ideas
• Because students do not regard food as scarce they do not consider competition for food. Even when they have been introduced to food chains and webs, students think in terms of linear chains and so do not consider competition between species for the same food resource. They think that a change in population will affect only those species directly related to it as predator or prey (Leach et. al. 1992). • Student's understanding of competition depends on the context they are considering; they can more readily understand competition between predators for a food source, than competition between prey in 'avoiding being eaten' (Leach et al. 1992). • Many students at all ages seemed unable to conceptualize organisms and their environments independent of human involvement (Leach et al. 1992). • From ages 9 to 16 there are few changes in knowledge about ecological crises. By age 16 students have a greater number of relevant concepts and meaningful connections between them. They believe that pollution can affect everything. Half the students held misconceptions including: anything natural is not pollution; biodegradable materials are not pollutants; the oceans are a limitless resource; solid waste in dumps is safe; and the human race is indestructible as a species (Brody 1992). • Biodiversity refers simply to all species living in a certain area and does not take into account species abundance and richness. The scope of plant biodiversity may be minimized. Not all species are important, so some have to be sacrificed. Extinction is not a "natural" phenomenon (Environmental Literacy Council 2007).	• Instruction needs to include that resources are finite globally and within any ecosystem they are scarce. Humans compete for resources. Making the step from 'finite resources' to 'scarcity' involves thinking proportionally in terms of supply and demand. The more organisms in an ecosystem the more demand there is on the supply of all resources. Introduce students to a number of examples of competition between organisms for resources (different than competition in human activities such as sports) (Leach et al. 1992). • Instruction needs to include one species being dependent on another with various levels of dependency (Leach et al. 1992). • The model of interdependence needs to be used for understanding cycling of specific elements, population dynamics and flow of energy in more advanced stages of study (Leach et al. 1992). • Instruction on biodiversity needs to support understanding of species, how scientists estimate number of species, and biodiversity in different ecosystems (Environmental Literacy Council 2007).

Table 6.5 (continued)

Learning Target #3. Some human activities enhance biodiversity while other activities decrease biodiversity. Humans depend on ecosystems for resources and other benefits provided through ecosystem biodiversity.

Misconceptions	Instructional Ideas
Biodiversity misconceptions related to humans include: • Biodiversity of other species is not relevant to humans. • Biodiversity loss on the Earth is inevitable. • A lot of biodiversity means the ecosystem is healthy. • Scientists know how many species exist at any given time (Environmental Literacy Council 2007). • Few students relate their ideas about feeding and energy to a framework of ideas about interactions of organisms. About a quarter of the students expressed views suggesting that other organisms exist for the benefit of humans (Brumby 1982).	• James Rutherford recommends that science content be taught from a real-world perspective that can include crosscutting themes such as scale, systems, constancy and change, and models. Environmental issues and concerns provide a particularly attractive context that can improve science learning. The learning of science can improve the ability of students to deal with environmental issues. Special emphasis should be put on learning about the various causes of biodiversity depletion (Environmental Literacy Council 2007). • The honeybee issue provides a real-world context that addresses biodiversity and ecosystem stability.

Learning Target #4. There are various, potential solutions to enhance biodiversity and ecosystem stability in a selected ecosystem that include constraints, benefits, and environmental impacts. These must be evaluated to determine the best possible solution.

Misconceptions	Instructional Ideas
Biodiversity misconceptions related to humans include: • Biodiversity of other species is not relevant to humans. • Biodiversity loss on the Earth is inevitable. • A lot of biodiversity means the ecosystem is healthy (Environmental Literacy Council 2007).	• As students evaluate solutions and consider trade-offs, they need to consider the impacts by humans on other species. • The human species has a major impact on other species in many ways: reducing the amount of the Earth's surface available to those other species, interfering with their food sources, changing the temperature and chemical composition of their habitats, introducing foreign species into their ecosystems, and altering organisms directly through selective breeding and genetic engineering (AAAS 1993). • Human activities, such as reducing the amount of forest cover, increasing the amount and variety of chemicals released into the atmosphere, and intensive farming, have changed the Earth's land, oceans, and atmosphere. Some of these changes have decreased the capacity of the environment to support life forms (AAAS 1993). • Provide more opportunities to explore alternative viewpoints through small group and class discussions (Engel Clough and Wood-Robninson 1985).

to determine what students know about ecosystems in general and how parts in the system are interrelated (Learning Target #1).

Clustering has students use ovals and words to generate initial ideas. It is a good idea to have students work alone to start the initial process. Each student places an oval with ecosystem components and interrelationships written inside at the center of their papers. Then they come up with as many components and interrelationships as they can that branch off the center oval. Students can use both words and drawings.

After students have generated their own thoughts, they work in small groups to discuss their different representations. From their discourse, each small group generates a new cluster that represents their best thinking.

Finally, each small group shares their thinking with the class. This can be done with each group's poster on the wall for a Gallery Walk where the small groups review and discuss each group's thinking. Students should be encouraged to ask questions about posters and changes they would recommend. When small groups make changes on their posters, these changes can be shown in a different color so group thinking can be tracked over time. As you look for misconceptions and misunderstandings from previous instruction, ask probing questions to elicit further responses. If the human species is missing, pose questions about where humans fit in and about human activities that impact the ecosystem and biodiversity. Misconceptions and depth of understanding will guide further instruction.

Stage VIII: Determine Strategies to Elicit and Confront Your Students' Preconceptions

The last part of the previous section begins to elicit and confront student preconceptions about ecosystems. Small- and large-group discourse helps students understand where they have still have misunderstandings. To further determine what preconceptions students have about biodiversity and human impact, I developed a probe with written questions that ask students to explain their thinking for each question. Refer to the probe example in Figure 6.8. The probe aligns with Learning Targets #1, #2, and #3 (Table 6.1, pp 203–204).

You can probe group thinking further by asking students to share explanations for each question. To model the process, one student shares an explanation. You then ask if there are alternate explanations and students share other explanations. The different explanations are posted on chart paper with a tally of the number of students who wrote the same or a similar explanation. The other nine questions are posted on chart paper around the room and student groups of 3 or 4 start at one question and post their different explanations. As the students in a group post their explanations, they need to ask questions and discuss their responses. Groups move to another chart and add new explanations or put a tally mark with ones where they wrote the same explanation. Groups continue around the room until all the charts are completed with the different explanations. You can then ask them how they will find out which expla-

Figure 6.8

Student Probe on Biodiversity and Human Impact

Think about biodiversity and human impact in an ecosystem. Answer each question with a T for true and a F for false. Below each answer give an explanation for your response.

1.____Biodiversity mostly looks at a count of the number of species in an ecosystem.
My explanation: _____
_____

2.____Ecosystems are static and unchanging so biodiversity remains stable.
My explanation: _____
_____

3.____Not all species are important so some have to be sacrificed.
My explanation: _____
_____

4.____Biodiversity of other species is relevant to humans.
My explanation: _____
_____

5.____Extinction is not a "natural" phenomena.
My explanation: _____
_____

6.____Human activities can have big impacts on biodiversity.
My explanation: _____
_____

7.____If there is a lot of one species, then the ecosystem must be healthy.
My explanation: _____
_____

8.____Many large empty spaces that are unmanaged are wastelands (e.g., abandoned city
lot, desert).
My explanation: _____
_____

9.____Biodiversity includes genetic diversity, species diversity, and ecosystem diversity.
My explanation: _____
_____

10.___Two major causes of biodiversity loss are human population growth and the increasing consumption of natural resources.
My explanation: _____
_____

Source: Adapted from Environmental Literacy Council 2007. *Biodiversity: Resources for Environmental Literary.* Arlington, VA: NSTA Press.

nations are plausible and which ones are not. This should lead them to talking about doing research to find out more about biodiversity and human impact.

An alternate strategy is using a modified Frayer Model (See Instructional Tool 3.7, p. 127) in lieu of the above probe or in combination with the probe. My past experience with students shows this to be a powerful approach. The graphic organizer that is illustrated for this chapter is a modification of the Frayer Model. This strategy is coupled with other strategies (small- and large-group discourse). The original model was developed by Frayer and colleagues (1969) to analyze and assess attainment of concepts. A new concept is defined giving its necessary attributes. A picture illustrating the concept can also be helpful. Vocabulary for concepts such as biodiversity and related vocabulary for understanding the concept are helpful for learners (Billmeyer 2003). I modified the four components in the Frayer Model for this chapter (See Figure 6.9).

Stage IX: Determine Sense-Making Strategies

Sense Making Part 1: Introduction to the Honeybee Real-World Issue

Students take what they found out about ecosystem stability and resilience, biodiversity and human impact and begin research to make sense of the decrease in the honeybee population and colony collapse disorder. Introduce the issue as a real-world problem to the students. An engaging way to start is to bring some food items that we eat that have their plant's flowers pollinated by honeybees. Some examples of plants that are pollinated mainly or totally by honeybees are the almond (100%), avocado (90%), apple (90%), broccoli (90%) blueberry (90%), onion (90%), cucumber (80%), celery (80%), cherry (80%), plum (65%), and watermelon (65%). During this introductory part, the questions can be discussed by small groups and then followed by large group conversation.

Introductory questions include:

- How are these fruits and vegetables connected to honeybees?

- How are these food items impacted by a decreasing honeybee population and colony collapse disorder (CCD)? If students don't know what CCD is about, they can do a quick internet search or they can learn about CCD when they do their research.

- What food do you eat that contains honey?

- How is the honeybee linked with other animals such as cows, sheep, and wild animals?

Figure 6.9

Modified Frayer Plus Discourse to Elicit Student Understanding

Overview of Modified Frayer Strategy

This strategy is used with small groups of students to analyze and assess understanding of the biodiversity concept. The model is set up with "Biodiversity" written in the center oval. The four quadrants around the biodiversity oval are definition (upper-left quadrant), essential characteristics (upper right quadrant), examples (lower-left quadrant), and human impact (lower right quadrant). Small groups complete the chart using words and drawings to illustrate biodiversity.

Steps in Process Using the Modified Frayer Strategy, Small- and Large-Group Discourse, and Gallery Walk

1. Each group sets up the Frayer diagram on a large sheet of chart paper. If students are not familiar with this strategy, the teacher should show a model on chart paper.

2. Small-group discourse should occur as each group completes the Frayer diagram. Each member in the group gives one idea so everyone has a turn. The group continues to get ideas from each member following the same format until they complete all four quadrants. It's important that every student speaks and contributes to the group's work. When the diagram is completed, each group posts their diagram on the wall for other groups to review.

3. The small groups look at the work done by other groups and write down similarities and differences with their group's diagram. Group members can write questions for other groups on post it notes and put the note on other group diagrams. When the process is finished, groups go back to their own diagrams and look at questions from the other groups.

4. The next step is large-group discourse to talk about what they observed from their review of all the groups' diagrams. What were the similarities? What were the differences? What questions surfaced? What changes would they make to their Frayer diagrams? Where can you look for additional information?

5. Small groups make changes using a different colored marker on their diagrams based upon the large group conversation. The posters are left on the wall so they can be referenced in later work and changes can be added.

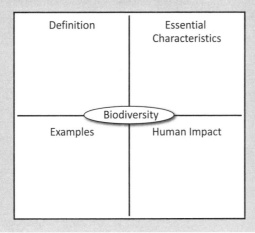

Sense Making Part 2: Early Investigation Into a Real-World Issue: The Honeybee Puzzle

Now challenge the students to become "citizen scientists" and find out what is happening to honeybees in the world. They will later focus on a local or regional ecosystem after their initial investigation and information search, mainly on the internet. However, if there is a college or university nearby, they may want to interview some scientists. The first task is to have students work individually and find out as much information as they can. Since they will share their information in small groups, they don't need to research all the questions.

Some initial questions for investigation (refer to the two keys at the end of all these questions for alignment with *NGSS* crosscutting concepts and *NGSS* science and engineering practices):

- Would it matter if the honeybees became extinct? Explain. (Learning Target #1, Crosscutting Concept CE, Science and Engineering Practices ECER)

- How does a decrease in the honeybee population impact an ecosystem? (Learning Target #1, Crosscutting Concepts MSI and CE, Science and Engineering Practices MR and ECER)

- Where is the greatest decline in honeybee population and CCD occurring in the world? Least decline? (Learning Target #1, Crosscutting Concept MSI, Science and Engineering Practices MR)

- Are human activities impacting the decline in honeybee populations? If so, how? (Learning Target #2, Crosscutting Concept CE, Science and Engineering Practices MR and ECER)

- What does the decrease in the honeybee population in the United States look like in a mathematical representation (graph, table, statistics)? (Learning Target #1, Crosscutting Concept MSI, Science and Engineering Practices MR)

- What evidence are scientists discovering about the decrease in honeybee populations and colony collapse disorder? What are the scientific explanations for the decrease in the honeybee populations and CCD? (Learning Targets #1 and #2, Crosscutting Concepts MSI, CE, Science and Engineering Practices MR and ECER)

- Are wild native bee populations also decreasing in population? If yes, what is the scientific evidence and explanations? (Learning Targets #1 and #2, Crosscutting Concepts MSI and CE, Science and Engineering Practices MR and ECER)

- How does the honeybee problem relate to biodiversity and ecosystem stability? (Learning Targets #1 and #2, Crosscutting Concepts MSI and CE, Science and Engineering Practices MR and ECER)

Key on Alignment With *NGSS* Crosscutting Concepts

- Models (e.g., physical, mathematical, computer models) can be used to simulate systems and interactions including energy, matter, and information flows-with and between systems at different scales (MSI)

- Much of science deals with constructing explanations of how things change and how they remain stable (CE)

Key on Alignment With *NGSS* Science and Engineering Practices

- Use mathematical representations of phenomena or design solutions to support and revise explanations (MR)

- Evaluate the claims, evidence, and reasoning behind currently accepted explanations or solutions to determine the merits of arguments (ECER)

After the initial class discussion on the introductory questions, students conduct individual research using the internet to gather some basic information on the honeybee population decline and CCD using the above questions as a springboard. This is an important part of "Sense Making."

You can have the students generate the questions to investigate rather than using the above list and fill in questions that were not included by the students. Question asking by students is a critical skill. As indicated in Instructional Tool 3.5 (p. 110) you can encourage deeper discussion and investigation with student-generated questions. Refer back to Instructional Tool 3.2 (p. 67) about Writing to Learn as an excellent approach for students as they conduct their individual research. The Science Notebook strategy (Instructional Tool 3.3, p. 93) is a way for students to organize their thinking so they can share information with small groups later. The information gathering can be set up for one selected question at a time about the honeybee issue. An example for recording on the left side full page of the notebook could include the following: general information on the honeybee, key vocabulary words (understood and not understood), mathematical models (statistics, tables, and graphs), and any other evidence on reasons for the honeybee population decline and CCD. On the full page right side of the notebook the following entries are recorded: scientific explanations and arguments, a brief summary with key points individuals want to share with small-group members, and additional questions to investigate.

There are many resources available on the honeybee population decline and CCD problem. Figure 6.10 (p. 222) gives some resources to help students get started.

Figure 6.10

Honey Bee Population Decline and Colony Collapse Disorder Resources

Internet Resources

U.S. Department of Agriculture Website: *www.nps.ars.usda.gov* (Note: This specific article by the Agricultural Research Service is a good place for students to start): U.S. Agricultural Research Service. 2012. Colony collapse disorder: An Incomplete Puzzle. 60 (6): 4–8.

Other research on honeybees can be found at:

www.ars.usda.gov/is/AR/archive/jul12/colony0712.htm.
At the top of the website page, search honeybee population or colony collapse disorder.

U.S. Environmental Protection Agency Website: *www.epa.gov*
There is an advanced search box in the upper-right corner to get started.

Natural Resources Defense Council Website: *www.nrdc.org*
There is a Search the Site box in the upper-right corner to get started.

BeeSpotter Website: *http://beespotter.mste.illinois.edu*
A University of Illinois Citizen Science Project that educates the public about pollinators by engaging them in a data collection effort. This has good basic information on honeybees and bumblebees.

Bee Health extension Website: *www.extension.org/bee_health*

COLOSS Network (Prevention of honey bee COlony LOSSes): *www.coloss.org*

Bee Informed Partnership: *http://beeinformed.org*

Horn, T. 2008. Honeybees: A history. *New York Times.*
topics.blogs.nytimes.com/2008/04/11/honey-bees-a-history

Kellar, B. n.d. Honey bees across America.
www.orsba.org/htdocs/download/HoneyBeesAcrossAmerica.html

University of Colorado Citizen Science Project on Wild Native Bees
http://beesneeds.colorado.edu

WatchKnowLearn: *WatchKnowLearn.org*

Print Resource
Walsh, B. 2013. The plight of the honeybee: Mass deaths in bee colonies mean disaster for farmers—and your favorite foods. *Time* 182 (8): 24–31.

Sense Making Part 3: Small-Group Discourse on Honeybee Research and Organizing Group Research

After individuals complete their research on the honeybee issue, form small groups to discuss and share their findings. Each student shares one question they researched and asks others in the group if they can add to the information, claims, evidence, and reasoning. The group compiles the key evidence and explanations that address the honeybee population decline and CCD. This includes models such as graphs and statistics. During this part of "Sense Making," groups also connect information. You should check to see if groups have come up with several big categories for CCD. Possibilities include the following:

- pathogens (example: virus, bacteria)

- parasites (example Varroa mite)

- environmental stressors (examples: pesticides or lack of nectar diversity)

- management stressors (example: how hives are managed)

Sense Making Part 4: Making Sense of Evidence and Scientific Explanations

The students' next challenge is to analyze the evidence and varying scientific explanations for the honeybee population decline and CCD. They need to address human impact on ecosystems that threaten survival of some species and decrease biodiversity (Learning Target 2). During this part of their work, small groups evaluate the claims, evidence and reasoning for several accepted scientific explanations. A reference that gives a "The Claim, Evidence, and Reasoning Framework for Talk and Writing" was developed by McNeill and Krajcik for grades 5–8 students but is applicable for high school (McNeill and Krajcik 2012).

You should guide groups to have a visual representation for the different explanations including mathematical models with statistics and trends. Groups might select a graphic organizer such as cause and effect to help make sense of their evidence. A note of caution to you: There doesn't seem to be one definitive cause for the honeybee population decline and CCD. There is considerable evidence that points to multiple reasons. It's important that human impacts are brought into the group discussions (e.g., agriculture practices, natural habitat loss, human population growth, loss of biodiversity in the ecosystem). It is suggested that small groups look at human population growth that shows graphic exponential growth. There are many resources available on this going back to Thomas Malthus who wrote the famous essay in 1798, "An essay on the principle of population" (UCMP website). One suggestion is for students to hear Al Bartlett's (deceased, University of Colorado physics professor) talk titled "arithmetic, population and energy" that he gave thousands of times since 1969 (Bartlett 1969).

Sense Making Part 5: Analysis of Arguments for Alternate Explanations

In this stage of sense making, students retrieve, extend, and apply information; using knowledge in relevant ways. Reflection is an important component. After student groups have discussed alternate explanations for the honeybee population decrease including CCD, the next step is to analyze arguments for the alternate explanations including mathematical models. The NRC *A Framework for K–12 Science Education* states the following about argumentation "As all ideas in science are evaluated against alternative explanations and compared with evidence; acceptance of an explanation is ultimately an assessment of what data are reliable and relevant and a decision about which explanation is most satisfactory. Thus knowing why the wrong answer is wrong can help secure a deeper and stronger understanding of why the right answer is right. Engaging in argumentation from evidence about an explanation supports students' understanding of the reasons and empirical evidence for that explanation, demonstrating that science is a body of knowledge rooted in evidence (NRC 2012, p. 44).

Sampson and Schleigh developed 30 biology classroom activities using a model built around scientific argumentation (Sampson et al. 2012). Their Stages of the Evaluate Alternatives Instructional Model are briefly summarized below.

- Stage 1: Introduce the phenomenon to investigate, the research question, and the alternative explanations (teacher selection).

- Stage 2: The generation of data. Small groups determine and implement a method to collect their data.

- Stage 3: The generation of tentative arguments and counterarguments. Small-group arguments need to include a claim that is supported by sufficient evidence and an adequate rationale. The group should develop a challenge for one alternative explanation. (Argument and challenge components include the research question, claim, evidence, justification of the evidence, an alternative claim, and challenge to the alternative claim.)

- Stage 4: An argumentation session. Groups share and critique each other's arguments.

- Stage 5: The reflection discussion. Original groups reconvene and discuss what they learned by interacting with individuals from other groups. Based upon their discussion, they modify their original argument or collect and analyze additional data. After the small-group discussion, the teacher guides the whole class through a reflective discussion on what was learned and ways to improve their arguments.

- Stage 6: The production of a final written argument. Each student writes an argument in support of one of the explanations including a challenge to an alternative explanation. Students write about what they know, how they know it, and why one explanation is more valid than the alternatives. Their evidence needs to include data, analysis of data, and interpretation of data.

In our honeybee example with the issue of why the population is decreasing including CCD, the small groups' task is to make sense of evidence and explanations from a variety of sources. Students are choosing the alternative explanations. They need to make sense of the data from their sources. The above stages 4–6 should be implemented. You might choose to have small groups rather than individuals write their final arguments.

Sense Making Part 6: Designing a Solution to Increase Biodiversity in a Local Ecosystem That Will Improve the Health of the Honeybee Population and Reduce Negative Human Impacts

This step has student groups address Learning Target #3. Refer to Table 6.2, p. 205 where the learning targets, crosscutting concepts, and science and engineering practices for ecosystem stability and resilience are defined. The major focus for this target is for student groups to "design a solution to a complex real-world problem, based on scientific knowledge, student-generated sources of evidence, prioritized criteria, and tradeoff considerations." The focus for student groups is designing a solution to increase biodiversity in a local or regional ecosystem that will also improve the health of the honeybee population and increase their numbers. You can keep the same small groups as for the previous five parts of sense making or have students form new groups to enhance further collaboration in the class.

It was suggested in Stage V: Identify Metacognitive Goals and Strategies that students use the "Options Diamond." Review the Stage V section (pp. 209–210). The "Options Diamond" routine focuses the student groups to look at trade-offs and solutions for each trade-off. Reflection is an important component for this routine addressing what the group has learned through the process.

Group solutions must be based upon previous class research on enhancing biodiversity and ecosystem stability including constraints, benefits, criteria, and environmental impacts. After small groups have designed their solutions, each group shares their solution with the other groups. They should include the use of technology to enhance their presentations.

Sense Making Part 7: Evaluating and Refining Solutions to Increase Biodiversity, Increase the Honeybee Population, and Reduce Negative Human Impact on a Local Ecosystem

This part of sense making has students use knowledge in relevant ways related to a real-world problem with a focus on Learning Target #4. Groups evaluate the different solutions giving strengths, weaknesses, and suggestions. Finally small groups go back to the drawing board to improve their solutions and share their final products with the class. Groups evaluate the final solutions.

Stage X: Determine Responsive Actions Based on Formative Assessment Evidence

Throughout the instructional sequence, you determine if your students understand the key concepts on ecosystem stability and resilience, biodiversity, and human impact. The products developed by the small groups are assessed. You can assign additional writing tasks for the student notebooks. Students write their understanding of vocabulary throughout the unit. For example, before groups design a solution for Part 6 sense making, individual students write their ideas first and then share their thinking with their small group. For Part 7 sense making, as students listen to other group's solutions, they can write notes, strengths, weaknesses, and questions for each group presentation. During the evaluation for each group solution, students should use their notebook thinking as a springboard to ask questions and critique each solution. Small groups use the same process to refine and evaluate final solutions for the honeybee issue. If students do not understand some concepts, you should provide additional instruction to support understanding. You can use probes to check for understanding. The *Uncovering Student Ideas* volumes through NSTA Press provide models for developing probes. You can learn more about these volumes at *www.uncoveringstudentideas.org*.

Time for Reflection (Teacher)

Now is the time for you to reflect on what you learned in this chapter. Think about the following questions.

- What did you learn about an instructional sequence that incorporates a real-world issue?

- What real-world issue related to ecosystem stability and resilience, biodiversity, and human impact might you select that is appropriate for a local or regional ecosystem? Why might this issue be important for your students to understand?

Other Real-World Issues and Resources

There are many compelling real-world issues on ecosystem stability and resilience, biodiversity, and human impact that you might select for students to research. You want to engage students in a problem that impacts their own environment. Some suggestions are:

- introduction of exotic or invasive species (non-native species)

- removal of predators from an ecosystem (e.g., mountain lion, wolf, black footed ferret)

- impact of drought on ecosystems (e.g., fire, beetle kill of lodgepole pine forests)

- habitat loss and degradation (e.g., agriculture, urban development)

- habitat fragmentation

- intersection of humans and wild animals (coyotes, bears, mountain lions) in an urban environment

- overharvesting some species

- climate change related to changes in an ecosystem (e.g., plants blooming earlier, less annual precipitation, increase in air temperature, increase in ocean temperature)

- pollution

- water quality

- endangered species

I highlight here two of the above issues. The first issue is invasive species. A key reason to select this issue comes from the following quote. "Introduced invasive species are the second major cause of biodiversity loss in North America" (Krasney et al. 2003, p. 9). *Invasive Species,* a book by Krasney et al. for high school teachers with a companion student book, introduces the issue and provides student protocols for ecology research. The WatchKnowLearn website has free videos on invasive species.

Another compelling issue is climate change with a focus on ecosystems. A current resource for biology teachers is the book *Climate Change from Pole to Pole: Biology Investigations* (Constible et al. 2008). The authors provide background on climate and six case studies with teacher and student pages.

Since ecosystems are very complex, this can lead scientists and others to misinterpret data and give explanations that don't take other possible explanations into account. When I first studied ecology at the university level, the Grand Canyon Kaibab Plateau research on what happened to the deer population when predators were removed from the ecosystem was used as an example of the negative impact of disrupting the predator-prey relationship in an ecosystem. Later, some scientists said the data used and the analysis of the data had flaws. Information on this research, the controversy, and more recent research on the issue can be found on the internet. The data showing the change in the deer population can be found on various internet sites and provides an interesting data set for students to graph and interpret.

Citizen science projects are a great way to engage students in scientific research with scientists and other citizens to collect data on a particular topic. There is a website for citizen science that gives information on many of these projects by a general topic area such as climate change, animals and plants. The website is *www.citizenscience.org*.

Examples of citizen science projects that provide support for classroom use is found in chapter 12, "Who poses the question?: Using citizen science to help K–12 teachers meet the mandate for inquiry" (Trautmann et al. 2012). The following citizen science projects are listed in the text.

- Project GLOBE (Global Learning and Observations to Benefit the Environment): *http://globe.gov*

- Journey North: *www.learner.org/north*

- Forest Watch: *www.forestwatch.sr.unh.edu*

- GREEN (Global Rivers Environmental Education Network): *www.earthforce.org/GREEN*

- Project BudBurst: *www.neoninc.org/budburst*

- Monarch Larva Monitoring Project: *www.mimp.org*

- ebird: *www.ebird.org*

Other excellent classroom resources are available on citizen science for student involvement. Trautmann et al. (2013) published a book, *Citizen Science: 15 Lessons That Bring Biology To Life, K–12*. It explores why to use citizen science in your teaching, implementation strategies, case studies, and 15 lessons. Another resource is the LearningScience website with free emerging tools to teach science. One of the selection links is "Help Scientists Do Science!" where many citizen science options are described.

You should incorporate student investigation on local ecosystems into any issue-based ecosystem unit. Ideally students ask a question related to an issue in a local ecosystem that leads to an investigation. Depending on your location, you need to consider the best time of year for fieldwork. Another option is to have students set up a closed ecosystem such as Bottle Biology to observe changes over time (University of Wisconsin). Students could further investigate what happens when pollutants are introduced.

Chapter 7

Variations of Traits

By Ravit Golan Duncan and Brian J. Reiser

"Genetics is to biology what atomic theory is to physics. Its principle is clear: that inheritance is based on particles and not on fluids. Instead of the essence of each parent mixing, with each child the blend of those who made him, information is passed on as a series of units. The bodies of successive generations transport them through time, so that a long-lost character may emerge in a distant descendant. The genes themselves may be older than the species that bear them."

—*John Stephen Jones 1999, p. 115*

About the Authors

By way of introduction, we would like to provide the reader with a sense of who we are and what expertise and relevant experiences we bring to the writing of this chapter. We are educational researchers who focus our research on the teaching and learning of science. Both of us have extensive expertise in developing inquiry-based curriculum materials in biology for the middle and high school grades. An integral part of our research involves working closely with teachers and students, to implement innovative instruction in the science classroom. Teachers nationwide from grades 4–12 have used the curriculum materials we have developed over the years.

In addition to expertise in curriculum design, Ravit Duncan also brings content expertise. She has a graduate degree in molecular biology and has studied the learning and teaching of genetics in grades 6–16 over the past decade. In collaboration with colleagues she has developed a learning progression in genetics that lays out a roadmap describing how students' knowledge of genetics should develop across grades 5–10.

Brian J. Reiser's research over the past two decades has involved the interweaving of scientific practices such as modeling, argumentation, and explanation with core content ideas in biology. Reiser was a member of the National Research Council committees authoring the reports *Taking Science to School* (2007) which provided research-based recommendations for improving K–8 science education; *A Framework for K–12 Science Education* (2012), which guided the design of the *Next Generation Science Standards;* and *Developing Assessments for the Next Generation Science Standards* (2014). Reiser co-led the development of IQWST (*Investigating and Questioning our World through Science and Technology*), a three-year middle school curriculum that supports students in science practices to develop disciplinary core ideas.

We believe that the *Next Generation Science Standards* hold the promise of a much brighter future for science education and a science-literate public in the United States.

Therefore, we are dedicated to helping teachers understand these standards and be able to employ effective teaching strategies to support their students in achieving them. We hope this chapter provides you with useful insights and tools for teaching genetics.

Overview

The disciplinary core ideas (DCIs) for heredity (LS3) focus on two fundamental aspects of genetics that are important to understand both from the perspective of scientific literacy for civic and personal engagement, and in preparation for STEM careers. The first aspect relates to the cellular and molecular mechanisms by which genes bring about our physical traits, in essence the gene to protein connection discussed in Chapter 3. This aspect deals with genetic mechanisms within an individual and explains how the genes we have (once we inherit them) result in our traits. The second aspect relates to the passing of genetic information across generations and the mechanisms involved in generating variations of genes and traits. In this chapter we will focus primarily on this second aspect, as described in the NRC *Framework* and NGSS- LS3B: Variation of Traits. Using the Instructional Planning Framework we will establish developmentally appropriate learning targets, figure out what performances can be expected of students who achieve these targets, and develop a coherent storyline and corresponding instructional sequence to support student engagement with the learning targets. There are many possible ways to develop instruction about the heredity DCI. We provide but one example of our own thinking process with the aim of making explicit the kinds of decisions and trade-offs involved in operationalizing the 10 stages of the framework in relation to this core idea.

Why This Topic?

The authors of the NRC *Framework* had good reasons for selecting heredity as a DCI. Scientific achievements, both theoretical and technical, in the field of genetics over the past few decades have been tremendous. These accomplishments have made headlines, spurred public debate, and resulted in many practical applications that citizens may encounter in their everyday lives. As a few examples: screening tests for a variety of genetic disorders are a fairly routine feature of obstetrics care; gene therapy methods have also been gaining some success with a few types of inherited cancers, retinal disease, and severe combined immunodeficiency; and sequencing of one's genetic material is now commercially available in many countries. What seemed like substance of science fiction 30 years ago is now within reach of ordinary citizens. We are wielding considerable power in our ability to manipulate genetic material, and with great power comes great responsibility.

The public needs to be educated about the meaning of these tremendous discoveries and technologies, as well as their shortcomings and the ethical issues that surround them (e.g., Billings et al. 1992; Feldman 2012). For example, research has shown that most high school graduates do not possess sufficiently sophisticated understandings in genetics to provide truly *informed* consent to the use of genetic technologies (Condit 2010; Miller 2004, Mills Shaw, Van Horne, Zhang, and Boughman 2008). As science educators, we must take part in the effort to help the next generation of citizens develop understandings in genetics that will be powerful and generative such that they can use them to reason about genetic phenomena and technologies that they may encounter in their lives now and in the future.

Phase I: Identifying Essential Content

Stage I: Identify DCI, Practices, and Crosscutting Concepts

We began our process of identifying the DCI, practices, and crosscutting concepts, with a close examination of the DCI as described in the NRC *Framework* and the relevant high school-level *NGSS* performance expectations (shown below).

> **By the end of grade 12.** The information passed from parents to offspring is coded in the DNA molecules that form the chromosomes. In sexual reproduction, chromosomes can sometimes swap sections during the process of meiosis (cell division), thereby creating new genetic combinations and thus more genetic variation. Although DNA replication is tightly regulated and remarkably accurate, errors do occur and result in mutations, which are also a source of genetic variation. Environmental factors can also cause mutations in genes, and viable mutations are inherited. Environmental factors also affect expression of traits, and hence affect the probability of occurrences of traits in a population. Thus the variation and distribution of traits observed depend on both genetic and environmental factors. (NRC 2012, p. 160)

The first point to note is that there are some important differences between the specifications in the NRC *Framework* and the PEs of the *NGSS* (see Figure 7.1). While the *Framework's* expectations for grade 12 are directly relevant to our topic of variation of traits, the PEs include some ideas that are not. Specifically, the PE HS-LS3-1 is a combination of several ideas, not all of which are relevant to the concept of genetic variation. This PE includes ideas from LS3-A, which deals with the connection between genes and proteins (see Chapter 3) as well as LS1-A, which includes two ideas: (a) all organisms contain genetic information in the form of DNA molecules, and (b) genes are regions in the DNA that code for the formation of proteins. The latter idea here (b) is also about the gene-protein connection. For our topic we will not tackle the gene-protein connection and the parts of the PE HS-LS3-1 that deal with this connection. Rather we will assume that content relevant to the *proteins and genes* topic will be a separate unit as described in Chapter 3. We assume that students will have studied that unit prior to this one, and will have figured out how genes determine the characteristics of organisms through proteins.

Figure 7.1

NGSS Performance Expectations for Heredity

HS-LS3   Heredity: Inheritance and Variation of Traits

Students who demonstrate understanding can:

HS-LS3-1. Ask questions to clarify relationships about the role of DNA and chromosomes in coding the instructions for characteristic traits passed from parents to offspring. [Assessment Boundary: Assessment does not include the phases of meiosis or the biochemical mechanism of specific steps in the process.]

HS-LS3-2. Make and defend a claim based on evidence that inheritable genetic variations may result from: (1) new genetic combinations through meiosis, (2) viable errors occurring during replication, and/or (3) mutations caused by environmental factors. [Clarification Statement: Emphasis is on using data to support arguments for the way variation occurs.] [Assessment Boundary: Assessment does not include the phases of meiosis or the biochemical mechanism of specific steps in the process.]

HS-LS3-3. Apply concepts of statistics and probability to explain the variation and distribution of expressed traits in a population. [Clarification Statement: Emphasis is on the use of mathematics to describe the probability of traits as it relates to genetic and environmental factors in the expression of traits.] [Assessment Boundary: Assessment does not include Hardy-Weinberg calculations.]

Examination of the DCI and PEs suggests that in terms of our topic (how variation in traits occurs) the focus is on four important sources of genetic variation that can arise between individuals even from the same parents: (a) independent assortment in sexual reproduction (HS-LS3-1); (b) chromosomal recombination (HS-LS3-2); (c) mutations due to random errors in DNA replication (HS-LS3-2); and (d) environmental sources of variation, namely mutation and influence on gene expression (HS-LS3-2). It is important to note here that that the idea of independent assortment (captured in the first bullet) was part of the expectations for the end of middle school:

> Sexual reproduction provides for transmission of genetic information to offspring through egg and sperm cells. These cells, which contain only one chromosome of each parent's chromosome pair, unite to form a new individual (offspring). Thus offspring possess one instance of each parent's chromosome pair (forming a new chromosome pair). Variations of inherited traits between parent and offspring arise from genetic differences that result from the subset of chromosomes (and therefore genes) inherited or (more rarely) from mutations. (NRC 2012, p. 159)

Thus, at the end of middle school, students should be able to explain how having two alleles (one from each parent) can explain how parents can have an allele for a trait that is not expressed in their own phenotype, but can pass that trait on to progeny.

At the high school level supplementary mechanisms for generating variation in traits are added to students' developing understanding of this idea. However, at both the middle and high school level the goal is to understand meiosis, the cellular process for generating sex cells, at the input-output level. That is, the focus is on understanding how meiosis helps explain how variation between siblings can arise, and how sexually reproducing species combine genetic material without doubling the amount of genetic material in subsequent generations. The NRC *Framework* and the *NGSS* do *not* call for a focus on the steps of the process. This is a really important point that we wish to emphasize. Much of current instruction, often driven by textbooks, includes an extensive focus on the steps of the process (Kurth and Roseman 2001). This focus on detail and the specific steps in meiosis does not result in a deeper understanding of what the process accomplishes; students still struggle to reason about the implications of this process for inheritance (Freidenreich, Duncan, and Shea 2011).

Another interesting point to note is the way the role of the environment is discussed. There are actually two ways in which the environment impacts traits. The first is by mutating the DNA. This is similar to the random errors generated in DNA replication, but in this case the errors are due to the involvement of mutagens—substances that can alter DNA and thus enhance the rate of mutation. The specific mechanisms by which mutagens act (for example, the generation of thymidine dimers by UV light) are not part of the expected understandings. Rather the idea here is that there are two types of mutations, spontaneous errors in replication and errors due to harmful mutagens in the environment.

The second way in which the environment influences our physical appearance is by impacting the expression of traits. Here too the story can get very complicated rather quickly and there are many different ways in which trait expression can be altered, these fall into the growing field of epigenetics. A simple example is the phenomenon of tanning. While almost all individuals can tan to some extent, the resulting skin color is a factor of both our genes and the extent to which we expose our skin to UV rays. Tanning is mediated by the activation of genes that ultimately enhance the production of melanin—the pigment that gives skin its brownish color. Similarly to the mutagen story, the NRC *Framework* expects students to be able to explain how, at the phenomena level, the environment can influence an organism's characteristics that are partially determined by genes. The actual mechanisms are not part of the story for even high school students. However, the focus on the role of the environment in genetics is emphasized to a greater extent in the NRC *Framework* and *NGSS* compared to prior standards (NRC 1996; AAAS 1993). This is a welcome and important shift given students' tendency towards genetic determinism (Dougherty 2009); more on this in the section on students' alternative conceptions and conceptual challenges.

The last PE stated in the *NGSS* for this core idea is about explaining the distribution of traits in a population. This PE does not directly relate to how variation in traits is generated, the focus of our topic. We have therefore chosen not to address it as part of our topic. Ideas about population genetics may be a better fit as part of a unit on evolution or as its own unit.

Now that we have a sense of the content for heredity, let's turn our attention to the practices and crosscutting concepts. The *NGSS* provide some suggestions for relevant practices: asking questions, defending claims with evidence, and developing explanations. These three practices are fairly complementary and dovetail nicely with each other. In addition, we plan to bring in the practice of developing and using models, since our ultimate goal is for students to develop a model for how variation occurs and use it to explain phenomena. We can develop our instructional activities (in the next phase) by engaging students with phenomena that will raise questions and require the development of an explanation. To support the development of the explanation, we will provide evidence. (More on this later in the chapter.)

In terms of crosscutting concepts there are two that seem most conducive to our topic: Patterns, and Cause and Effect. Since we are trying to explain relationships between genes and phenotypes, in particular patterns of association between these, the "patterns" concept is pertinent. While we noted there is a limited emphasis on the biochemical and molecular mechanisms involved in the generation of variation, there is still a sufficient role for mechanisms at the cellular level to merit attending to the *mechanism and explanation* concept. Structure and Function is another crosscutting concept that we considered. However, this concept seemed to fit better with the first part of LS3 detailing how genes bring about their effects in the organism. If our focus is on sources of variation, then structure and function relationships are less crucial, although they are still relevant and may gain some attention in the instructional sequence. Energy, Systems, Scale, and Change were also less clearly relevant to our topic.

We now have our relevant ideas, practices, and crosscutting concepts identified and we are ready to move on to the next stage: unpacking the DCIs into specific learning targets that include the practices and crosscutting concepts and developing the storyline for instruction.

Stage II: Deconstruct DCIs, Create a Storyline, and Align With Practices and Crosscutting Concepts

In some ways we have already started the unpacking process in the prior section in terms of identifying critical boundaries, namely, what is not included in the expectations for learning. We feel that it is just as crucial to identify the limits of what we should be teaching as the core, or essential understandings. Thus our topic is defined by both what is at the center and what is out of bounds.

We unpacked the sources of variation into the following learning targets:

1. DNA is the information-encoding molecule in our body that makes up chromosomes. The molecular structure of DNA enables it to encode instructions for making proteins in the sequence of nucleotides that make up the DNA molecule. Changes to this sequence can alter the instructions and result in substantive changes to traits (mostly detrimental but occasionally beneficial). The molecular structure also affords accurate replication so that the information can be passed on to future generations.

a. DNA is made up of two strands, each composed of smaller building blocks called nucleotides of which there are four types (ATGC).

b. The A-T nucleotides can bind to each other, as can G and C, thus the two strands are connected by bonds between A-T and G-C on opposite strands.

c. DNA is packed into structures called chromosomes. It is these chromosomes that are segregated into sex cells in the process of meiosis.

2. During the process of meiosis, chromosomes can swap parts. This results in additional shuffling of alleles and more variation (unlinking of allele variants of genes that were on the same chromosome).

3. Mutations due to random errors in DNA replication also generate genetic variation that leads to variation in traits.

a. DNA is replicated by splitting the double strand and building complementary strands to each of the existing strands. This process thus makes use of the DNA strand as a template, increasing the accuracy of the process.

b. However, errors do still occur and these can result in substantial changes to the information encoded (recipes for proteins).

c. There are proofreading mechanisms in the cells that catch some of the errors of replication, but not all of them.

4. The environment can also impact genetic variation:

a. Either by mutating the DNA (chemical mutagens found in the environment),

b. or by influencing which genes are activated in the cell.

The first learning target builds on what students learned about genes in middle school, specifically, the idea that individuals have pairs of chromosomes each with many genes and particular alleles of those genes, and that these influence traits by determining which proteins are made.

At the high school level students delve further into the molecular level to understand how genes influence traits, and we want them to understand that DNA is the information-carrying molecule that makes up the chromosomes they learned about before, and to be able to reason about how the structure of that molecule enables it to encode information. In terms of boundaries the *NGSS* and NRC *Framework* do not expect students to know the detailed structure of the DNA molecule. The focus is on how that structure of DNA affords encoding of information that influences the organism's traits. This presents a significant departure from current teaching in most high schools. It seemed unrealistic to us that teachers would be willing to forgo teaching about the structure of DNA entirely, we therefore surmised that some compromise is needed. In the *Framework* and *NGSS*, the key ideas about DNA structure are those needed for understanding how the genetic code is "read," and how DNA is replicated in ways that minimize errors; namely, the complementary base pairing of the double stranded molecule. We do not think that additional details about DNA structure, such as the phosphate backbone, the 3 and 5 prime ends, or the chemical structure of the nucleotides, are necessary. This instructional design decision involved a critical trade-off—balancing the addition of ideas beyond what is specified in the *NGSS* while not derailing the intent of the *NGSS* to focus on a few core ideas. We strongly agree with the NRC *Framework's* position regarding the need to substantially prune existing instruction; however, it is important to remember that the DCIs and PEs state the minimum requirements and that there is an acknowledgement that teachers may choose to go beyond those specifications. The key here is to ensure that the big ideas do not get buried and lost in a minutia of details. We have chosen to add some detail to the core idea and PEs that elaborate what the core ideas can explain, but only after we justified to ourselves the need for the addition.

The second learning target also clearly builds on what students learn at the middle school level: Chromosomes come in pairs and children inherit one member of the pair from each of their two parents, and this can be used to explain variation in traits between siblings. At the high school level we provide a mechanism for the laws of segregation and independent assortment—meiosis. Here we are also in complete agreement with the NRC *Framework's* boundaries and do not believe there is a need to teach the details (i.e., stages) of the process. For one thing, meiosis is a rather confusing process, with its two cellular divisions. Students do not see a need for two divisions given that the "goal," so to speak, of the process is to halve the genetic content in the cell. From that standpoint it would be more logical to simply divide the material and generate two sex cells rather than first duplicating all of it and then halving it twice to generate four sex cells. Even understanding recombination does not really require an understanding of the steps of meiosis, merely the existence of homologous pairs of chromosomes that swap parts before segregating. Note that the argument is *not* that these ideas are too complicated for students to learn; rather it is that these extra details about the process, while making a more complete story, do not add to the utility of what meiosis can explain. Learning the stages can easily become an exercise in learning the process as an outcome by itself, rather than using the understanding of the process to reason about and explain phenomena, such as how sexually reproducing species combine genetic information from two parents to produce offspring.

The third learning target builds on the first, in that we added the idea of complementary base pairing to our expectations of what students should know about DNA structure. In our unpacking here we have specified what is meant in the DCI by the statement "DNA replication is tightly regulated and remarkably accurate." This refers to the idea that DNA replication is semi-conservative, i.e., each strand serves as a template for the construction of a new strand. Thus after cell division the parent and daughter cells each end up with a DNA molecule that is made up of one "old" strand and one "new" strand. The use of a template increases the accuracy of the process. It is this latter idea that is at the core of why understanding that DNA replication is semi-conservative (terminology is far less important that the idea itself) as it allows one to explain the how accuracy in replication is achieved. Had the process been less accurate we would probably not have multicellular organisms because the rate of mutation would not sustain the delicate complexity involved. However, errors do occur and while they are often fixed by the proofreading process, some errors are still missed. We do not expect students to learn about the details of the proofreading machinery, these are highly involved molecular complexes, just the idea that there are some "black-boxed" ways in which the cell can proofread and catch errors in the process.

We have already discussed the ideas captured by the fourth learning target (the role of environment in generating variation) in the previous section. We will therefore simply reiterate that this learning target captures the two ways by which the environment impacts trait variation—modifying the DNA and modifying gene expression. The mechanisms of gene expression remain black-boxed in the *Framework* and *NGSS*. We do not advocate that teachers delve into the complex, albeit interesting, signal transduction pathways by which cells alter the profile of genes they express (for example, the famous lac-operon). Rather, students can simply learn that genes can be turned on or off and that environmental factors (chemical and physical) can trigger activating or inactivation of genes (i.e., they can influence which genes are translated into proteins by the cell).

The key to helping students develop understandings of sources of trait variation is pushing them to develop causal mechanistic models of these sources. Often this will entail problematizing incomplete or nonmechanistic explanations. Problematizing involves leading students to see how their current ideas cannot explain a new case. For example, at the middle school level, the need for two alleles per trait is not readily obvious. Left to their own devices, students are likely to develop models of inheritance that link single alleles to phenotype. One kind of phenomenon that can call that model into question consists of cases in which a trait appears to "skip" a generation. That is, a trait apparent in a parent is not apparent in its progeny, but appears again in the third generation. This leads to the idea that it is possible to have genetic information that does not affect the traits of the individual but can be passed on to progeny. This need can motivate investigation of the process of sexual reproduction to trace how genetic information is physically transmitted from parent to offspring, and ultimately to the construction of the idea that there are two copies of genetic information for each trait. In this way, the skipped generation case uncovers a problem in the single allele to trait model, motivating learners to revise and extend the model to encompass these more complex phenomena (Stewart, Cartier, and Passmore 2005).

Now that we have unpacked the DCIs and PEs into a set of four learning targets, the question becomes: What is the storyline that will allow students to develop these understandings in a coherent manner? Our overall approach here is one of problematizing students' existing models of genetic phenomena. These existing models were likely developed in middle school (some of which we alluded to above) and while they work well for the phenomena students are expected to explain at that level, they may break down in trying to explain more complicated phenomena. These shortfalls of the models serve as the motivation for revising and refining the models such that they provide explanations of a wider array of genetic phenomena.

Let's take for example the model that *NGSS* targets students to construct in middle school, in which each parent contributes half the genetic information by contributing half the chromosomes. This model works well until one starts tracking which information is encoded on which chromosome, and encounters phenomena that reveal a lack of expected linkage (correlation) between alleles of genes on the same chromosome. If sex cells simply contain either one chromosome or the other from each pair, at random, how can specific alleles stored on the same chromosome fail to travel together? Pushing to explain this problematizes the simple model of random segregation and introduces the need for recombination. From a pedagogical stance the "game" is to problematize students' initial models such that they continue to develop more sophisticated and complex models that account for more aspects of the phenomenon at hand (variation).

Using this general approach of problematizing will be helpful in thinking about the phenomena we will want to engage students with as we develop the unit (in the second phase of the instructional framework). However, this approach does not dictate a particular order of which models or aspects of students' cognitive models we should problematize first. We identified at least two viable starting points: investigating unexpected patterns of inheritance (i.e., problematizing for recombination), or investigating sources of mutations. If we begin with the former, we will be problematizing students' conceptions about how genes are passed on from one generation to the next and what they would expect to see in terms of patterns of correlation between traits on the same chromosome. We would then have to "switch gears" to problematize changes to traits due to mutations (caused by either replication errors or mutagens).

In contrast, if we begin with sources of mutations, we can problematize students' notions of the permanency of alleles, for example, how is it possible that a child would have a recessive trait if both parents are homozygote dominant? Or how can we get a completely new variant of a trait? We can begin with replication errors, which happen very infrequently, and then move to mutagens, which can significantly increase mutation rate (and the appearance of new trait variants). We would then switch gears again and tackle genetic recombination. In either case, regardless of the starting point, the second aspect of environmental influences on gene expression is not tightly related to either starting points and would have to be addressed as a third separate piece.

Since the mechanisms involved are rather distinct, we would have to "switch gears" more than once no matter where we start. We ended up selecting the first option with the following rationale: If we start with this option, we are not introducing entirely new variants right off the bat. Rather we are introducing a mechanism for reshuffling existing variants, that is, introducing variation by creating new combinations of traits rather than new traits. We can then problematize students' revised model (that now included recombination) further with phenomena in which new variations of the traits appear. This will entail revising the model of meiosis by examining changes to the actual genes (the instructions) rather that how genes are shuffled and sorted into sex cells. To us this revision of the model seemed less of a "jump" or switch in gears than the alternative sequence in which students first revise a model of mutations and then see phenomena that require recombination to explain. We will develop this instructional sequence in somewhat more detail in the second phase. Now we want to revisit the issue of practices and crosscutting concepts and their alignment with our current storyline.

Recall we selected the practices of: Asking Questions, Defending Claims With Evidence, Developing and Using Models, and Developing Explanations. The problematizing existing models approach is certainly well aligned with these practices. Problematizing involves introducing phenomena carefully selected to reveal the limitations of students' current models, leading them to construct questions that motivate further investigation. This driving question leads to additional questions that are more specific to the phenomenon itself that students can generate (e.g., can alleles change from dominant to recessive or vise versa?). The entire basis of this approach is in revising existing models and explanations; we can support students in their revision attempts by proving them with evidence that can help them pinpoint the troublesome aspects of their models and can give them clues about the genetic mechanisms at play.

We selected patterns and mechanism as our two crosscutting concepts. These also fit nicely with our general approach. Students are trying to explain anomalies in patterns of inheritance and traits by developing their existing mechanistic accounts of these. The idea of patterns suggests consistency and predictability; we capitalize on these properties in our problematizing approach. We deliberately introduce a phenomenon that "breaks" these patterns in unforeseen ways, thus problematizing the existing cognitive models that can explain the regular pattern but not the new counterexample. The way to solve this conundrum for students is by examining the mechanisms underlying their existing models and modifying or elaborating them.

Stage III: Determine Performance Expectations and Identify Criteria to Determine Student Understanding

NGSS specifies learning targets as the integration of content (disciplinary core ideas and crosscutting concepts) and practices. The *NGSS* performance expectations establish performances that students should be able to accomplish if they understand the content and can engage with the practice. We anticipate the students who have mastered the learning targets will be able to develop model-based explanations of the key processes involved that lead to the sources of variation in the evidence. These explanations should include a causal mechanism and be accurate. Below we provide an example of an assessment task and the criteria one can use to evaluate student performance on this task.

For our assessment task we use the context of a common phenomenon of bacteria developing resistance to drugs. The question posed to students is how is it possible that population of bacteria that were once susceptible to particular antibiotics can now tolerate them? As evidence students can be shown the proportion of resistant bacteria developing in a two populations: one that is irradiated with UV light (more resistant cells), and one that is not disturbed in this way (less resistant cells). Although natural selection can explain why preexisting variation would increase over time, it cannot provide an explanation for the difference between the two conditions. For that, students need to make a claim about the source of the new trait of resistance and defend it using the evidence provided. We would expect students' responses to include reference to environmental mutagens as the cause, in this case UV, and to explain how mutations, when not corrected, can result in new traits. Students should be able to explain why the non-irradiated bacteria population still have some resistant cells, but fewer by invoking the idea that some random errors occur during replication and that these can also generate the new trait. Thus this task can address two of our four learning targets. Students should be able to provide both mechanisms (mutagens and replication errors), they should be able to link the evidence to both mechanisms thus defending their claim about the source of variation in this interesting phenomenon.

Stage IV: Determine Nature of Science (NOS) Connections

There are three nature of science ideas that are most relevant to the modeling and explanation students need to engage in to develop these disciplinary core ideas:

- Scientific Knowledge Is Based on Empirical Evidence

- Scientific Knowledge Is Open to Revision in Light of New Evidence

- Scientific Models, Laws, Mechanisms, and Theories Explain Natural Phenomena

The strategy of problematizing models to motivate investigations relies, of course, on the understanding that knowledge building in science needs to be guided by empirical evidence. Understanding that models need to explain natural phenomenon is also relevant, and realizing that the process of building models is incremental. That is, models provide the best explanation of phenomena we have figured out so far, but new evidence may push us to revise existing models so they can handle new findings. Thus students need to see scientific knowledge as open to revision as we obtain new evidence, and see the role of investigation as uncovering evidence that may help us decide between competing models or may call an existing model into question. Thus, the problematizing of the model from middle school and its extension to handle new cases of variation are a great example of how science knowledge can be refined over time. Models are partial explanations that leave some open questions. With new investigations of phenomena, we can revise models so that they explain everything they explained earlier, but also address new phenomena or unpack steps that had been black-boxed earlier.

We envision incorporating these nature of science ideas as they are relevant to the model-building work we would do with students, rather than attempting to "front-load" the unit with discussion of these issues. For example, we might start with phenomenon that can be explained by the simple two-allele model, and then bring in phenomenon that cannot be explained by that model. In discussion about how to proceed, the importance of revising models in light of new evidence could then become an explicit focus, as students strategized about how to respond to the problematic evidence.

Stage V: Identify Metacognitive Goals and Strategies

The key metacognitive goal we have identified is self-regulated thinking. This goal is key given the incremental model building and model revision we are targeting for this unit. Students will continually need to be monitoring what we can explain and what evidence poses problems for the current model, and what are our current questions that will help us resolve these problems. There are several specific strategies we see as important to draw on in supporting this type of self-regulated thinking.

The strategy "identify what you know and what you don't know" will be important for helping students keep track of the model they are building incrementally. We envision that we will want to keep track of progress and open questions in an ongoing chart as we progress through the unit. At each point, we will need to discuss the current assumptions in the model (there are two alleles, genes determine traits, genes are located on chromosomes, independent assortment, and so on), the evidence we have collected that led us to these parts of the model, and the open questions (why don't traits that are on the same chromosome "travel together" as we would expect from independent assortment?). We will have students periodically revisit the chart to record new evidence and the questions it motivates, and revisions to the model as we figure out other ways that variation can arise.

The strategy "talking about thinking" will also be a part of students' work in this area. It will be important to push students to talk through their reasoning as they make arguments about how the current model can account for the patterns in data or fails to do so. Self-evaluation strategies will also be a key part of the work. We will need students to review the progress in explaining the variation data, and to identify what questions they have about these patterns. Our goal is for students to take responsibility for figuring out which parts of the phenomena cannot be explained with the current model and articulating research questions to guide the next steps in the investigation, and which parts of the model are problematic and need to be revised.

Phase II: Planning for Responsive Action

Stage VI: Research Student Misconceptions Common to This Topic That Are Documented in the Research Literature

Having identified the core ideas, practices, and crosscutting concepts as well as the beginning of a coherent storyline for teaching them, we next turn to the research to identify the cognitive challenges students may encounter when learning these ideas, practices, and concepts. We review the research as it relates to our four learning targets.

The domain of genetics is rife with terminology and scientific representations. A quick search of Google images related to chromosomes can give you a sense of the many ways in which we represent these structures in science as well as in instructional texts and tools (see Figure 7.2, p. 244). The abundance of terms and representations do not help students disentangle these structures and their role in genetics (Bahar, Johnstone, and Hansell 1999; Lewis and Wood-Robinson 2000). We know that students do not always see a connection between DNA, genes, and chromosomes (Lewis and Wood-Robinson 2000; Venville, Gribble and Donovan 2005). Students tend to think of DNA as a unique personal identifier found in the blood, determining our traits and found in cells (Venville et al. 2005). As can be seen in Figure 7.2 (p. 244), some of these representations make the connection between DNA and chromosomes relatively clear, while the others do not. Thus helping students integrate their conceptions of DNA and chromosomes into a coherent explanatory model is a non-trivial task we will need to undertake as we design instruction to address our first learning target.

The ideas captured in our second learning target involving the process of meiosis are fairly well researched in genetics education. We have known for decades now that students can develop algorithmic understandings of meiosis and even accurately use tools like Punnett squares without truly understanding what this process accomplishes. For example, students may be able to recount the phases of meiosis or calculate Punnett squares correctly, but they cannot use the process of meiosis to explain how the system of two copies of information for each trait can lead to passing on information for a trait that the parent doesn't exhibit (Stewart and Dale 1989). For example, we have shown that while students can correctly determine the genotype of a homozygote recessive parent with a genetic disorder, when asked what proportion of the parent's sex cells will carry the allele for the disorder, they do not always realize that all the sex cells will have this allele. Thus the correlation between a parent's genotype and the makeup of their sex cells is not obvious even after instruction about meiosis (Freidenreich et al. 2011). The process itself is also confusing for students. Research suggests that students struggle to distinguish between mitosis and meiosis (i.e., students state that these are the same cell division processes), and often are not be able to accurately describe how meiosis leads to the generation of genetic diversity among offspring (Lewis and Wood-Robinson 2000; Williams et al. 2012). Again, we feel the

Figure 7.2

Images of Chromosomes

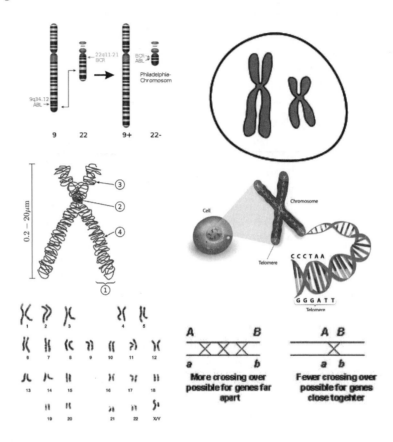

key here is to focus on the input and output of the process with the aim of explaining genetic variation.

In regard to our third learning target, mutations during DNA replication, there is not much research about this specific idea. However, we know that students tend to hold a view of mutations as being harmful (Mills Shaw et al. 2008) and an unwanted consequence. However, mutations form the basis of genetic variation and the fodder of natural selection. While most are probably harmful, some low rate of mutation is critical for the survival of populations and species. We therefore need to help students recognize the important role of mutations in generating genetic diversity and provide them with some examples of both neutral and beneficial mutations.

As noted earlier, the NRC *Framework* and *NGSS* have placed much greater emphasis on the role of the environment in influencing trait expression, our fourth learning target. This is not an idea that has been extensively researched. However, we do know that students have little understanding of the ways in which the environment can influence our genes and traits (Mills Shaw et al. 2008). Overall students tend to hold a rather deterministic view of genetics (Dougherty 2009) as exemplified by ideas like each trait is controlled by a single gene, genes determine all traits, and there is a lack of environmental influence. This view is highly problematic given that most traits are under the heavy influence of the environment and complex genetic interactions. For example, students tend to think that if a parent has cancer the child is very likely to have cancer as an adult (de Vries, Mesters, van de Staag and Honing 2005). However, most cases of cancer are not due to inherited mutations related to that cancer (like mutations BRCA 1 and 2 genes). Moreover, while students likely understand that exposure to some environmental factors like UV rays can cause cancer, it is less clear that they understand the underlying mechanism as involving mutations in our DNA. Thus students' prior conceptions here seem to miss the mark on both causal mechanisms by which the environment influences our traits.

In this section we highlighted only those alternative conceptions and conceptual challenges that we deem are most directly relevant to the topic at hand. However, instruction may surface alternative ideas that we did not discuss. For example, we know that students do not always presume a two-allele model, and may not even assume equal contribution of genetic material by both parents. For the purposes of this chapter we assumed that students have developed relatively sophisticated and accurate understandings of the content expected for the middle school level, and thus do not continue to hold such alternative ideas.

Stage VII: Determine Strategies to Elicit and Confront Your Students' Preconceptions

Now that we have identified key alternative ideas that students may bring with them to learn our focus topic, we need to find ways of identifying which of these are held by students in a particular context. The most obvious way is to use a written or interview-based preassessment. We would argue for a two-tiered approach, giving all students a written survey and then interviewing a handful of them regarding ideas that are unclear on the written survey. Russ and Sherin (2013) present helpful guidance for using interviews to tap students' prior conceptions. The survey and interviews should be done relatively close to the beginning of instruction and after other relevant concepts have been taught. That is, we would recommend teaching the topic of genes and proteins first and then surveying students. This is because the former topic may change students' ideas in ways that are relevant to the current topic. An alternative to surveying individual students is to conduct a brainstorming discussion in which students are asked to speculate about how to explain challenging cases drawing on the target ideas.

The written preassessment, or survey, need not be long and time-consuming; however, it should focus on the four learning targets and be designed to elicit potential alternative ideas that students may have. One could do this by using a combination of forced choice items and short open-ended items. The forced choice items can be designed to pinpoint specific alternative that we suspect exist (as described in the prior section), whereas the open-ended items leave room for identifying other alternative conceptions that we did not anticipate. An example of such items is shown in Figure 7.3.

Figure 7.3

Preassessment Item Formats

Alternative Items for Learning Target I

Forced choice: T/F format	Open-ended short response
Determine whether each of the following statements is True or False.	What is the connection, if any, between genes and DNA?
1. Genes are in our cells and DNA is in our blood.	
2. Genes determine our traits but DNA does not.	
3. Genes are part of the DNA, they are what holds the two strands of DNA together.	
4. Genes are segments of DNA that code for proteins.	
5. Everyone's DNA is unique, we do not share DNA with other family members.	

An alternative approach to identifying student conceptions we considered was to merge this stage with the next and to periodically engage students with an activity designed to elicit their naïve ideas prior to instruction about those concepts. For example, before beginning the instructional sequence for Learning Target #1 and #2, having a class discussion about how the egg and sperm are formed in the parent and what is the nature of the genetic material in these specialized cells. Then once that sequence is completed engaging in another group or whole-class activity to elicit students' understandings about mutations or the role of the environment in determining our traits. The drawback to this piecemeal approach is that students' alternative ideas about one concept may be related or influenced by their ideas about another (e.g., ideas about DNA, and ideas about mutations). Having a sense of the broader gamut of ideas students have about the different aspects of the topic, that is seeing the whole landscape, may impact design decisions about specific instructional strategies and activities that a more piecemeal view would not afford.

Stages VIII and IX: Strategies for Eliciting and Addressing Preconceptions; Sense-Making Strategies

As we discussed earlier, the general approach we have chosen is to problematize exiting models. We have combined stages VIII and IX from the instructional framework, weaving our strategies for eliciting and addressing prior conceptions into an instructional sequence to support students' development of the target models. In this section we discuss some potential instructional strategies and activities for doing so, with the aim of helping students see the gaps in their understanding and motivating the need to revise their existing conceptual models. We also describe the sense-making strategies we use to help students develop these revised models. The overall sequence is summarized in Table 7.1

Table 7.1

Instructional Sequence and Strategies

Part 1: How can variation happen?
• Elicit current ideas about genes and traits (Concept Mapping)
• Reconstruct middle school model of simple dominance to explain variation
• Draw out predictions of original model
• Present problematizing cases that violate predictions
• Help students revise the independent assortment model to include recombination in meiosis
Part 2: How does new variation occur?
• Elicit current ideas about the outcomes of mutations (Concept Cartoon)
• Provide evidence to help students understand that whether a mutation is beneficial or disadvantageous depends on context
• Provide competing models for source of mutations
• Provide evidence to reason about models
• Develop combined model that includes both sources of mutations

We suggest an instructional approach such as project-based science in which there is an overarching question, sometimes called a driving question (Krajcik and Czerniak 2014) that teachers develop with students, anchored in phenomena from the classroom and their daily lives. This driving question provides the coherence for the unit, and motivates specific subquestions to explore and develop pieces of the story leading to the four learning targets. The driving question for student inquiry should focus on the general issue of how variation between individuals in a species can occur; what makes us look different from each other? It is probably wise to focus the exploration on humans and then expand and generalize to other organisms. This is because students have more experiences, and prior knowledge to draw on, with regard to humans. Their understandings of genetics are more robust for mammals than for other organisms.

Part I: How Can Variation Happen?

We have chosen concept mapping as a general strategy for getting students to think about and discuss their ideas in ways that would help them see differences and potential gaps in their understanding. While vocabulary is not the learning focus of the unit, there are many terms and concepts that students bring into the classroom. We felt that it would be helpful to get students to articulate their initial ideas and definitions, and identify problematic vocabulary and open questions right at the start. Concept maps are akin to brainstorming webs (Tool 3.7) except that students are provided with a set of concepts they need to think about and connect. For this activity students can work in small groups or pairs and receive a large piece of paper and cards with the following words on them (one per card): genes, traits, DNA, chromosomes, nucleotides, cells, parents, blood (and others can be added). Students are asked to place these cards on the large paper and draw labeled arrows to connect the terms.

Students can share their concept maps in various ways. We like the idea of a gallery walk as a means of sharing ideas because it can help students understand the importance of explaining their thinking and it allows students to provide constructive feedback to peers. In a gallery walk students walk around and review other students' concept maps and provide comments and questions (on sticky notes). Students do not have to view all the maps, commenting on three or four is sufficient. A whole-class discussion can follow, in which the teacher can attempt to develop a class concept map that captures students' ideas as well as their questions and uncertainties. In essence this class discussion can combine a concept mapping and KWL (Know, Want to know, and Learn) thus engaging students in the practices of defending their claims and questioning. Overall, the goal of this activity would be to have students share their ideas, note differences in what they believe to be true, and highlight aspects of their understandings that are incomplete.

Following the concept mapping would be a sense-making activity in which students use evidence to figure out hierarchical organization of the genetic information (such as genes as segments of DNA on a chromosome). The evidence is this case can include historically important experiments that contributed to our understanding of the genetic material and its structure. For example, students can explore patterns of diffraction made by shining a laser pointer at by various structures (like a rod, a circle, a spring) and then they can compare the patterns they have with photo 51, the famous x-ray diffraction image made by Rosalind Franklin. The pattern in the picture can be replicated by shining a laser pointer through a spring from a ballpoint pen. A more detailed description of this activity and how it can be used with students can be found in an article in *The Physics Teacher* by Braun, Tierney, and Schmitzer (2011). Additional evidence can be provided to help students understand the relationship between DNA and chromosomes (for example, simplified descriptions of the experiments of Boveri and Sutton). Using these kinds of evidence pieces, students can construct a more complete and accurate model of DNA structure and organization (from nucleotide to chromosomes), which addresses the first learning target. This model can be used as the basis for activities that address the second learning target.

The next step is to help students articulate their current models about how traits are inherited, as developed at the middle school level. It is key to have all students on the same page, being able to use the model to reason about how genetic information is passed on, how variation can occur between siblings due to having two alleles that are segregated and randomly assorted, and how the information in the two alleles are used to determine the trait. To do this, we would present students with phenomena that need to be explained (e.g., a specific genetic trait or disorder), and have them use their current model to develop an explanation for the pattern of inheritance as evidenced in pedigrees of families with the specific trait. The pedigrees would show multiple siblings across three generations, and would include cases of inheritance where the recessive trait skips a generation. Students would have to explain how it is possible for a child to have a trait that neither of her parents have. In framing this as a task in which students are asked to explain using a model, students need to do more than simply say, "traits can skip generations when they are recessive alleles." They will need to provide an explanatory account that connects the observation above with the underlying mechanism of meiosis and Mendel's Laws. Such an account would have to show what alleles must have been passed to each of the parents in order for them to have produced a child that has two recessive alleles, which is what the child would have to possess in order to show the recessive phenotype. The account would also have to explain how each combination of alleles for each individual determined the individual's trait.

Once we have helped students reconstruct this model from middle school, we introduce new phenomena that will problematize their current conceptual models. The second learning target requires some setup first. It is unlikely that students have thought about all the consequences that are derived from their model of genes as segments on chromosomes. One would have to walk students through some predictions that can be made based on their current model, namely, that genes located on the same chromosome would show a pattern of association for their alleles. The prior problems students have solved (in middle school and the activities described above) only dealt with one gene at a time, and so are completely handled by students' model from middle school. To push that model, we present problems tracing two genes at once, and ask students to analyze possible genotypes for the phenotypes in a pedigree. In doing this, we ask students to identify which chromosomes the alleles must have come from. We establish the process first with traits on different chromosomes, which present no challenge to the existing model. Then we provide a case in which the genes (and therefore their alleles) are on the same chromosome, and a resulting child has what appears to be an impossible combination of traits (drawn from two different chromosomes). We do not tell the students the difference in the case, the goal is for them to try and figure out how to explain this unexpected result.

The case of two genes on the same chromosome can more easily be observed with X-linked traits like color blindness and hemophilia since only one chromosome is involved and the allele on that chromosome is always the one expressed in males (i.e., recessive phenotype is always shown). One would expect that in families with color blindness and hemophilia these traits would go together. Indeed, according to students' current model it is not possible for a male to have one trait and not the other. Once students come to understand this prediction they can be presented with a discrepant event, the existence of males who only have one trait and not the other (a real phenomenon). We will then use these cases to pose a problem for the model, first getting students to realize why the current model cannot explain these cases. This establishes the need to revise the model in a way that it can handle how a sex cell could end up with traits drawn from two different chromosomes of the same pair.

As a class, we discuss and summarize (a) the problem for our model (b) our current investigation goal, namely how can a sex cell end up with one allele from one chromosome and another allele from the other chromosome of the same pair. We then ask students to brainstorm different types of modifications to the meiosis model that could potentially explain the anomalous data. We do not expect students to solve the problem without additional evidence and support. However, the experience of trying to solve it and failing is still highly productive for learning as it allows students to notice important features of a potential solution (Kapur and Bielaczyc 2012). The approach of letting students struggle with a problem they are likely unable to solve is called productive failure, and it has been shown to be very effective when coupled with direct instruction following the problem-solving activity (Kapur 2008).

As noted above we can provide students with some relevant evidence to support students in their sense making around this phenomenon and their attempts to come up with a mechanism to explain it. The evidence, again, can be drawn from historical experiments conducted by Barbara McClintock, and her student Harriet Creighton, on traits in maize and their relationships to specific alleles on chromosomes. McClintock and Creighton mapped crossing over events they observed under a microscope to resultant traits in maize plants to show that homologous chromosomes were swapping sections. Once the class has established what the evidence shows us must be happening, we can then present a mechanism that can explain it—the recombination as a part of meiosis. Recall, there is no need to delve into much detail here, it is sufficient that students understand that chromosomes can swap sections. Given their attempt to account for the anomalous data it is very likely that the idea of swapping parts will be seen as way to resolve the conundrum as soon as it is introduced. Students' struggles earlier will prepare them to see the fruitfulness of this solution more readily. Moreover, the understanding of meiosis and recombination will satisfy a need-to-know (Edelson

2001) that the prior activity established; students are likely to be motivated to learn the solution given their attempts to come up with one.

Part II: How Does New Variation Occur?

We expect the concept of mutations to arise in part 1 of this unit as a possible explanation for how unexpected outcomes could occur. If the idea of mutations arises, it provides a natural transition to the second half of the unit. If it does not arise from the students, the teacher can bring in this idea as something that we know can influence inheritance in particular ways. We will then focus students' attention on trying to figure out what actually happens in mutations and why it matters.

To elicit students' alternative conceptions about mutations, we chose the strategy of concept cartoons (Tool 3.8, p. 134). Concept cartoons essentially pit two or more competing ideas/perspectives about a phenomenon and the idea is to have students choose the position they agree with and argue with students who have chosen one of the alternative positions. The claims or alternative positions presented in the concept cartoon often reflect known alternative conceptions that students have (for examples and explanation about concept cartoons go to *www.conceptcartoons.com*). One can construct a concept cartoon for practically any topic for which students may harbor alternative conceptions that are nonaligned with canonical understandings.

In Figure 7.4 we present an example of a potential concept cartoon that addresses our third learning target. Students select which of the three perspectives they agree with the most and then engage in a whole-class argumentation discussion about these ideas. Much like the concept mapping activity, the goal of this activity is to help students see gaps in their understanding by highlighting alternative viewpoints that may "shake the foundations" of their existing conceptual models, thus problematizing them. It is also likely that such a class discussion would raise questions for students, such as, what are mutations, and how are they caused? Following this activity one could, again, provide students with evidence to help them revise their existing model.

One approach to doing this is to present some evidence about actual cases of known mutations and their consequences. A good example for this purpose is sickle-cell anemia, because the mutation itself is simple to explain and its consequences vary depending on whether the individual inherits one or two mutated copies. The mutation in

Figure 7.4.

Concept Cartoon for the Third Learning Target

sickle cell is a change to the hemoglobin proteins that makes it "sticky" and causes hemoglobin proteins to clump together, altering the shape of the cell and preventing proper transportation of oxygen. Overall, this change is a bad thing and the resulting phenotype is a painful and dangerous anemia. However, in heterozygous (individuals with only one mutated copy) there is some advantage because sickle cells are not as susceptible to malaria. In a complex mechanism the sickled hemoglobin makes the host tolerant to the parasite that causes malaria. This type of evidence can help students understand that "beneficial" and "harmful" are rather context dependent and what might be harmful in one context can be beneficial in another. There are many other examples that can be used to bolster this understanding: (a) a mutation in the CCR5 protein that confers resistance to HIV, (b) a mutation in MC1R that results in fair skin was advantageous when our ancestors moved to colder northern climates because fair skin allows more vitamin D production, and (c) a mutation in the lactase gene that stays turned "on" and expressed throughout an individual's life allows us to drink milk (earlier humans were lactose intolerant from the age of 4 or 5). For all of these examples, students can compare the mutated and normal genes to identify the change. The exact nature of the mutation is less important that having students see that errors can arise in the genetic sequence, and how that could then affect the role of the gene in determining the phenotype.

After students understand what a mutation is, we will raise the central question of how mutations arise. Again our goal is to help students work out two different mechanisms for mutation—spontaneous errors in replication and changes to the DNA due to harmful mutagens in the environment. The question of where novel variation comes from could also be introduced using a concept cartoon. There could be just two alternative perspectives provided in this cartoon: (a) mistakes made when copying DNA, and (b) changes to the DNA due to toxin in the environment. We chose to juxtapose two correct options in the cartoon this time (and a third incorrect option) in order to help students later understand that often there are multiple causes for a particular outcome. Students are likely to initially pick one side of the cartoon; however, with appropriate evidence we hope they will come to see that both claims are true and can account for the new phenomenon. The kinds of evidence that can be used here include studies about the rate of changes to DNA when single-celled organisms are exposed to mutagens in the environment; the disorder xeroderma pigmentosum, in which there is a mutation in a DNA repair protein causing patients to have higher incidence of mutations and cancer; and studies of the effects of UV rays on DNA structure. Through analysis of such evidence and engagement in sense making regarding what they suggest about potential causes of mutations, students can develop models that include the two mechanisms for mutations—random replication errors and the effects of mutagens. These activities would conclude the unit and address all four learning targets. We next discuss some useful assessment strategies that can help tailor the instruction to students' developing understandings.

Stage X: Determine Responsive Actions Based on Formative Assessment Evidence

As we have described earlier, the science practice that frames most of the students' activity will be constructing an explanatory model and using their models to construct explanations of the particular phenomena. In the sequences in stages VIII and IX, we have already mentioned some of the strategies we will use to respond to formative assessments in order to support sense making and revealing understanding. The models we ask students to construct and use can be seen as visual models involving diagrams and verbal models of metaphors (Instructional Tool 3.6, p. 119). We envision diagrams would be key in the first learning target, explaining how two genes on the same chromosome could become swapped during meiosis. The model would have to track the location of the genes on each allele in each of the two parents, then the resulting sex cells, and finally the alleles in the offspring that exhibit the recombination. We will also ask students for written explanations using the model (Instructional Tool 3.3, p. 93), in which they walk through the cause-and-effect sequence and tie observed data (e.g., what we know about the location of the two genes on the chromosomes, and the observed combination of phenotypes in the parents and the offspring). By combining the diagrammatic models with written explanations that draw on the models, we will ensure that students are pushed to make their reasoning explicit, tracing the genetic information from parent to child, and using the information to determine traits.

We envision that large- and small-group discussions (Instructional Tool 3.5, p. 110) will play a key role. To engage in scientific modeling, students need to talk through their ideas, use their models to try to account for evidence in test cases, and attempt to reach consensus as a class when there are multiple candidate models or candidate explanations of cases proposed by students (Reiser, Berland, and Kenyon 2012). Teachers will need to manage classroom discussion to attempt to push continually for mechanism in students' diagrammatic models and written explanations. Questions such as "But why does that happen?" and "What is going on with the chromosome that would allow that to happen?" will be key in pushing students to go beyond simply labeling things and attempting to reason through how the mutation could happen or why it would have the effect it does.

Conclusion

The general subject matter of heredity is a great example of the way that the shifts in *NGSS* have real implications for classroom instruction. Too often through traditional instruction students learn the details of the structure of DNA, and learn about the process of transcription and translation at the algorithmic level, without engaging in developing explanatory models that use these structures and processes to explain how and why various heredity phenomena occur. For example, many students learn to work out Punnett squares to calculate probabilities of various trait combinations and learn to name and describe the steps of meiosis, and yet they are not able to explain how it is possible for a parent to pass on genetic information to a child for a trait that the parent does not possess. Students may learn the steps of translation and transcription or the names of particular types of mutations (such as deletion, substitution), without being able to explain how mutations can occur or why these types of mutations would lead to effects on the organism.

The strategies presented in this chapter are intended to help students achieve the type of learning targeted by *NGSS*, in which students tackle the core explanatory ideas by engaging in the science and engineering practices. The focus on models and explanations is intended to help students develop these explanatory accounts so that they can use the disciplinary core ideas of LS3 to explain how variation arises across generations, how new variations can appear, and ultimately how genetics determines our traits.

Resources

- Genetic Science Learning Center: *http://learn.genetics.utah.edu*

- The Concord Consortium: Digital learning for science, math, and engineering: *http://concord.org*

- The "Tools for Ambitious Science Teaching" website (*http://tools4teachingscience. org*) provides helpful guidance and tools for managing classroom discourse involving scientific models.

- The Talk Science Primer can be found online at TERC (*http://inquiryproject.terc. edu/shared/pd/TalkScience_Primer.pdf*). The primer, written by Sarah Michaels and Cathy O'Connor, presents a set of talk moves teachers can use to help students articulate their thinking and to help students build on each others' ideas in science discussions.

Chapter 8

The Role of Adaptation in Biological Evolution

By Sue Whitsett

"Nothing in biology makes sense except in the light of evolution."

—*Dobzhansky 1973*

I was asked to contribute to this book by Susan Koba and Anne Tweed, colleagues and mentors. I was overwhelmed when asked to take on probably the most controversial science topic in life science: evolution. Personally, I do not consider this topic to be a "hard- to-teach" topic, at least when thinking of the content associated with the topic. It is hard to teach because of what students bring to the classroom—misconceptions and beliefs. I began my career as a Catholic school teacher, and learned early in my career that science is about evidence, not about beliefs. I had to learn how to become a teacher of science. I did not take a specific course in evolutionary biology during my undergraduate degree. It was not until I had taught for 30 years that I took a formal course in evolution; prior to that time I took advantage of many presentations at NSTA and NABT conferences to learn about activities to help teach the topic in my life science classes.

Why This Topic?

Students come into life science with both their own experiences and those they have learned through direct instruction. Depending on a student's background and the schools the students have attended, the range of information about evolution can be very diverse, more so than many other topics. As life science teachers, we need to acknowledge the diversity of opinions, beliefs, and factual information a student brings to the discussion of evolution. "Acknowledge" does not mean bring religion into the conversation, but rather help a student understand the difference between a belief system and the nature of science.

As I begin my unit on evolution, I use a survey that I developed as an introductory lesson. This is given immediately following the previous unit's summative exam. I purposefully avoid any mention of the "E" (Evolution) word until after the survey has been handed out. The evolution preassessment contains short-answer questions covering multiple topics that I plan to address in the unit of study. The first question is very open-ended: "When you hear the word evolution, what do you think of?" Why do we need to address this topic in this book? I still had at least 10% of my students every year answer "man coming from monkeys!" I had straight-A students sit in class with a note from their parents stating their child may not participate in any of the activities nor the assessments during the unit on evolution. This one example of a misconception and many others that appear in the preassessment, along with parents that will not let their son or daughter participate, are reasons to include a chapter on evolution again in the second edition of this book.

Evolution is a major theory that ties together many concepts in life science and gives meaning to those other concepts (Mayr 1991, 1997). It can help the student understand why there is the diversity of life, why there has been extinction of life, why there is concern about the overuse of prescription medications, why there are still some viruses that do not have a vaccination, and why there is hope to help cure some diseases in the future. According to Mead and Branch, "By providing a unifying principle for biology, evolution provides a powerful framework for investigating the living world, enabling us to develop models, frame theories, and test hypothesis" (2011). A report was published based on a national conference held in 2009, titled Vision and Change in Undergraduate Biology Education: A Call to Action. The core concept of that report states that the diversity of life evolved over time by processes of mutation, selection, and genetic change. The explanation gives a reason why evolution should be part of any biology course:

> Darwin's theory of evolution by natural selection was transformational in scientist's understanding of the patterns, processes, and relationships that characterize the diversity of life. Because the theory is the fundamental organizing principle over the entire range of biological phenomena, it is difficult to imagine teaching biology of any kind without introducing Darwin's profound ideas. Inheritance, change, and adaptation are recurring themes supported by evidence drawn from molecular genetics, developmental biology, biochemistry, zoology, agronomy botany, systematics, ecology, and paleontology. A strong preparation in the theory of evolution remains essential to understanding biological systems at all levels. (AAAS 2011)

Phase I: Identifying Essential Content

Stage I: Identify Disciplinary Core Ideas, Practices, and Crosscutting Concepts

Following the design process found in earlier chapters, Stage I identifies the disciplinary core ideas from the *Next Generation Science Standards* that are the focus for this unit. The standard that is the focus for this chapter is HS-LS4 Biological Evolution: Unity and Diversity. I have chosen the disciplinary core idea (DCI) of Adaptation, LS4.C, as the major focus of this chapter.

Under DCI LS4.C, the first, second, and third bullets focus on the specific content for this chapter (see Figure 8.1, pp. 258–259). The first bullet looks at the process of evolution. The second bullet focuses on natural selection that leads to adaptation. The third bullet focuses on the distribution of adaptations in a given population after conditions in an environment change.

Figure 8.1

HS-LS4 Biological Evolution: Unity and Diversity

HS-LS4 Biological Evolution: Unity and Diversity

HS-LS4	Biological Evolution: Unity and Diversity

Students who demonstrate understanding can:

HS-LS4-1. Communicate scientific information that common ancestry and biological evolution are supported by multiple lines of empirical evidence. [Clarification Statement: Emphasis is on a conceptual understanding of the role each line of evidence has relating to common ancestry and biological evolution. Examples of evidence could include similarities in DNA sequences, anatomical structures, and order of appearance of structures in embryological development.]

HS-LS4-2. Construct an explanation based on evidence that the process of evolution primarily results from four factors: (1) the potential for a species to increase in number, (2) the heritable genetic variation of individuals in a species due to mutation and sexual reproduction, (3) competition for limited resources, and (4) the proliferation of those organisms that are better able to survive and reproduce in the environment. [Clarification Statement: Emphasis is on using evidence to explain the influence each of the four factors has on number of organisms, behaviors, morphology, or physiology in terms of ability to compete for limited resources and subsequent survival of individuals and adaptation of species. Examples of evidence could include mathematical models such as simple distribution graphs and proportional reasoning.] [Assessment Boundary: Assessment does not include other mechanisms of evolution, such as genetic drift, gene flow through migration, and co-evolution.]

HS-LS4-3. Apply concepts of statistics and probability to support explanations that organisms with an advantageous heritable trait tend to increase in proportion to organisms lacking this trait. [Clarification Statement: Emphasis is on analyzing shifts in numerical distribution of traits and using these shifts as evidence to support explanations.] [Assessment Boundary: Assessment is limited to basic statistical and graphical analysis. Assessment does not include allele frequency calculations.]

HS-LS4-4. Construct an explanation based on evidence for how natural selection leads to adaptation of populations. [Clarification Statement: Emphasis is on using data to provide evidence for how specific biotic and abiotic differences in ecosystems (such as ranges of seasonal temperature, long-term climate change, acidity, light, geographic barriers, or evolution of other organisms) contribute to a change in gene frequency over time, leading to adaptation of populations.]

HS-LS4-5. Evaluate the evidence supporting claims that changes in environmental conditions may result in: (1) increases in the number of individuals of some species, (2) the emergence of new species over time, and (3) the extinction of other species. [Clarification Statement: Emphasis is on determining cause and effect relationships for how changes to the environment such as deforestation, fishing, application of fertilizers, drought, flood, and the rate of change of the environment affect distribution or disappearance of traits in species.]

HS-LS4-6. Create or revise a simulation to test a solution to mitigate adverse impacts of human activity on biodiversity.* [Clarification Statement: Emphasis is on designing solutions for a proposed problem related to threatened or endangered species, or to genetic variation of organisms for multiple species.]

The performance expectations above were developed using the following elements from the NRC document *A Framework for K–12 Science Education*.

Science and Engineering Practices

Analyzing and Interpreting Data
Analyzing data in 9–12 builds on K–8 experiences and progresses to introducing more detailed statistical analysis, the comparison of data sets for consistency, and the use of models to generate and analyze data.
- Apply concepts of statistics and probability (including determining function fits to data, slope, intercept, and correlation coefficient for linear fits) to scientific and engineering questions and problems, using digital tools when feasible. (HS-LS4-3)

Using Mathematics and Computational Thinking
Mathematical and computational thinking in 9–12 builds on K–8 experiences and progresses to using algebraic thinking and analysis, a range of linear and nonlinear functions including trigonometric functions, exponentials and logarithms, and computational tools for statistical analysis to analyze, represent, and model data. Simple computational simulations are created and used based on mathematical models of basic assumptions.
- Create or revise a simulation of a phenomenon, designed device, process, or system. (HS-LS4-6)

Constructing Explanations and Designing Solutions
Constructing explanations and designing solutions in 9–12 builds on K–8 experiences and progresses to explanations and designs that are supported by multiple and independent student-generated sources of evidence consistent with scientific ideas, principles, and theories.
- Construct an explanation based on valid and reliable evidence obtained from a variety of sources (including students' own investigations, models, theories, simulations, peer review) and the assumption that theories and laws that describe the natural world operate today as they did in the past and will continue to do so in the future. (HS-LS4-2),(HS-LS4-4)

Engaging in Argument from Evidence
Engaging in argument from evidence in 9–12 builds on K–8 experiences and progresses to using appropriate and sufficient evidence and scientific reasoning to defend and critique claims and explanations about the natural and designed world(s). Arguments may also come from current or historical episodes in science.
- Evaluate the evidence behind currently accepted explanations or solutions to determine the merits of arguments. (HS-LS4-5)

Obtaining, Evaluating, and Communicating Information
Obtaining, evaluating, and communicating information in 9–12

Disciplinary Core Ideas

LS4.A: Evidence of Common Ancestry and Diversity
- Genetic information provides evidence of evolution. DNA sequences vary among species, but there are many overlaps; in fact, the ongoing branching that produces multiple lines of descent can be inferred by comparing the DNA sequences of different organisms. Such information is also derivable from the similarities and differences in amino acid sequences and from anatomical and embryological evidence. (HS-LS4-1)

LS4.B: Natural Selection
- Natural selection occurs only if there is both (1) variation in the genetic information between organisms in a population and (2) variation in the expression of that genetic information—that is, trait variation—that leads to differences in performance among individuals. (HS-LS4-2),(HS-LS4-3)
- The traits that positively affect survival are more likely to be reproduced, and thus are more common in the population. (HS-LS4-3)

LS4.C: Adaptation
- Evolution is a consequence of the interaction of four factors: (1) the potential for a species to increase in number, (2) the genetic variation of individuals in a species due to mutation and sexual reproduction, (3) competition for an environment's limited supply of the resources that individuals need in order to survive and reproduce, and (4) the ensuing proliferation of those organisms that are better able to survive and reproduce in that environment. (HS-LS4-2)
- Natural selection leads to adaptation, that is, to a population dominated by organisms that are anatomically, behaviorally, and physiologically well suited to survive and reproduce in a specific environment. That is, the differential survival and reproduction of organisms in a population that have an advantageous heritable trait leads to an increase in the proportion of individuals in future generations that have the trait and to a decrease in the proportion of individuals that do not. (HS-LS4-3),(HS-LS4-4)
- Adaptation also means that the distribution of traits in a population can change when conditions change. (HS-LS4-3)
- Changes in the physical environment, whether naturally occurring or human induced, have thus contributed to the expansion of some species, the emergence of new distinct species as populations diverge under different conditions, and the decline—and sometimes the extinction—of some species. (HS-LS4-5),(HS-LS4-6)

Crosscutting Concepts

Patterns
- Different patterns may be observed at each of the scales at which a system is studied and can provide evidence for causality in explanations of phenomena. (HS-LS4-1),(HS-LS4-3)

Cause and Effect
- Empirical evidence is required to differentiate between cause and correlation and make claims about specific causes and effects. (HS-LS4-2),(HS-LS4-4),(HS-LS4-5),(HS-LS4-6)

Connections to Nature of Science

Scientific Knowledge Assumes an Order and Consistency in Natural Systems
- Scientific knowledge is based on the assumption that natural laws operate today as they did in the past and they will continue to do so in the future. (HS-LS4-1),(HS-LS4-4)

*The performance expectations marked with an asterisk integrate traditional science content with engineering through a Practice or Disciplinary Core Idea.

The section entitled "Disciplinary Core Ideas" is reproduced verbatim from A Framework for K-12 Science Education: Practices, Cross-Cutting Concepts, and Core Ideas. Integrated and reprinted with permission from the National Academy of Sciences.

May 2013 ©2013 Achieve, Inc. All rights reserved. 1 of 2

258

Figure 8.1 (continued)

HS-LS4 Biological Evolution: Unity and Diversity

builds on K–8 experiences and progresses to evaluating the validity and reliability of the claims, methods, and designs. • Communicate scientific information (e.g., about phenomena and/or the process of development and the design and performance of a proposed process or system) in multiple formats (including orally, graphically, textually, and mathematically). (HS-LS4-1) ------------------------------------ *Connections to Nature of Science* **Science Models, Laws, Mechanisms, and Theories Explain Natural Phenomena** • A scientific theory is a substantiated explanation of some aspect of the natural world, based on a body of facts that have been repeatedly confirmed through observation and experiment and the science community validates each theory before it is accepted. If new evidence is discovered that the theory does not accommodate, the theory is generally modified in light of this new evidence. (HS-LS4-1)	• Species become extinct because they can no longer survive and reproduce in their altered environment. If members cannot adjust to change that is too fast or drastic, the opportunity for the species' evolution is lost. (HS-LS4-5) **LS4.D: Biodiversity and Humans** • Humans depend on the living world for the resources and other benefits provided by biodiversity. But human activity is also having adverse impacts on biodiversity through overpopulation, overexploitation, habitat destruction, pollution, introduction of invasive species, and climate change. Thus sustaining biodiversity so that ecosystem functioning and productivity are maintained is essential to supporting and enhancing life on Earth. Sustaining biodiversity also aids humanity by preserving landscapes of recreational or inspirational value. (HS-LS4-6) *(Note: This Disciplinary Core Idea is also addressed by HS-LS2-7.)* **ETS1.B: Developing Possible Solutions** • When evaluating solutions, it is important to take into account a range of constraints, including cost, safety, reliability, and aesthetics, and to consider social, cultural, and environmental impacts. *(secondary to HS-LS4-6)* • Both physical models and computers can be used in various ways to aid in the engineering design process. Computers are useful for a variety of purposes, such as running simulations to test different ways of solving a problem or to see which one is most efficient or economical; and in making a persuasive presentation to a client about how a given design will meet his or her needs. *(secondary to HS-LS4-6)*

Connections to other DCIs in this grade-band: **HS.LS2.A** (HS-LS4-2),(HS-LS4-3),(HS-LS4-4),(HS-LS4-5); **HS.LS2.D** (HS-LS4-2),(HS-LS4-3),(HS-LS4-4),(HS-LS4-5); **HS.LS3.A** (HS-LS4-1); **HS.LS3.B** (HS-LS4-1),(HS-LS4-2) (HS-LS4-3),(HS-LS4-5); **HS.ESS1.C** (HS-LS4-1); **HS.ESS2.D** (HS-LS4-6); **HS.ESS2.E** (HS-LS4-2),(HS-LS4-5),(HS-LS4-6); **HS.ESS3.A** (HS-LS4-2),(HS-LS4-5),(HS-LS4-6); **HS.ESS3.C** (HS-LS4-6); **HS.ESS3.D** (HS-LS4-6)

Articulation across grade-bands: **MS.LS2.A** (HS-LS4-2),(HS-LS4-3),(HS-LS4-5); **MS.LS2.C** (HS-LS4-5),(HS-LS4-6); **MS.LS3.A** (HS-LS4-1); **MS.LS3.B** (HS-LS4-1),(HS-LS4-2),(HS-LS4-3); **MS.LS4.A** (HS-LS4-1); **MS.LS4.B** (HS-LS4-2),(HS-LS4-3),(HS-LS4-4); **MS.LS4.C** (HS-LS4-2),(HS-LS4-3),(HS-LS4-4),(HS-LS4-5); **MS.ESS1.C** (HS-LS4-1); **MS.ESS3.C** (HS-LS4-5),(HS-LS4-6)

Common Core State Standards Connections:

ELA/Literacy –

RST.11-12.1	Cite specific textual evidence to support analysis of science and technical texts, attending to important distinctions the author makes and to any gaps or inconsistencies in the account. *(HS-LS4-1)*,(HS-LS4-2), *(HS-LS4-3)*,(HS-LS4-4)
RST.11-12.8	Evaluate the hypotheses, data, analysis, and conclusions in a science or technical text, verifying the data when possible and corroborating or challenging conclusions with other sources of information. (HS-LS4-5)
WHST.9-12.2	Write informative/explanatory texts, including the narration of historical events, scientific procedures/ experiments, or technical processes. *(HS-LS4-1)*,(HS-LS4-2), *(HS-LS4-3)*,(HS-LS4-4)
WHST.9-12.5	Develop and strengthen writing as needed by planning, revising, editing, rewriting, or trying a new approach, focusing on addressing what is most significant for a specific purpose and audience. *(HS-LS4-6)*
WHST.9-12.7	Conduct short as well as more sustained research projects to answer a question (including a self-generated question) or solve a problem; narrow or broaden the inquiry when appropriate; synthesize multiple sources on the subject, demonstrating understanding of the subject under investigation. (HS-LS4-6)
WHST.9-12.9	Draw evidence from informational texts to support analysis, reflection, and research. *(HS-LS4-1)*,(HS-LS4-2), *(HS-LS4-3)*,(HS-LS4-4),(HS-LS4-5)
SL.11-12.4	Present claims and findings, emphasizing salient points in a focused, coherent manner with relevant evidence, sound valid reasoning, and well-chosen details; use appropriate eye contact, adequate volume, and clear pronunciation. *(HS-LS4-1)*,(HS-LS4-2)

Mathematics –

MP.2	Reason abstractly and quantitatively. *(HS-LS4-1)*,(HS-LS4-2),(HS-LS4-3), *(HS-LS4-4)*,(HS-LS4-5)
MP.4	Model with mathematics. *(HS-LS4-2)*

*The performance expectations marked with an asterisk integrate traditional science content with engineering through a Practice or Disciplinary Core Idea.

The section entitled "Disciplinary Core Ideas" is reproduced verbatim from A Framework for K-12 Science Education: Practices, Cross-Cutting Concepts, and Core Ideas. Integrated and reprinted with permission from the National Academy of Sciences.

May 2013   ©2013 Achieve, Inc. All rights reserved.   2 of 2

Looking at the crosscutting concepts in the foundation box in Figure 8.1, both concepts "Patterns" and "Cause and Effect" should be included as part of the unit of study. One kind of pattern found in evolution can be looking at different adaptations, which can help a student understand that change has occurred. Cause and effect is incredibly powerful when trying to understand evolutionary change. "One goal of instruction about cause and effect is to encourage students to see events in the world as having understandable causes, even when those causes are beyond human control (NRC 2012, p. 88)." Natural selection is the cause that leads to adaptations, which is the effect that organisms over time may experience in their environment.

Again referring to Figure 8.1 and looking at the first column, I chose the following science and engineering practices: Analyzing and Interpreting Data, Constructing Explanations and Designing Solutions, and Engaging in Argument from Evidence. During the unit, students will be able to participate in some scientific investigations in which they will be able to produce data that they then can analyze and help them understand how change can occur. Students may find patterns that emerge as they interpret their data. Taking that one step further, the students will focus on constructing an explanation from evidence they obtain throughout the course of study. "Asking students to demonstrate their own understanding of the implications of a scientific idea by developing their own explanations of phenomena, whether based on observations they have made or models they have developed, engages them in an essential part of the process by which conceptual change can occur" (NRC 2012, p. 68). By the end of the unit, students should be able to engage in argument from evidence that has been introduced, collected, and discussed. "Becoming a critical consumer of science is fostered by opportunities to use critique and evaluation to judge the merits of any scientifically based argument" (NRC 2012, p. 71). This practice may come with the most resistance given the controversial nature of the topic of evolution.

Stage II: Deconstruct DCIs, Create a Storyline, and Align Practices and Crosscutting Concepts

A Framework for K–12 Science Education: Practices, Crosscutting Concepts, and Core Ideas (NRC 2012, pp. 164–166) has information to help deconstruct LS4.C Adaptations. Table 8.1 shows the chosen bullets, the essential understandings, the learning targets, and information that can be left out of the instruction for this unit.

Most often, I taught my unit on evolution after I taught units on genetics and ecology. Students had an understanding of DNA and how a change in base sequence could lead to a change in a particular trait. This progression from a unit on genetics to a unit on evolution made it easier for my students to understand the changes that could occur both by mutation and by reproduction. Students were therefore ready to understand that changes in DNA sequence could cause a variation in traits. For my students, they could see a concrete example of the changes (for example, showing the DNA sequence of hemoglobin versus the DNA sequence of sickle cell anemia along with physical 3-D models of both molecules). When students had these concrete examples in previous units of study, it provided them with a stepping stone to help understand the large abstract concept of evolution. Additionally, during the study of ecology earlier in the year, students had learned about limited resources in an environment, which leads to competition for those resources. Learning Target #2 is a concept that is reinforced in this unit of study.

The sequence of learning targets was chosen to begin with concrete examples that students are able to visualize and understand. These concepts should have been taught in the years prior to my course, but based on information gained from the preassessment, almost

Table 8.1

Deconstructing DCIs Into Learning Targets for DCI HS-LS4.C

Bullets Selected From LS4.C	Essential Understanding	Learning Targets	Pruning
Evolution is a consequence of the interaction of four factors: (1) the potential for a species to increase in number, (2) the genetic variation of individuals in a species due to mutation and sexual reproduction, (3) competition for an environment's limited supply of the resources that individuals need in order to survive and reproduce, and (4) the ensuing proliferation of those organisms that are better able to survive and reproduce in that environment.	Environment can influence organisms over multiple generations. Species can increase in number over time if conditions in a given environment are favorable to those individuals. Some individuals may have traits caused by either mutation or sexual reproduction that allow them to survive and then reproduce, thus passing on that favorable trait to the next generation. In an environment, there will be competition for the limited amount of resources necessary for survival. Those individuals that have traits that are not suited for a given environment will not be able to survive, and thus cannot reproduce to pass on their traits. Those individuals that are the most fit will have an advantage at obtaining these resources, surviving, and reproducing. Eventually over multiple generations, there will be more organisms with the traits that are best suited for the environment, and fewer individuals with traits that are not suited. If conditions change in an environment, the distribution of traits may also change.	1. Individuals of a given species have variation of heritable traits. Genetic mutation and/or sexual reproduction cause the variations. 2. There are limited resources in an environment that can lead to competition for those resources. The individuals with access to these resources will survive and can then reproduce leading to more organisms with the same traits.	Do not include other mechanisms of evolution, such as genetic drift (including but not limited to "the bottleneck effect" or "the founder effect"), gene flow from immigration or emigration, and co-evolution.

Table 8.1 (continued)

Bullets Selected From LS4.C	Essential Understanding	Learning Targets	Pruning
Natural selection leads to adaptation, that is, to a population dominated by organisms that are anatomically, behaviorally, and physiologically well suited to survive and reproduce in a specific environment. That is, the differential survival and reproduction of organisms in a population that have an advantageous heritable trait leads to an increase in the proportion of individuals in the future generations that have the trait and to a decrease in the proportion of individuals that do not.		3. An adaptation is a trait an individual is born with that allows the individual to survive in its environment. Not all traits are adaptations. Adaptations can be anatomical, behavioral or physiological. Structures perform functions that allow organisms to survive. #4. Individual organisms with certain traits are more likely than others to survive and pass their traits on to their offspring. Survival of the fittest means best suited to survive and reproduce. Over successive generations, individuals with an advantageous heritable trait may increase in number and those without that trait may decrease in number. 5. Natural selection provides a mechanism for species to adapt to changes in their environment.	Do not include allele frequency calculations (i.e.. Hardy Weinberg equation)
Adaptation also means the distribution of traits in a population can change when conditions change.		6. If an environment changes, the population living in that environment may have a different distribution of traits over time.	Do not include allele frequency calculations (i.e., Hardy Weinberg equation)

every year I need to reinforce Learning Target #1. Once students understand the basic idea of variations, then I can move forward and expand that idea into Learning Target #2. Again, some of this material should have been covered in previous grades, and some I have covered in earlier units of my course. Both Learning Target #1 and #2 may be eliminated if your students have a sound understanding of these concepts. Learning Target #3 begins the foundation of the LS4.C Adaptations with a basic understanding of the terminology and the concept of what an adaptation is and what it is not. From this point, the students then move to Learning Target #4, the idea that adaptations allow organisms to survive, reproduce and increase that characteristic in a population. If organisms in that population do not have the adaptation, future generations may have fewer individuals with other traits that are not advantageous. Another disciplinary core idea covers natural selection, which is not discussed in this chapter, but the concept is very important for the last two learning targets. In the course of teaching evolution, it is necessary to include information about natural selection, but for the DCI I chose, there are no learning targets specifically for the understanding of natural selection. Once students understand Learning Target #5, then the final learning target for this unit follows, forming the idea that if an environment changes, over time the traits found in that environment will also have a different distribution (see Figure 8.2, p. 264).

Correct scientific vocabulary is critical when discussing evolution and the misuse of terms can lead to many of the misconceptions that students have. "Misunderstanding the distinction between individuals and species underpins many alternative conceptions of evolutionary processes. We argue for the use of clear, unambiguous, and consistent language in this and future documents dealing with evolutionary education" (Catley, Lehrer and Reiser 2005, p. 60). The authors continue:

> In particular, organisms—individuals that comprise a species or population (part of a species), be restricted to this usage. Further, when reference is made to collections of organisms; i.e., populations, species, or higher taxa, that these terms be consistently used. As the species and not the organism is the unit of evolutionary change it is particularly important to make this clear." (Catley, Lehrer, and Reiser 2005, p. 60)

Students particularly have struggled in my classes with the words *adapt*, *adapted*, and *adaptation*. They have used the terms interchangeably on many occasions leading to confusion and misconceptions of how evolution occurs.

As the unit of study progresses following the storyline, it is helpful to infuse the practices and crosscutting concepts into the curriculum and not as separate entities. The practices and crosscutting concepts are identified in Table 8.2 (pp. 265–266).

Figure 8.2.

Adaptations Learning Sequence

Target #1	Target #2	Target #3	Target #4	Target #5
Individuals of a given species have variation of heritable traits. Genetic mutation and/or sexual reproduction cause the variations.	There are limited resources in an environment that can lead to competition for those resources. The individuals with access to these resources will survive and can then reproduce leading to more organisms with the same traits.	An adaptation is a trait an individual is born with that allows the individual to survive in its environment. Not all traits are adaptations. Adaptations can be anatomical, behavioral or physiological. Structure perform functions that allow organisms to survive.	Individual organisms with certain traits are more likely than others to survive and pass their traits on to their offspring. Survival of the fittest means best suited to survive and reproduce. Over successive generations, individuals with an advantageous heritable trait may increase in number and those without that trait may decrease in number.	Natural selection provides a mechanism for species to adapt to changes in their environment.

Target #6:
If an environment changes, the population living in that environment may have a different distribution of traits over time.

Essential Understandings

Environment can influence organisms over multiple generations. Species can increase in number over time if conditions in a given environment are favorable to those individuals. Some individuals may have traits caused by either mutation or sexual reproduction that allow them to survive and then reproduce, thus passing on that favorable trait to the next generation. In an environment, there will be competition for the limited amount of resources necessary for survival. Those individuals that have traits that are not suited for a given environment will not be able to survive, and thus cannot reproduce passing on their traits. Those individuals that are the most fit will have an advantage at obtaining these resources, surviving, and reproducing. Eventually over multiple generations, there will be more organisms with the traits that are best suited for the environment, and fewer individuals with traits that are not suited. If conditions change in an environment, the distribution of traits may also change.

Table 8.2.

Developing a Storyline

Learning Targets	Crosscutting Concept Addressed in Learning Target	Science and Engineering Practices to Include	Possible Aligned Representations or Activities to Include
1. Individuals of a given species have variation of heritable traits. Genetic mutation and/or sexual reproduction cause the variations.	Patterns: Different patterns may be observed at each of the scales at which a system is studied and can provide evidence for causality in explanations of phenomena.	Analyzing and Interpreting Data: Apply concepts of statistics and probability…	Class data collection on some common human traits.
2. There are limited resources in an environment that can lead to competition for those resources. The individuals with access to these resources will survive and can then reproduce, leading to more organisms with the same traits.	Patterns: Different patterns may be observed at each of the scales at which a system is studied and can provide evidence for causality in explanations of phenomena. Cause and Effect: Empirical evidence is required to differentiate between cause and correlation and make claims about specific causes and effects.	Analyzing and Interpreting Data: Apply concepts of statistics and probability… Construct an explanation based on valid and reliable evidence obtained from a variety of sources…	Develop an explanation of how a population can have a higher frequency of certain traits in a given environment.
3. An adaptation is a trait an individual is born with that allows the individual to survive in its environment. Not all traits are adaptations. Adaptations can be anatomical, behavioral, or physiological. Structures perform functions that allow organisms to survive.	Patterns: Different patterns may be observed at each of the scales at which a system is studied and can provide evidence for causality in explanations of phenomena.	Engage in Argument from Evidence: Evaluate the evidence behind currently accepted explanations or solutions to determine the merits of arguments. Construct an explanation based on valid and reliable evidence obtained from a variety of sources…	Class activity that identifies an adaptation Simulation about camouflage in an environment Historical summary of data from Peppered Moth. Historical summary of how Lamarck describes why giraffes have long necks as compared to Darwinian explanation and modern-day examples of Lamarckian thinking Develop an initial explanation of an adaptation that an organism has today that has allowed it to survive (for example, snakes with no legs)

Table 8.2 (continued)

Learning Targets	Crosscutting Concept Addressed in Learning Target	Science and Engineering Practices to Include	Possible Aligned Representations or Activities to Include
4. Individual organisms with certain traits are more likely than others to survive and pass their traits on to their offspring. Survival of the fittest means best suited to survive and reproduce. Over successive generations, individuals with an advantageous heritable trait may increase in number and those without that trait may decrease in number.	Patterns: Different patterns may be observed at each of the scales at which a system is studied and can provide evidence for causality in explanations of phenomena.	Construct an explanation based on valid and reliable evidence obtained from a variety of sources... Engage in Argument From Evidence	Fishy frequency activity Revise and expand the explanation about adaptations and how it affects organisms in successive generations
5. Natural selection provides a mechanism for species to adapt to changes in their environment.	Patterns: Different patterns may be observed at each of the scales at which a system is studied and can provide evidence for causality in explanations of phenomena. Cause and Effect: Empirical evidence is required to differentiate between cause and correlation and make claims about specific causes and effects.	Construct an explanation based on valid and reliable evidence obtained from a variety of sources...	Computer simulation of natural selection Develop an explanation of how organisms adapt to their environment
6. If an environment changes, the population living in that environment may have a different distribution of traits over time.	Scientific Knowledge Assumes an Order and Consistency in Natural Systems: Scientific knowledge is based on the assumption that natural laws operate today as they did in the past and will continue to do so in the future Cause and Effect: Empirical evidence is required to differentiate between cause and correlation and make claims about specific causes and effects.	Construct an explanation based on valid and reliable evidence obtained from a variety of sources...	Revise and expand the explanation of how organisms adapt to their environment. Argumentation activity to justify claim based on evidence Class discussion of the adaptations of bacteria to antibiotics

Stage III: Determine Performance Expectations and Identify Criteria to Determine Student Understanding

As we begin Stage III, it is important to know what you are going to teach. Many times a curriculum document with a list of objectives or standards is given to teachers. You need to turn these into lesson plans that make sense, not only to you but also to your students. In the previous section, I put together a cohesive storyline that built upon ideas and continued to build until Learning Target #6, which in my opinion, pulls all the other learning targets together. My students would have had a very hard time comprehending Learning Target #6 without the scaffolding of ideas leading up to it. Without knowing the criteria you want to achieve, it is possible to skip some of the important steps along the way to understanding. In Table 8.3 I have identified the criteria I expected my students to know and be able to do to show me their understanding of the learning targets.

Table 8.3.

Criteria to Determine Student Understanding of Adaptations

Learning Target	Criterion
1. Individuals of a given species have variation of heritable traits. Genetic mutation and/or sexual reproduction cause the variations.	Identify patterns of variations in a population, analyze and interpret data about such variations.
2. There are limited resources in an environment that can lead to competition for those resources. The individuals with access to these resources will survive and can then reproduce leading to more organisms with the same traits.	Construct an explanation of how a population can have a higher frequency of certain traits in a given environment.
3. An adaptation is a trait an individual is born with that allows the individual to survive in its environment. Not all traits are adaptations. Adaptations can be anatomical, behavioral or physiological. Structures perform functions that allow organisms to survive.	Give an example of an adaptation that an organism has today and explain why that has allowed the organism to survive.
4. Individual organisms with certain traits are more likely than others to survive and pass their traits on to their offspring. Survival of the fittest means best suited to survive and reproduce. Over successive generations, individuals with an advantageous heritable trait may increase in number and those without that trait may decrease in number.	Engage in argument from evidence presented and construct an explanation why certain traits (adaptations) can lead to more individuals with that trait over successive generations and that some traits will decrease over successive generations.
5. Natural selection provides a mechanism for species to adapt to changes in their environment.	Construct an explanation of how organisms adapt to their environment.
6. If an environment changes, the population living in that environment may have a different distribution of traits over time.	Refine and expand the explanation for Learning Target 5 to include why an environment over time may have different distribution of traits.

Now, since I have established the criteria for the unit, it is time to connect the learning targets to the performance expectations. One DCI of HS-LS4 Biological Evolution: Unity and Diversity is Adaptations. Eugenie Scott states, "I believe that there are three genetics-related concepts that, if taught properly, would greatly improve the biological literacy of our fellow citizens. The first is evolution …. This brings us to the second basic concept, adaptation" (Scott 2013). The performance expectations that match the criteria as determined come from DCI Adaptations and are: HS-LS4-2, HS-LS4-3, and HS-LS4-4 (see Table 8.4).

Table 8.4.

Alignment of Assessment Criteria and Selected NGSS Performance Expectations

Developed Criteria for Unit of Study	Aligned HS-LS4 Performance Expectations
Identify patterns of variations in a population, analyze and interpret data about such variations.	Construct an explanation based on evidence that the process of evolution primarily results from four factors: (1) the potential for a species to increase in number, (2) the heritable genetic variation of individuals in a species due to mutation and sexual reproduction, (3) competition for limited resources, and (4) the proliferation of those organisms that are better able to survive and reproduce in the environment.
Construct an explanation of how a population can have a higher frequency of certain traits in a given environment.	
Give an example of an adaptation that an organism has today and explain why that has allowed the organism to survive.	Construct an explanation based on evidence for how natural selection leads to adaptation of populations.
Construct an explanation of how organisms adapt to their environment.	
Engage in argument from evidence presented and construct an explanation why certain traits (adaptations) can lead to more individuals with that trait over successive generations and that some traits will decrease over successive generations.	Apply concepts of statistics and probability to support explanations that organisms with an advantageous heritable trait tend to increase in proportion to organisms lacking this trait.
Refine and expand the explanation for Learning Target 5 to include why an environment over time may have different distribution of traits.	

Stage IV: Determine Nature of Science (NOS) Connections

In this unit of study there is one crosscutting concept that connects to the nature of science: Scientific knowledge assumes an order and consistency in natural systems. Scientific knowledge is based on the assumption that natural laws operate today as they did in the past and will continue to do so in the future. This connection links to performance expectation HS-LS4-4.

I found that before I begin a unit on evolution, if my students have a better understanding of the nature of science there was typically less negativity toward the topic of evolution once we began the formal study. I also noticed when I switched the nature of science activity from the first unit of the year to the day before the evolution unit there was again less negativity toward the topic. The activity used is Activity 1, Introducing Inquiry and the Nature of Science (NAS 1998, pp. 66–73). There are many reasons for choosing this activity, as it introduces both basic procedures for inquiry and concepts that deal with the Nature of Science. I emphasized the following concepts found in the science background for teachers, which is included in the activity:

- Technology used to gather data enhances accuracy and allows scientists to analyze and quantify results and investigations.

- Scientific explanations emphasize evidence, have logically consistent arguments, and use scientific principles, models, and theories.

- Science distinguishes itself from other ways of knowing and from other bodies of knowledge through the use of empirical standards, logical arguments, and skepticism, scientists thrive for the best possible explanations about the natural world. (NAS 1998, p. 66)

Additionally, I wanted to stress that we were going to approach our next unit of study from the viewpoint of scientific explanations. According to the background for teachers in the activity, the following conditions need to be met to make explanations scientific:

- Scientific explanations are based on empirical observations or experiments.
- Scientific explanations are made public.
- Scientific explanations are tentative.
- Scientific explanations are historical.
- Scientific explanations are probabilistic.
- Scientific explanations assume cause-effect relationships.
- Scientific explanations are limited. (NAS 1998, p. 67)

Students were put into groups of three for the activity. Following the directions in the book, students practiced the skills of scientists. They made observations, worked together and not in isolation, discussed their findings, hypothesized, collected data, and came up with a conclusion based on their evidence. The students became frustrated when I would not tell them or show them the answer on the bottom of the cube. All of the above concepts were discussed throughout the activity in a variety of ways, including individual reflection, group discussion, and class discussion. We had a discussion about the parts of the activity that reinforced the scientific practices and the tentativeness of science. I stressed both that scientists do not always find an answer and the importance of scientists communicating their findings. We talked about what students could have done to find more data that are factual. Did "technology" help them propose an answer? Did it help to hear other students' thinking? I wrapped up the discussion with the idea that our next unit of study still has a lot of unanswered questions; that we will look at evolution through the lens of science as a way of knowing and that we will do some hands-on, minds-on activities and talk about the history of discoveries various scientists have made. I stressed to the students that we need to be open minded as we begin this unit; beliefs are not science and need to be set aside for this unit. I encouraged them to hold onto their beliefs, since we were in a science class and we use the practices of science to better understand evolution. With the goals mentioned above achieved in the classroom, my students were ready to begin a unit of study on evolution. To reinforce the nature of science concepts, this activity was referenced numerous times throughout the unit on evolution.

Stage V: Identify Metacognitive Goals and Strategies

Students did not know when I gave the preassessment for the unit on evolution immediately after the previous unit's summative assessment, that I was using it as a self-regulated thinking activity (see Figure 8.3).

The students turned in the preassessment; I did an item analysis, and kept the preassessments until the end of the unit of study. Two to three days prior to the summative assessment for the unit, I handed back the unmarked sheet and asked them to make any corrections that they could to their original answers using a different color writing utensil. All concepts for the unit of study had been covered by this point of time. If students did not know an answer, I encouraged them to study that evening or to come in for some extra help. If there were many students that did not understand a particular concept at this point, I would do some re-teaching of the idea to help clear up any problems.

The use of a study guide sheet was another self-regulated thinking activity used at the beginning of every unit of study throughout the year. The study guide sheet listed the standards we were going to cover, a daily list of the activities for the unit, a list of vocabulary words important to the understanding of the standards, a list of any homework for the unit, and a mini self-assessment of the standards. The mini self-assessment was set up as a table listing the standard number in column 1, preassessment

Figure 8.3

Preassessment on Unit #8

Name _____

Date _____

Please answer all of the following questions. If you do not know an answer, please write I do not know. Please do not include any religious beliefs in your answers. Thank-you!

1. When you think of evolution, what comes to your mind?

2. Give an example of a trait and list a variation for that trait.

3. List an example of an adaptation an organism has and explain why it is an adaptation.

4. Are all adaptations traits, or are all traits adaptations.? Explain your answer.

5. Can an individual adapt to a changing environment during its lifetime? If yes, give an example. If no, explain why not.

6. Lamarck was a scientist who believed that the reason a giraffe has a long neck was that during its lifetime it had to stretch its neck to reach the leaves that were higher up in the branches of trees to survive. The long neck giraffe then passed this acquired characteristic onto its offspring.

7. Is it possible for an organism to pass on an acquired characteristic to their offspring?

knowledge in column 2, midassessment knowledge in column 3, and postassessment knowledge in column 4. Before beginning the unit, the students needed to self-assess each standard in the appropriate column of the table using the following statements:

0. No clue, never heard of it
1. Heard of it, cannot tell you anything
2. Heard of it, some ideas
3. Can explain most of the idea
4. Could teach to others

I would collect these and do an item analysis, and then return the study guides the following day. About midway through the unit, the students had to pull out the sheet and again do a self-assessment and turn it in so I could do a quick analysis and return it the following day; two days prior to the summative unit assessment there was one more self-assessment. Again, I would collect and analyze where the students were at, and if needed, re-teach or do additional activities to help the students with any lack of understanding. An example of what the study guide would look like using learning targets is shown in Figure 8.4.

Throughout the course of study, the students would engage in many activities to help them understand what they knew and did not know. Opportunities were given for students to add to their own body of knowledge and demonstrate that knowledge to themselves, to others in the class and to me.

One example of having students think about what they knew was a KWL activity. Students set up the KWL chart in their notes and began on day 1 of the unit about what they knew and what they wanted to know about evolution. As specific topics came up during the unit, students would add to their KWL chart—for example, adding the concept of *adaptation* to the chart and another day adding *natural selection* to the chart. I would then use this chart as a reflection for the students and a type of formative assessment for myself throughout the unit. Students would draw a line every time we took out the KWL charts, add a new date, and write down what they knew about the new topic, what they still wanted to learn and what they learned about previous topics. I would collect these KWL charts to inform me about where the students were with their learning and then return the charts when I wanted them to add more. Throughout the unit, students identified what they learned as we added new topics.

In my classes, I used whiteboards on a regular basis. Sometimes they were used as a quick reflection, and other times they helped students formulate their ideas and share with partners. An example of this critical thinking and learning metacognition technique that I used in my class during the study of adaptations was Think, Share, Pair. I would ask the students to list on a whiteboard up to five adaptations organisms have and explain why each is an adaptation. The students shared their list with their assigned partners and determined which were truly adaptations, and then generated one list of adaptations to share with the class. Another example that coincides with one of the practices chosen for this chapter, "Construct an explanation," was also a critical thinking and learning metacognition activity using whiteboards. After a class discussion and note taking on the adaptation of long necks on giraffes, students were asked to explain the current anatomy of a snake— without legs. Students were given information that there was fossil evidence of snakes with leg bones. They were to construct an explanation of how over time, the adaptation of no legs may have occurred. This activity was done on a whiteboard with the choice of either drawing their scenario or writing the scenario. The students did a Think, Share, Pair again and then presented their explanation to the class.

A full listing of all the strategies identified in this chapter can be found in Table 8.5 (pp. 274–275).

Figure 8.4.

Unit Sheet for Adaptations

Learning Target #1. Individuals of a given species have variation of heritable traits. Genetic mutation and/or sexual reproduction cause the variations.

Learning Target #2. There are limited resources in an environment that can lead to competition for those resources. The individuals with access to these resources will survive and can then reproduce, leading to more organisms with the same traits.

Learning Target #3. An adaptation is a trait an individual is born with that allows the individual to survive in its environment. Not all traits are adaptations. Adaptations can be anatomical, behavioral, or physiological. Structures perform functions that allow organisms to survive.

Learning Target #4. Individual organisms with certain traits are more likely than others to survive and pass their traits on to their offspring. Survival of the fittest means best suited to survive and reproduce. Over successive generations, individuals with an advantageous heritable trait may increase in number and those without that trait may decrease in number.

Learning Target #5. Natural selection provides a mechanism for species to adapt to changes in their environment.

Learning Target #6. If an environment changes, the population living in that environment may have a different distribution of traits over time.

On the following table, place a number in the appropriate column for your understanding of the concept as of today. Use the following numbers:

0. No clue, never heard of topic
1. Heard of the topic, cannot tell you anything about it
2. Heard of the topic, know about it, some ideas
3. Can explain parts of the topic to someone else
4. Could teach the class about the topic

	Beginning of unit	**Mid-unit**	**End of unit**
Learning Target #1			
Learning Target #2			
Learning Target #3			
Learning Target #4			
Learning Target #5			
Learning Target #6			

Table 8.5

Completed Strategy Selection Template for Learning Target #1

	Learning Target #1	Learning Target #2	Learning Target #3	Learning Target #4	Learning Target #5	Learning Target #6
	Individuals of a given species have variation of heritable traits. Genetic mutation and/or sexual reproduction cause the variations.	There are limited resources in an environment that can lead to competition for those resources. The individuals with access to these resources will survive and can then reproduce, leading to more organisms with the same traits.	An adaptation is a trait an individual is born with that allows the individual to survive in its environment. Not all traits are adaptations. Adaptations can be anatomical, behavioral, or physiological. Structures perform functions that allow organisms to survive.	Individual organisms with certain traits are more likely than others to survive and pass their traits on to their offspring. Survival of the fittest means best suited to survive and reproduce. Over successive generations, individuals with an advantageous heritable trait may increase in number and those without that trait may decrease in number.	Natural selection provides a mechanism for species to adapt to changes in their environment.	If an environment changes, the population living in that environment may have a different distribution of traits over time.

Table 8.5 (continued)

Strategies for:	Learning Target #1	Learning Target #2	Learning Target #3	Learning Target #4	Learning Target #5	Learning Target #6
Identifying Preconceptions	Preassessment (Figure 8.3 on p. 271) and KWL					
Eliciting and Confronting Preconceptions	Student discourse; informational text strategies; physical movements/ gestures	Student discourse; Informational text strategies	Student discourse; Informational text strategies; Probe: "Changing Environment", Physical movements/gestures	Student discourse; Informational text strategies; Probe: "Is it Fitter?"	Student discourse; Informational text strategies; Probe: "Adaptation"	Student discourse; Informational text strategies
Sense Making	Informational Text Strategies	Informational text strategies	Informational text strategies; scientific explanations; Think, pair, share	Informational text strategies; Scientific explanations; hands-on experiments	Informational text strategies; scientific explanations, dynamic models	Informational text strategies; hands on experiments, scientific explanations
Demonstrating Understanding				Scientific argumentation and explanations; Graphic organizers; questions and probes during discussions	Scientific argumentation and explanations; Graphic organizers; questions and probes during discussions	Scientific argumentation and explanations; Graphic organizers; questions and probes during discussions
Selected Metacognitive Strategy	Self-Regulated Thinking: KWL Critical Thinking and Learning: Think, Pair, Share					

Phase II: Planning for Responsive Action

Stage VI: Research Student Misconceptions Common to This Topic That Are Documented in the Research Literature

In a recent study of 181 middle school science teachers and almost 10,000 of their students, researchers found that the teachers who could predict their students' wrong answers on a standardized test, which were misconceptions, helped their students learn the most. "If teachers are to help students change their incorrect beliefs, they first need to know what those are" (Harvard University 2013).

Table 8.6 lists some of the common misconceptions that students may have for each of the learning targets. This is not a complete listing of misconceptions found in the literature. The citations listed after the misconceptions will provide other misconceptions. Learning Target #2 was removed from the chart as it did not pertain directly to adaptations, the focus of this chapter. One common instructional idea throughout the table is the reinforcement of correct terminology when teaching concepts in evolution. In the article "Correcting Some Common Misrepresentations of Evolution in Textbooks and the Media" (Padian 2013), the author gives examples of terms to avoid when teaching about evolution and suggests terms that would help students better understand the concepts.

An example of an instructional idea that students in my class enjoyed and that helped correct the misconception of acquired traits being passed on was a humorous scenario that I described to the students. I asked students to picture a Caucasian couple about to have a baby. I mentioned that the woman decided upon learning about her pregnancy that she would treat herself to many sessions at the tanning parlor until it was time to deliver. We discussed the idea that a tan is an acquired trait. Before the due date, the woman had a very dark tan. I asked the students, would the baby be born with a tan? Of course, when the students laughed and said no, I pointed out that this would be an example of Lamarckian thinking about inheritance of traits. Having examples that the students can relate to easily does help them in developing their own thinking.

To help overcome the misconception listed for Learning Target #6, The Frayer Model of Vocabulary Development (Barton and Jordan 2001, pp. 53–57) is an excellent tool to reinforce the vocabulary that is very often misused. Using the vocabulary words *species* and *population* will reinforce the concept that evolution occurs to a population, not to an individual organism. Using the word *adaptation* will help students develop a better understanding of what an adaptation is and is not, and provide students another chance to show their understanding with examples.

Table 8.6

Misconceptions and Instructional Ideas for Adaptations

Learning Target #1:
Individuals of a given species have variation of heritable traits. Genetic mutation and/or sexual reproduction cause the variations.

Misconceptions	**Instructional Ideas**
"Students invariably attribute observable variation to environmental factors alone." (Driver et al. 1994)	• Concrete examples of heritable traits that students can measure (e.g., eye color, attached or free ear lobes. Avoid traits that can change due to environmental conditions, such as hair that lightens up during the summer, height (nutrition, smoking during pregnancy can play a role in overall height).

Learning Target #3:
An adaptation is a trait an individual is born with that allows the individual to survive in its environment. Not all traits are adaptations. Adaptations can be anatomical, behavioral, or physiological. Structures perform functions that allow organisms to survive.

Misconceptions	**Instructional Ideas**
• All traits are adaptations. • "Students confused an individual's adaptation during its lifetime with inherited changes in a population over time: they appeared to believe in the inheritance of acquired characteristics." (Driver et al. 1994) • The inheritance of acquired traits (BSCS 2005)	• Examples of traits that are not adaptations. (hand activity) • Attention should be paid to the terms *adapt, adaptation*. (vocabulary development—Frayer model) • Camouflage lab "When the Chips are Down" to address adaptations • Probe • Examples of Lamarckian thinking about acquired characteristics. • "Middle-school and high-school students may have difficulties with the various uses of the word *adaptation*. In everyday usage, individuals adapt deliberately. But in the theory of natural selection, populations change or *adapt* over generations, inadvertently. Students of all ages often believe that adaptations result from some overall purpose or design, or they describe adaptation as a conscious process to fulfill some need or want. Elementary- and middle-school students also tend to confuse non-inherited adaptations acquired during an individual's lifetime with adaptive features that are inherited in a population." (National Science Digital Library n.d.)

Note: Learning Target #2 was removed from the chart as it did not pertain directly to adaptations, the focus of this chapter.

Hard-to-Teach Biology Concepts, Revised 2nd Edition

Table 8.6 (continued)

Learning Target #4:
Individual organisms with certain traits are more likely than others to survive and pass their traits on to their offspring. Survival of the fittest means best suited to survive and reproduce. Over successive generations, individuals with an advantageous heritable trait may increase in number and those without that trait may decrease in number.

Misconceptions	Instructional Ideas
• "The fittest organisms in a population are those that are strongest, healthiest, fastest, and/or largest" (University of California Museum of Paleontology 2013).	• Students need to understand the term "fitness" as meaning the ability to get its genes into the next generation, not relating the term to health" (University of California Museum of Paleontology 2013). • Use of probes, and "Fishy Frequencies" lab

Learning Target #5:
Natural selection provides a mechanism for species to adapt to changes in their environment.

Misconceptions	Instructional Ideas
• Natural selection gives organisms what they need. (University of California Museum of Paleontology 2013) • "Natural selection involves organisms trying to adapt" (University of California Museum of Paleontology 2013).	• "Natural selection leads to the adaptation of species over time, but the process does not involve effort, trying, or wanting" (University of California Museum of Paleontology 2013). • "Natural selection acts on the genetic variation in a population, and this genetic variation is generated by random mutation – a process that is unaffected by what organisms in the population need" (University of California Museum of Paleontology 2013). • Use the example of antibiotic resistance in bacteria.
• "Pupils tend to see adaptation in a naturalistic or teleological sense: undertaken to satisfy the organism's need or desire to fulfil some future requirement" (Driver et al. 1994).	• Computer simulation of natural selection where the students can manipulate the environmental changes.

Learning Target #6:
If an environment changes, the population living in that environment may have a different distribution of traits over time.

Misconceptions	Instructional Ideas
"Most pupils appear to regard adaptation in terms of individuals changing in major ways in response to their environment in order to survive" (Driver et al. 1994).	Terminology needs to be stressed as to the difference between an individual and a population, and the difference between the words *adapt* and *adaptation*, to help students understand this misconception. (Use of vocabulary strategies)

Stage VII: Determine Strategies to Identify Your Students' Preconceptions

There are a variety of ways to elicit information from our students to find out what they know, and more importantly what they do not know. When I began my unit on evolution, I always began with an open-ended preassessment. I let my students know that it was not for a grade, but was going to be used to help me teach our next unit. This preassessment was given as the students were finishing their summative assessment on the previous unit. The only other direction I gave the students was they might not include any religious beliefs as part of their answers. I included questions that were based on our district's curriculum and the learning targets for my unit. These questions could very easily be modified to fit your district's curriculum. The activity is meant to take less than 10 minutes, and was purposefully given before we began any discussion on the topic. I had students from at least three different middle schools and some from home schools enrolling in public education for the first time. I was looking for knowledge of the topic, lack of knowledge, misconceptions, and beliefs. I would score the students' papers and do an item analysis pertaining to the items I was looking for. I did not make any marks on the student paper except to circle questions the students did not answer. This information was used to guide my instruction for the unit.

Referring back to Figure 8.3 (p. 271), I listed some of the questions I used that are specific to the learning targets discussed in this chapter. I have included Question #1, though it is not directly related to the DCI "Adaptation," to show how open-ended the questions were. I saved these preassessments, and after the students finished the unit, they were used before the summative assessment for the unit. The students received their original preassessment, used a different-colored writing utensil, and chose any questions they wanted to correct. Since there were no marks on their paper, they did not know which were incorrect from the preassessment. This was used to help the students identify what they still needed to study. I corrected the formative assessment and returned it the next day. I used phrases such as "expand your answer," "clarify your answer," and "check this answer." The students then had about two days to seek additional help for any concepts that were still confusing, and gave me an idea of what I still needed to clarify for the students.

A second method used to identify preconceptions was a unit study guide, described in Stage V with the identification of metacognition activities found in Figure 8.4 (p. 273).

One other strategy, which has been previously mentioned in Stage V of this chapter, is the use of a KWL.

Stage VIII: Determine Strategies to Elicit and Confront Your Students' Preconceptions

I have used my textbook as a resource for information, as a tool for formative assessment, and for vocabulary building. During this unit of study, students were assigned specific parts of the text to read that matched the learning targets. The students were assigned specific questions to answer that helped guide their reading. In the instructional tool box in Table 8.5, this would fall under informational text strategies.

I used these assigned readings as a formative assessment. I collected and analyzed how the students answered the questions, and then used this to lead a class discussion with both small-group and large-group discourse. It is often through these assignments that I find the lack of understanding or the misconceptions the students have about specific concepts. The class discussions were based on the questions the students answered; the students were asked to add to their answers with correct information if their answer is not accurate. These two instructional strategies are used with all of the learning targets.

As I began the unit, I assumed that Learning Target #1 was covered in previous years and in the previous unit of genetics, including a hands-on activity that looked at individual traits. Depending on the results of the preassessment and the study guides, some reinforcement of this concept might be required. I have used a very easy kinesthetic activity to reinforce this idea of variation; students physically measure their heights and we plot the data for the class to see. We discuss the variation found in our class and relate it to other variations that organisms may have.

Learning Target #2 is also a concept that should have been taught in earlier grades and I also cover it during a unit on ecosystems. Again, if the concept needs reinforcement, a quick review is provided. Learning Targets #3–6 address new concepts which have not been previously taught and require additional instruction to elicit and confront preconceptions. For Learning Target #3 I have used another kinesthetic activity to help students understand the concept of an adaptation. I have students take out a piece of paper and a writing utensil. The students set both on the desk. I assign every other row in my class the trait of an open hand, and the opposite row the trait of a closed fist (thumb inside the fist). I set up a scenario that in this environment, those that will survive and reproduce are those that can write their full name in cursive in a given amount of time. The students may only use the trait that they are born with to do so. I have them get ready and say "go." If you give students only five seconds, those with the open hand will be able to accomplish the task, those with a closed fist will not. This leads to a discussion of the difference between adaptations and traits, one of the misconceptions identified with Learning Target #3.

I follow this activity with the probe "Changing Environment" (Keeley 2011). This probe addresses ideas about individual organisms adapting to changes in the environment and Lamarckian ideas. It also helps with the misconceptions identified in Learning Target #3, especially with the terms of *adaptation* and *adapt*.

As I move through the unit of study, I follow with sense making and demonstrate understanding for each individual learning targets before proceeding to the next learning target. For this section of this chapter, I will elaborate on the strategies for eliciting and confronting preconceptions only for each learning target, and I will discuss the strategies for the sense making and demonstrating understanding for each learning target later in the chapter.

Learning Target #4 brings with it some vocabulary and misconceptions that students have problems with (identified in Stage IV of this chapter). Through the use of the probe, "Is It Fitter?" (Keeley and Tugel 2009), I am able to address the misconception of survival of the fittest terminology.

For Learning Target #5, I use one more probe, "Adaptation" (Keeley and Tugel 2009), which addresses the concept of whether organisms intentionally adapt to a change in the environment. The use of this probe helps with the misconceptions of organisms trying to adapt and organisms getting what they need to survive from natural selection.

Stage IX: Determine Sense-Making Strategies

As I move into this stage, I try to find activities that both reinforce the learning targets and help with the misconceptions that have been identified, as well as tie together the concepts and weave in the practices and crosscutting ideas. I try to do as many hands-on, minds-on activities to help students with the concepts. As stated above, these activities take place after eliciting and confronting preconceptions. I typically try to scaffold the learning for the students by the learning target unless an activity covers multiple learning targets. I did not identify any sense-making strategies for Learning Targets #1 and 2, as these should be only review, unless individual students need additional help. If they do need help, I typically use some kind of informational text strategy that breaks down the learning target further than I had during the classroom activities and discussion.

For Learning Target #3, I begin with a whiteboard activity after there has been some class discussion. The activity is a Think, Pair, Share to help students self-reflect on their learning up to this point. I give each student a whiteboard and ask them to make two columns: the left is labeled "adaptation," and the right is labeled "how it helps the organisms survive." Students are asked to write any five adaptations they can think of along with the explanation. When all are finished, I ask them to share with their partners. They decide which five they want to present to the class—that possibly no one else considered—and then fill out one whiteboard for the two of them to share with the class. This activity serves as a formative assessment for me to find out if students are on their way to understanding adaptations. The simple question of "how it helps organisms survive" helps students to begin to explain their thoughts scientifically without a lot of writing at this point of time.

I follow this activity with a discussion of acquired traits versus adaptations. I bring in background of Lamarckian ideas with a story about how Lamarck explained why giraffes have long necks. We have some class discussion and use some "modern ideas" to reinforce the concepts. Class discussion is not lecture and notes, but rather information and student discourse. Classes are not quiet, since exchange of ideas between students with other students and students with the teacher is encouraged. I have students both verbalize and write their ideas to help them develop their own understanding of concepts.

To help the kinesthetic- and mathematical-minded students, I use an activity that looks at the adaptation of camouflage, "The Chips are Down: A Natural Selection Simulation" (DiGiovanni 1999). This lab activity uses colored chips and cloth with a floral background. Students collect data on the adaptation of camouflage. The students are given 100 paper dots of 10 different colors. They distribute the dots onto material with a floral background and collect data over five successive "generations." Using the same dots, the students follow the same procedure on a different piece of material with a different color background and collect the data. The students use graphs to help analyze their data. The graphs are color-coded to match the color of the dots, and to make a bigger impression, I run off graph paper on colored paper to match the floral background. After the activity, the students write answers to specific questions based on the evidence they collected. The color visualization and the use of math helped students begin to explain how an adaptation can have an effect on a population.

Now that the students have experienced an activity to connect to their thinking, we discuss one more example of adaptations. We have a historical discussion about the evolution of coloration of the peppered moths that were light colored and evolved over time to be a much darker coloration. We talk about the age of the Industrial Revolution in England and how the change in the environment due to pollution led to light-colored moths not surviving, and those with a variation of darker colors surviving. Student learning is enhanced by providing multiple examples of evolutionary change, both hands-on and in verbal discussions, followed up by activities and student writing about adaptations.

At this point, I have students do some type of graphic organizer, for example a concept definition web. An example of this type of graphic organizer can be found in *Teaching Reading in Science* (Barton and Jordan 2001). This activity can "help students to explore connections, explain relationships, and elaborate on what they have learned. Throughout the lesson or unit, a teacher can evaluate students' understanding and check for misconceptions" (Barton and Jordan 2001, p. 81).

Continuing the unit, Learning Target #4 adds the idea of the survival of the fittest. Another hands-on activity that my students always enjoyed is called "Fishy Frequencies (or How Selection Affects the Hardy-Weinberg Equilibrium)" (Jones and Stanhope 1994). This activity can be modified to fit the level of mathematical abilities of your students. I used it with my remedial biology class, having students only collect the data and then graph results. My pre-AP Biology students did the calculations with the Hardy-Weinberg equation. Students in general biology who were strong in math also did the mathematical calculations. This activity helped with the misconception of relating the size of an individual to survival of the fittest since both "fish" in the sea were the same size, they just had a different variation of color. This activity also provided questions to help students explain their thinking and then had students develop their own activities.

Learning Target #5 then expands the idea of adaptations to include the concept of natural selection as a mechanism for species to adapt to their environment. To add other types of opportunities for students to learn, I used a computer simulation activity to reinforce this concept, "Natural Selection Simulation Lab" (Smith 2012). Students were able to manipulate the environment to see how natural selection can occur. This is a great activity to reinforce the scientific practice of using models. It would be very difficult in a biology class to provide an experience for students to observe natural selection causing a change with real organisms due to time. The use of a computer simulation that generates data for the students to analyze and explain is one option available to help with student understanding. Having students explain their results will reinforce the scientific practices of obtaining, evaluating, and communicating information and constructing explanations, as well as cover the crosscutting concept of Cause and Effect.

A class discussion of a current problem, antibiotic resistance in bacteria, helps bring the idea of adaptation and evolution to modern times. The discussion becomes very lively when I can relate a personal story of MRSA with the students. We use this discussion to emphasize the idea that organisms cannot just simply change due to an environmental change, but rather have adapted to changes in their environment. Using an organism that is not part of the animal kingdom gave students a different perspective on species adapting. This discussion was used during this learning target to help students with one of the misconceptions identified earlier in Table 8.6 (pp. 277–278). At the end of the discussion, to make sure the students are making the connections amongst all of the learning targets, I again use some kind of graphic organizer to pull together the concepts.

The last activity for this unit is "Color Variation in Venezuelan Guppies" (Sampson and Schleigh 2013) and is used to pull all the concepts and practices together, to challenge the students with the practice of scientific argumentation, and to cover Learning Target #6. Students work in teams to analyze data, develop a claim to a given question, state their evidence, and state their justification for their evidence in order to share the information with others on a whiteboard. They then use a round-robin format to listen and critique other groups' ideas, eventually coming back to their own group to modify their own claim. Sampson and Schleigh expain:

> The purpose of this activity is to help students understand how natural selection, sexual selection, and the interplay between these two mechanisms can shape the traits of a population found in different habitats over time. This activity also

helps students learn how to engage in practices such as constructing explanations, arguing from evidence, and communicating information. This activity is also designed to give students an opportunity to learn how to write in science and develop their speaking and listening skills, which are important goals for literacy in science. (2013, p. 24)

Additionally, this activity covers the scientific practice of using mathematics and computational thinking along with the crosscutting concepts of Patterns, and Cause and Effect: Mechanism and Explanation.

Conclusion

I began this chapter with a statement that evolution itself is not a "hard-to-teach" topic because of the content, but rather it is hard to teach because of the multitude of misunderstandings, preconceptions, and misconceptions that students bring into the classroom. As a teacher of science, I wanted my students to walk away from my unit on evolution with an understanding of science so they could understand the content and make future decisions based on scientific evidence, not on beliefs. In order to bring about this understanding, I had to provide learning that made the students' thinking visible to me so I could guide their learning. I needed to know what they knew, and what they did not know, in order to bring about conceptual change. Every activity in this chapter helped me see what the students were thinking, which led to a unit that would change slightly every year based on what background the students brought to class. The need for being able to "see" what the students were thinking may be more crucial with this topic because of the controversy that surrounds the topic. The strategies identified in this chapter from the preassessment to the scientific argumentation are scaffolded to provide the needed content and the needed scientific practices to really make conceptual change happen. It is my hope in writing this chapter that all teachers of science can have success in bringing conceptual change in the understanding of evolution with their students.

Resources

For more excellent resources and concepts to further student's learning on the roles of adaptation in biological evolution visit the links provided below:

- ENSI/SENSI (Evolution and Nature of Science Institutes): *www.indiana. edu/~ensiweb/main.fr.html*

- Evolution: Education and Outreach: *http://aibs.us1.list-manage1.com/track/click?u =a2886d199362c2554974f78af&id=3e402cef34&e=3c6770a9a0*

- NSTA (National Science Teachers Association): *www.nsta.org*

- NABT (National Association of Biology Teachers): *www.nabt.org*

- NCSE (National Center for Science Education): *www.ncse.com*

- NESCent (National Evolution Synthesis Center): Education and Outreach *www. nescent.org*

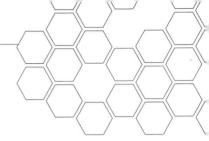

Reflections and Next Steps

Part I of this book provided all the information you might need to develop a unit of study on your own or in collaboration with your colleagues. It included our (1) Instructional Planning Framework, (2) process to implement the framework in the context of *NGSS*, and (3) Instructional Tools that provide a rich set of strategies, as well as research that supports the use of those strategies in the instructional setting. Chapters 1 and 2 introduced the framework and process, Chapter 3 modeled the implementation of the framework and use of the tools while focused on the topic proteins and genes, and Chapter 4 described various other connections that enhance the possibility that all students can learn and become more competent in your biology classroom.

We asked contributing authors to provide additional models of implementation of the framework, process, and tools. You found these examples in Part II of the book. Five authors contributed four different chapters, each interpreting Part I of the book and applying it to a particular science topic. There are variations among these chapters based on the author's interpretation of the framework and tools as well as their experiences and work contexts. These chapters, added to the model proteins and genes unit of study, provide instructional ideas for five different topics—one from each of the five *NGSS* disciplinary core ideas for life science.

Now it is time for you to consider how to best use what you are learning as you read. You should also determine whether or not you want to use any of the five model units of study. If you choose to use any of them, you might implement them as they are or modify them using other activities or instructional strategies. In addition, there are many topics beyond these five that you must teach. You can use this book as you develop or enhance various units of study.

You may have chosen to work on a unit of study as you read. Or you may have decided to wait until after you read either Part I of the book or the entire book to apply what you learned. You may have worked your way through the book individually or, as we suggested, with colleagues. Regardless of how you approached the book, the following sections outline some steps you might take.

Consider the following suggestions if you have already completed a unit of study:

1. Reflect on the framework, process, and tools. What was easy for you to do and what was most challenging? Why were some aspects easier and why were some more challenging?

2. Review the steps of the process (Table 2.1, p. 39) and the hints and resources found in Appendix 2 (p. 311). See if this review helps you better understand the most challenging aspects of the work. Think about ways in which you might alter your unit of study based on what you learned by revisiting the process and Appendix 2.

3. Consider the aspects that were most challenging and review again the way in which those aspects were addressed in Chapter 3 or in one or several of the contributed chapters in Part II of the book. How did the authors of those chapters address the same aspects that you found challenging? Think about ways in which you might alter your unit of study based on what you learned during review of the contributed models.

4. Work with a colleague and try to explain the choices you made in development of your unit of study.

5. Implement your unit of study and determine which aspects are most successful and which did not work as well. Revisit the instructional tools and consider different strategies that might work better with your students.

Consider the following suggestions if you have not completed a unit of study:

1. If at all possible, work with a colleague. Review the framework, process, and tools. Which aspects are easiest for you to understand and which are most challenging? Why are some aspects easier and why are some more challenging?

2. Identify a unit of study you find most difficult to teach. Review the steps of the process (Table 2.1, p. 39) and the hints and resources found in Appendix 2 (p. 311). Begin the process to develop your unit.

3. As you develop the unit of study, determine which aspects are the most difficult to complete. Review those aspects in either Chapter 3 or some of the contributed chapters in Part II of the book. How did the various authors address those aspects? How might you incorporate their thinking as you plan?

4. Implement your unit of study and determine which aspects are most successful and which did not work as well. Revisit the instructional tools and consider different strategies that might work better with your students.

Ideas for Further Learning

1. Choose one of the instructional tools—ideally one about which you would like to learn more. Review the various strategies in the tools, paying special atten-

tion to the research and the resources. Review some of the research in the original source. Also access some of the resources and learn more about the strategy. Try out the strategy in your classroom.

2. Reflect upon how well you are keeping student thinking evident in your classroom. Which aspect of the Instructional Planning Framework is your greatest challenge? Eliciting student thinking? Confronting student thinking? Sense making? Formative assessment? Pay special attention to specific strategies that work well for this aspect of your instruction and learn more about those strategies.

3. Work with a colleague to analyze one of the contributed chapters. Consider implementing one of contributing authors' unit of study as it is and then reflect upon effectiveness. Also consider modifying the unit of study to better align with your instructional style or to better address the needs of your students.

4. Modify a unit of study you have used in the past, enhancing the unit with a variety of strategies found in the Instructional Tools or by including some of the connections described in Chapter 4.

5. Use this book in a Professional Learning Community both as you read and study it and as you develop various instructional units.

We are all in different places in the professional science-teaching continuum. Regardless of where we are there is always room to learn more and to enhance our instruction. We hope that the Instructional Planning Framework, the planning process and the Instructional Tools help you as you work to enhance your science curriculum and instruction.

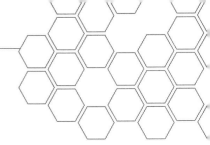

References

Adeniyi, E. O. 1985. Misconceptions of selected ecological concepts held by some Nigerian students. *Journal of Biological Education* 19 (4): 311–316.

Ambron, J. 1987. Writing to improve learning in biology. *Journal of College Science Teaching* 16 (4): 263–266.

American Association for the Advancement of Science (AAAS). 1990. *Science for all Americans.* New York: Oxford University Press.

American Association for the Advancement of Science (AAAS). 1993. *Benchmarks for science literacy.* Washington, DC: AAAS.

American Association for the Advancement of Science (AAAS). 2001a. *Atlas of science literacy.* Washington, DC: AAAS.

American Association for the Advancement of Science (AAAS). 2001b. *Designs for science literacy.* New York: Oxford University Press.

American Association for the Advancement of Science (AAAS). 2011. *Vision and change in undergraduate biology education: A call to action.* Washington, DC: AAAS.

Anderson, K. M. 2007. Tips for teaching: Differentiating instruction to include all students. *Preventing School Failure* 51 (3): 49–53.

Andersson, B. 1990. Pupils' conceptions of matter and its transformation (age 12–16). *Studies in Science Education* 18 (1): 53–85.

Annenberg. n.d. *The habitable planet: Unit 9: Biodiversity decline. www.learner.org/courses/envsci/unit/text.php?unit=9&secNum=0*

Association for Supervision and Curriculum Development (ASCD). 2008. Analyzing classroom discourse to advance teaching and learning. *Education Update* 50 (2): 1, 3, and 7.

Atkinson, H., and S. Bannister. 1998. Concept maps and annotated drawings. *Primary Science Review* 51: 3–5.

Atkin, J. M., and J. E. Coffey, eds. 2003. *Everyday assessment in the science classroom.* Arlington, VA: NSTA Press.

Avraamidou, L., and J. Osborne. 2009. The role of narrative in communicating science. *International Journal of Science Education* 31 (q2): 1685–1707.

Bahar, M., A. H. Johnstone, and M. H. Hansell. 1999. Revisiting learning difficulties in biology. *Journal of Biological Education* 33 (2): 84–86.

Baker, L. 2004. Reading comprehension and science inquiry: Metacognitive connections. In *Crossing borders in literacy and science instruction*, ed. E. W. Saul, 239–257. Newark, DE: International Reading Association and Arlington, VA: NSTA Press.

Banilower, E., K. Cohen, J. Pasley, and I. Weiss. 2010. *Effective science instruction: What does research tell us?* 2nd ed. Portsmouth, NH: RMC Research Corporation, Center on Instruction.

Banilower, E. R., P. S. Smith, I. R. Weiss, K. A. Malzahn, K. M. Campbell, and A. M. Weis. 2013. *Report of the 2012 national survey of science and mathematics education.* Chapel Hill, NC: Horizon Research.

Barker, M. 1985. *Where does wood come from? An introduction to photosynthesis for 3rd and 4th formers.* Hamilton, New Zealand: University of Waikato. Referenced in Wahanga Mahi Rangahau, *Curriculum, learning and effective pedagogy: A literature review in science education* (Wellington, New Zealand: Ministry of Education).

Barker, M. 1986. The description and modification of children's ideas of plant nutrition. PhD diss., University of Waikato. Referenced in Wahanga Mahi Rangahau, *Curriculum, learning and effective pedagogy: A literature review in science education* (Wellington, New Zealand: Ministry of Education).

Bartlett, A. 1969. *Arithmetic, population and energy: A talk by Al Bartlett. www.albartlett.org*

Barton, M. L., and D. L. Jordan. 2001. *Teaching reading in science: A supplement to teaching reading in the content areas teacher's manual.* 2nd ed. Aurora, CO: McREL.

Bee Health extension. *www.extension.org/bee_health*

Bee Informed Partnership. *http://beeinformed.org*

BeeSpotter. *beespotter.mste.illinois.edu*

Billings, P. R., M. A. Kohn, M. de Cuevas, J. Beckwith, J. S. Alper, and M. R. Natowicz. 1992. Discrimination as a consequence of genetic testing. *American Journal of Human Genetics* 50(3): 476–482.

Billmeyer, R. 2003. *Strategies to engage the mind of the learner.* Omaha, NE: Dayspring Press.

Black, P. J., and C. Harrison. 2004. *Science inside the black box: Assessment for learning in the science classroom.* London: NFER-Nelson.

Black, P., C. Harrison, C. Lee, B. Marshall, and D. Wiliam. 2003. *Assessment for learning: Putting it into practice.* London: Open University Press.

Black, P., and D. Wiliam. 1998. Assessment and classroom learning. *Assessment in Education* 5 (1): 7–74.

Blakey, E., and S. Spence. 1990. *Developing megacognition.* Syracuse, NY: ERIC Clearinghouse on Information Resources (ERIC Digest no. ED327218) *www.ericdigests.org/pre-9218/developing.htm.*

Block, C. C., and M. Pressley, eds. 2002. *Comprehension instruction: Research-based best practices.* New York: Guilford.

Bransford, J., A. Brown, and R. Cocking, eds. 1999. *How people learn: Brain, mind, experience, and school.* Washington, DC: National Academies Press.

Braun, G., D. Tierney, and H. Schmitzer. 2011. How Rosalind Franklin discovered the helical structure of DNA: Experiments in diffraction. *Physics Teacher* 49 (3): 140–143.

Briggs, B., T. Mitton, R. Smith, and T. Magnuson. 2009. Teaching cellular respiration and alternate energy sources with a laboratory exercise developed by a scientist-teacher partnership. *The American Biology Teacher* 71 (3): 164.

Brody, M. J. 1992. Student science knowledge related to ecological crises. Paper presented at AERA, San Francisco.

Brown, D. S. 2003. High school biology: A group approach to concept mapping. *The American Biology Teacher* 65 (3): 192–197.

Brumby, M. N. 1982. Students' perceptions of the concept of life. *Science Education.* 66 (4): 613–622.

Brunsell, E., ed. 2012. *E + S: Integrating engineering and science in your classroom.* Arlington, VA: NSTA Press.

BSCS. 2005. *The nature of science and the study of biological evolution.* Arlington VA: NSTA Press.

Burke, K. A., T. J. Greenbowe, and B. M. Hand. 2005. The process of the science writing heuristic. Iowa State University. *http://avogadro.chem.iastate.edu/SWH/homepage.htm*

Burke, K. A., B. Hand, J. Poock, and T. Greenbowe. 2005. Using the science writing heuristic: Training chemistry teaching assistants. *Journal of College Science Teaching* 35 (1): 36–41.

Buxton, C., and O. Lee. 2010. Fostering scientific reasoning as a strategy to support science learning for ELLs. In *Teaching science with Hispanic ELLs in K–16 classrooms,* ed. D. Senal, C. Senal, and E. Wright, 11–36. Charlotte, NC: Information Age Publishing.

Bybee, R. W. 2006. How inquiry can contribute to the prepared mind. *The American Biology Teacher* 68 (8): 454–457.

Bybee, R. W. 2013. *The case for STEM education: Challenges and opportunities*. Arlington, VA: NSTA Press.

Bybee, R. W., J. Taylor, A. Gardner, P. Van Scotter, J. C. Powell, A. Westbrook., and N. Knapp. 2006. *The BSCS 5E instructional model: Origins and effectiveness*. Colorado Springs, CO: BSCS.

Campbell, B., and L. Fulton. 2003. *Science notebooks: Writing about inquiry*. Portsmouth, NH: Heinemann.

Campbell, N. A., J. B. Reece, L. A. Urry, M. L. Cain, S. A. Wasserman, P. V. Minorsky, and R. B. Jackson. 2008. *Biology*. 8th ed. Upper Saddle River, NJ: Pearson.

Campbell, T., D. Neilson, and P. S. Oh. 2013. Developing and using models in physics. *The Science Teacher* 80 (6): 35–41.

Cañas, A. J., and J. D. Novak. 2006. Re-examining the foundations for effective use of concept maps. Cmap Tools. *http://cmc.ihmc.us/cmc2006Papers/cmc2006-p247.pdf*

Cartier, J. 2000. Research report: Using a modeling approach to explore scientific epistemology with high school biology students. Wisconsin Center for Education Research. *www.wcer.wisc.edu/NCISLA/publications/reports/RR991.pdf*

Catley, K., R. Lehrer, and B. Reiser. 2005. *Tracing a prospective learning progression for developing understanding of evolution*. Paper commissioned by the National Academies Committee on Test Design for K–12 Science Achievement, Vanderbilt University and Northwestern University. Washington, DC: National Academy of Sciences.

Cawelti, G. C. 1999. *Handbook of research on improving student achievement*. 2nd ed. Arlington, VA: Educational Research Service.

Center for Applied Special Technology (CAST). 2011. Universal Design for Learning (UDL) Lesson Builder. 2011. *http://lessonbuilder.cast.org*.

Centre for Educational Research and Innovation (CERI). 2008. Assessment for learning: Formative assessment. OECD/CERI International Conference: Learning in the 21st century: Research, innovation and policy.

Century, J. R., J. Flynn, D. S. Makang, M. Pasquale, K. M. Roblee, J. Winokur, and K. Worth. 2002. Supporting the science-literacy connection. In *Learning science and the science of learning*, ed. R. W. Bybee, 37–49. Arlington, VA: NSTA Press.

Chamberlain, K., and C. C. Crane. 2009. *Reading, writing, and inquiry in the science classroom, grades 6–12: Strategies to improve content learning*. Thousand Oaks, CA: Corwin Press.

Chattopadhyay, A., and B. S. Mahajan. 2006. Students' understanding of DNA and DNA technologies after "fifty years of DNA double helix." Electronic proceedings of the International Conference to Review Research on Science, Technology, and Mathematics Education (epiSTEME1).

Cheuk, T. 2012. *Relationships and convergences among the mathematics, science, and ELA practices*. Refined version of diagram created by the Understanding Language Initiative for ELP Standards. Palo Alto, CA: Stanford University.

Chi, M. T. H. 1992. Conceptual change within and across ontological categories: Examples from learning and discovery in science. In *Cognitive models of science: Minnesota studies in the philosophy of science*, ed. R. Giere, 129–186. Minneapolis, MN: University of Minnesota Press.

Chinn, C. A., and W. F. Brewer. 2001. Models of data: A theory of how people evaluate data. *Cognition and Instruction* 19 (3): 323–392.

Chinn, C., D. E. Brown, and C. B. Bertram. 2002. Student-generated questions: A meaningful aspect of learning in science. *International Journal of Science Education* 24 (5): 521–549.

Chubb, J. E., and T. Loveless. *Bridging the achievement gap.* Washington, DC: Brookings Institution Press.

Citizen Science Central. *www.citizenscience.org*

Clark, D. 2000. Scaffolding knowledge integration through curricular depth. Unpublished PhD diss. University of California, Berkeley.

Clark, D. B. 2006. Longitudinal conceptual change in students' understanding of thermal equilibrium: An examination of the process of conceptual restructuring. *Cognition and Instruction* 24 (4): 467–563.

COLOSS Network. *www.coloss.org*

Coffey, J., R. Douglas, and C. Stearns, eds. 2008. *Assessing science learning: Perspectives from research and practice.* Arlington, VA: NSTA Press.

Colbert, J. T., J. K. Olson, and M. P. Clough. 2007. Using the web to encourage student-generated questions in large-format introductory biology classes. *CBE Life Sciences Education* 6 (1): 42–48.

Coley, J. D., and K. D. Tanner. 2012. Common origins of diverse misconceptions: Cognitive principles and the development of biology thinking. *CBE-Life Science Education* 11: 209–215.

Committee on Undergraduate Science Education. 1997. *Science teaching reconsiders: A handbook.* Washington, DC: National Academies Press.

Concord Consortium. 2013. Molecular workbench. *www.mw.concord.org*

Condit, C. M. 2010. Public understanding of genetics and health. *Clinical Genetics* 77 (1): 1–9.

Constible, J., L. Sandro, and R. E. Lee, Jr. 2008. *Climate change from pole to pole: Biology investigations.* Arlington, VA: NSTA Press.

Costa, A. 2008. The thought-filled curriculum. *Educational Leadership* 65 (5): 20–24.

Cothron, J. H., R. N. Giese, and R. J. Rezba. 2000. *Students and research: Practical strategies for science classrooms and competitions.* Dubuque, IA: Kendall Hunt.

Darling-Hammond, L., B. Barron, P. D. Pearson, A. H. Schoenfeld, E. K. Stage, T. D. Zimmerman, G. N, Cervetti, and J. L. Tilson. 2008. *Powerful learning: What we know about teaching for understanding.* San Francisco, CA: Jossey-Bass.

Davies, A. 2003. Learning through assessment: Assessment for learning in the science classroom. In *Everyday assessment in the science classroom,* ed. J. M. Atkin and J. E. Coffey, 13–25. Arlington, VA: NSTA Press.

Dean, C. B., E. R. Hubbell, H. Pitler, and B. Stone. 2012. *Classroom instruction that works: Research-based strategies for increasing student achievement.* 2nd ed. Alexandria, VA: ASCD.

de Vries, H., I. Mesters, H. van de Staag, and C. Honing. 2005. The general public's information needs and perceptions regarding hereditary cancer: An application of the integrated change model. *Patient Education and Counseling* 56 (2): 154–165.

Diamond, J., C. Zimmer, E. M. Evans, L. Allison, and S. Disbrow. 2006. *Virus and the whale: Exploring evolution in creatures big and small.* Arlington VA: NSTA Press.

Dickinson, J. L., and R. Bonney, eds. 2012. *Citizen science: Public participation in environmental research*. Ithaca, NY: Comstock Publishing Associates.

DiGiovanni, N. 1999. ENSI (Evolution & the Nature of Science Institutes). *www.indiana.edu/~ensiweb/lessons/ns.chips.html*

diSessa, A. A. 2006. A history of conceptual change research: Threads and fault lines. In *The Cambridge handbook of the learning sciences*, ed. K. Sawyer, 265–281. Cambridge, UK: Cambridge University Press.

diSessa, A. A., N. Gillespie, and J. Esterly. 2004. Coherence versus fragmentation in the development of the concept of force. *Cognitive Science* 28: 843–900.

Dobzhansky, T. 1973. Nothing in biology makes sense except in the light of evolution. *The American Biology Teacher* 35: 125–129.

Donovan, S., and J. Bransford, eds. 2005. *How students learn: Science in the classroom*. Washington, DC: National Academies Press.

Donovan, S., J. Bransford, and J. Pellegrino, eds. 1999. *How people learn: Bridging research and practice*. Washington, DC: National Academies Press.

Dougherty, M. J. 2009. Closing the gap: Inverting the genetics curriculum to ensure an informed public. *American Journal of Human Genetics* 85 (1): 6–12.

Driver, R., A. Squires, P. Rushworth, and V. Wood-Robinson. 1994. *Making sense of secondary science: Research into children's ideas*. London: Routledge.

Duncan, R. G., and B. J. Reiser. 2003. Students' reasoning about phenomenon generated by complex systems: The case of molecular genetics. Paper presented at the annual international conference of the National Association for Research in Science Teaching, Philadelphia, PA.

Duncan, R. G., and B. J. Reiser. 2005. Designing for complex system understanding in the high school biology classroom. Paper presented at the annual international conference of the National Association for Research in Science Teaching, Dallas, TX.

Duschl, R. A., and J. Osborne. 2002. Supporting and promoting argumentation discourse in science education. *Studies in Science Education* 38 (1): 39–72.

Duschl, R, A., H. A. Schweingruber, and A. W. Shouse, eds. 2007. *Taking science to school: Learning and teaching science in grades K–8*. Washington, DC: National Academies Press.

Edelson, D. C. 2001. Learning-for-use: A framework for the design of technology-supported inquiry activities. *Journal of Research in Science Teaching* 38 (3): 355–385.

Edens, K. M., and E. F. Potter. 2003. Using descriptive drawings as a conceptual change strategy in elementary science. *School Science and Math Journal* 103 (3): 135–144.

Ekici, F., E. Ekici, and F. Aydin. 2007. Utility of concept cartoons in diagnosing and overcoming misconceptions related to photosynthesis. *International Journal of Environmental and Science Education* 2 (4): 111–124.

Eklund, J., A. Rogat, N. Alozie, and J. Krajcik. 2007. Promoting student scientific literacy of molecular genetics and genomics. Paper presented at the annual international conference of the National Association for Research in Science Teaching, New Orleans, LA.

Elrod, S. n.d. Genetics concepts inventory. Bioliteracy. *www.bioliteracy.net*

Engel Clough, E., and C. Wood-Robinson. 1985. How secondary students interpret instances of biological adaptation. *Journal of Biological Education* 19 (2): 125–128.

Enger, S. K., and R. E. Yager. 2009. *Assessing student understanding in science*. 2nd ed. Thousand Oaks, CA: Corwin Press.

Environmental Literacy Council and National Science Teachers Association (NSTA). 2007. *Biodiversity: Resources for environmental literacy*. Arlington, VA: NSTA Press

Eunice Kennedy Shriver National Institute of Child Health and Human Development, NIH, DHHS (NICHD). 2010. *What content-area teachers should know about adolescent literacy*. Washington, DC: U.S. Government Printing Office.

Feldman, E. A. 2012. The genetic information nondiscrimination act (GINA): Public policy and medical practice in the age of personalized medicine. *Journal of General Internal Medicine* 27 (6): 743–746.

Fisher, K. M. 1985. A misconception in biology: Amino acids and translation. *Journal of Research in Science Teaching* 22 (1): 53–62.

Fisher, K. M., J. H. Wandersee, and D. E. Moody. 2000. *Mapping biology knowledge*. Boston: Kluwer.

Flick, L. B. 1997. Understanding a generative learning model of instruction: A case study of elementary teacher planning. *Journal of Science Teacher Education* 7 (2): 95–122.

Frayer, D., W. C. Frederick, and H. J. Klausmeier. 1969. *A schema for testing the level of cognitive mastery*. Madison, WI: Wisconsin Center for Education Research.

Freedman, R. L. H. 1994. *Open-ended questioning: A handbook for educators*. Menlo Park, CA: Addison-Wesley.

Freidenreich, H. B., R. G. Duncan, and N. Shea. 2011. Promoting middle school students' understanding of three conceptual models in genetics. *International Journal of Science Education* 33 (17): 2323–2350.

Friedrichsen, P. M., and B. Stone. 2004. Examining students' conceptions of molecular genetics in an introductory biology course for non-science majors: A self study. Paper presented at the annual international meeting of the National Association for Research in Science Teaching, Vancouver, WA.

Gilbert, J., and M. Kotelman. 2005. Five good reasons to use science notebooks. *Science and Children* 43 (3): 28–32.

Gilbert, S. W. 2011. *Models-based science teaching*. Arlington, VA: NSTA Press.

Gilbert, S. W., and S. W. Ireton. 2003. *Understanding models in Earth and space science*. Arlington, VA: NSTA Press.

Gill, C. J. 1987. A new perspective on disability and its implications for rehabilitation. In *Sociocultural implications in treatment planning in occupational therapy*, ed. F. S. Cromwell, 49–55. New York: Haworth Press.

Glynn, S. M., and R. Duit. 1995. Learning science meaningfully: Constructing conceptual models. In *Learning science in the schools: Research reforming practice*, ed. S. M. Glynn and R. Duit, 3–33. Mahwah, NJ: Lawrence Erlbaum Associates.

Gore, M. C. 2004. *Successful inclusion strategies for secondary and middle school teachers: Keys to help struggling learners access the curriculum*. Thousand Oaks, CA: Corwin Press.

Gregory, G. H., and E. Hammerman. 2008. *Differentiated instructional strategies for science, grades K–8*. Thousand Oaks, CA: Corwin Press.

Grosslight, L., C. Unger, E. Jay, and C. L. Smith. 1991. Understanding models and their use in science—Conceptions of middle and high school students and experts. *Journal of Research in Science Teaching* 28 (9): 799–822.

Hahn, H. 1988. The politics of physical differences: Disability and discrimination. *Journal of Social Issues* 44 (1): 39–47.

Hale, M. S., and E. A. City. 2006. *The teacher's guide to leading student-centered discussions: Talking about texts in the classroom.* Thousand Oaks, CA: Corwin Press.

Hand, B., L. Hockenberry, K. Wise, and L. Norton-Meier, eds. 2008. *Questions, claims and evidence: The important place of argument in children's science writing.* Portsmouth, NH: Heinemann.

Hargrove, T. Y., and C. Nesbit. 2003. *Science notebooks: Tools for increasing achievement across the curriculum.* Columbus, OH: ERIC Clearinghouse for Science, Mathematics, and Environmental Education.

Harland, D. 2011. *STEM student research handbook.* Arlington, VA: NSTA Press.

Harlen, W. 2000. *Teaching, learning and assessing science 5–12.* 3rd ed. London: Paul Chapman.

Harlen, W. 2001. *Primary science: Taking the plunge.* 2nd ed. Portsmouth, NH: Heinemann.

Harlen, W. 2013. *Assessment and inquiry-based science education: Issues in policy and practice.* Trieste, Italy: Global Network of Science Academies (IAP) Science Education Programme (SEP).

Harrison, A. G., and R. K. Coll. 2008. *Using analogies in middle and secondary science classrooms: The FAR guide—An interesting way to teach with analogies.* Thousand Oaks, CA: Corwin Press.

Harrison, A. G., and D. F. Treagust. 2000. A typology of school science models. *International Journal of Science Education* 22 (9): 1011–1026.

Hartley, L. M., J. Momsen, A. Maskiewicz, and C. D'Avanzo. 2012. Energy and matter: Differences in discourse in physical and biological sciences can be confusing for introductory biology students. *BioScience* 62 (5): 488–496 .

Hartman, H. J., and N. A. Glasgow. 2002. *Tips for the science teacher: Research-based strategies to help students learn.* Thousand Oaks, CA: Corwin Press.

Harvard-Smithsonian Center for Astrophysics, Science Education Department, Science Media Group. 1987. *A private universe. www.learner.org/resources/series28.html*

Harvard-Smithsonian Center for Astrophysics, Science Education Department, Science Media Group. 1997. *Minds of our own. www.learner.org/resources/series26.html*

Harvard University. 2013. Understanding student weaknesses. *ScienceDaily,* May 2. *www.sciencedaily.com/releases/2013/05/130502131936.htm*

Harvey, S., and A. Goudvis. 2007. *Strategies that work: Teaching comprehension for understanding and engagement.* 2nd ed. Portland, ME: Stenhouse.

Hattie, J., and H. Timperly. 2007. The power of feedback. *Review of Educational Research* 77: 81–112.

Hazen, R. M., and J. S. Trefil. 1991. *Science matters: Achieving scientific literacy.* New York: Doubleday.

Heritage, M. 2007. Formative assessment: What do teachers need to know and do? *Phi Delta Kappan* 89 (2): 140–146.

Heritage, M. 2008. *Learning progressions: Supporting instruction and formative assessment.* Washington, DC: Council of Chief State School Officers.

Heritage, M. 2010. *Formative assessment and next-generation assessment systems: Are we losing an opportunity?* Washington, DC: Council of Chief State School Officers.

Hershey, D. R. 2004. Avoid misconceptions when teaching about plants. *www.actionbioscience.org/education/hershey3.html*

Hewson, P. W. 1992. *Conceptual change in science teaching and teacher education*. Madrid, Spain: National Center for Educational Research, Documentation and Assessment.

Hipkins, R., R. Bolstad, R. Baker, A. Jones, M. Barker, B. Bell, R. Coll, B. Cooper, M. Forret, A. Harlow, I. Taylor, B. France, and M. Haigh. 2002. *Curriculum, learning, and effective pedagogy: A literature review in science education*. Wellington, NZ: New Zealand Council for Educational Research.

Horn, T. 2005. *Bees in America: How the honeybee shaped a nation*. Lexington, KY: University Press of Kentucky.

Horn, T. *New York Times*. 2008. Honeybees: A history. April 11.

Horton, P., A. McConney, M. Gallo, A. Woods, G. Senn, and D. Hamelin. 1993. An investigation of the effectiveness of concept mapping as an instructional tool. *Science Education* 77 (1): 95–111.

Hyerle, D. 2000. *A field guide to using visual tools*. Alexandria, VA: Association for Supervision and Curriculum Development.

Ingram, M. 1993. *Bottle biology: An idea book for exploring the world through soda bottles and other recyclable materials*. Dubuque, IA: Kendall/Hunt.

International Centre for Development Oriented Research in Agriculture (ICRA). n.d. Systems diagrams: Guidelines. ICRA. *www.icra-edu.org/objects/anglolearn/Systems_Diagrams-Guidelines1.pdf*

Jensen, E. 1998. *Teaching with the brain in mind*. Alexandria, VA: Association for Supervision and Curriculum Development.

Jensen, J. E. 2008. *NSTA tool kit for teaching evolution*. Arlington, VA: NSTA Press.

Johnson, D. W., and R. T. Johnson. 1986. *Learning together and alone*. 2nd ed. Englewood Cliffs, NJ: Prentice Hall.

Johnson, L., S. Adams Becker, M. Cummins, V. Estrada, A. Freeman, and H. Ludgate. 2013. *NMC Horizon Report: 2013 K–12 edition*. Austin, TX: The New Media Consortium.

Jones, J., and J. Stanhope. 1994. Access Excellence. *Fishy Frequencies. www.accessexcellence.org/AE/AEPC/WWC/1994/fishfreq.php*

Jones, J. S. 1999. *Almost Like a Whale: The Origin of Species Updated*. New York: Doubleday.

Jones, M. G., M. R. Falvo, A. R. Taylor, and B. P. Broadwell. 2007. *Nanoscale science: Activities for grades 6–12*. Arlington, VA: NSTA Press.

Jones, S. R. 1996. Toward inclusive therapy: Disability as social construction. *NASPA Journal* 33: 347–354.

Joyce, B., and E. Calhoun. 2012. *Realizing the promise of 21st-century education: An owner's manual*. Thousand Oaks, CA: Corwin Press.

Kagan, S. 1989. The structural approach to cooperative learning. *Educational Leadership* 47 (4): 12–15.

Kamil, M. L. 2003. *Adolescents and literacy: Reading for the 21st century*. Washington, DC: Alliance for Excellent Education.

Kaplan, J. K. 2012. Colony collapse disorder: An incomplete puzzle. *Agricultural Research* 60 (6): 4–8.

Kapur, M., and K. Bielaczyc. 2012. Designing for productive failure. *Journal of the Learning Sciences* 21 (1): 45–83.

Kapur, M. 2008. Productive failure. *Cognition and Instruction* 26 (3): 379–424.

Keeley, P. 2008. *Science formative assessment: 75 practical strategies for linking assessment, instruction, and learning.* Thousand Oaks, CA: Corwin Press and Arlington, VA: NSTA Press.

Keeley, P. 2011. *Uncovering student ideas in life science, vol. 1: 25 new formative assessment probes.* Arlington, VA: NSTA Press.

Keeley, P., and F. Eberle. 2008. Using standards and cognitive research to inform the design and use of formative assessment probes. In *Assessing science learning,* ed. J. Coffey, R Douglas, and C. Stearns, 206–207. Arlington, VA: NSTA Press.

Keeley, P., and C. M. Rose. 2006. *Mathematics curriculum topic study: Bridging the gap between standards and practice.* Thousand Oaks, CA: Corwin Press.

Keeley, P., and J. Tugel. 2009. *Uncovering student ideas in science, vol. 4: 25 new formative assessment probes.* Arlington, VA: NSTA Press.

Keeley, P., J. Tugel, and C. Gabler. Forthcoming. *Science curriculum topic study: Bridging the gap between standards and STEM teaching and learning.* 2nd ed. Thousand Oaks, CA: Corwin Press.

Kellar, B. n.d. Honey bees across America *www.orsba.org/htdocs/download/HoneyBeesAcross America.html.*

Keogh, B., and S. Naylor. 1999. Concept cartoons, teaching and learning in science: An evaluation. *International Journal of Science Education* 21 (4): 431–446.

Keogh, B., and S. Naylor. 2007. Talking and thinking in science. *School Science Review* 88 (324): 85–90.

Kindfield, A. C. H. 1994. Understanding a basic biological process: Expert and novice models of meiosis. *Science Education* 78 (3): 255–283.

Koba, S., and A. Tweed. 2009. *Hard-to-teach biology concepts: A framework to deepen student understanding.* Arlington, VA: NSTA Press.

Köse, S. 2008. Diagnosing student misconceptions: Using drawings as a research method. *World Applied Science Journal* 3 (2): 283–293.

Krajcik, J., and C. Czerniak. 2014. *Teaching science in elementary and middle school: A project-based approach.* 4th ed. New York: Routledge.

Krajcik, J., P. C. Blumenfeld, R. W. Marx, K. M. Bass, and J. Fredricks. 1998. Inquiry in project-based science classrooms: Initial attempts by middle school students. *The Journal of the Learning Sciences* 7 (3 and 4): 313–350.

Krasny, M. E., and The Environmental Inquiry Team. 2003. *Invasion ecology.* Arlington, VA: NSTA Press.

Kurth, L. A., and J. Roseman. 2001. Findings from the high school biology curriculum study: Molecular basis of heredity. Paper presented at the Annual Meeting of the National Association for Research in Science Teaching, St. Louis, MO.

Layman, J., G. Ochoa, and H. Heikkinen. 1996. *Inquiry and learning: Realizing science standards in the classroom.* New York, NY: National Center for Cross Disciplinary Teaching and Learning.

Lazear, D. 1991. *Seven ways of teaching: The artistry of teaching with multiple intelligences.* Palatine, IL: IRI/Skylight Publishing.

Leach, J., R. Driver, P. Scott, and C. Wood-Robinson. 1992. *Progression in conceptual understanding of ecological concepts by pupils aged 5–16.* Leeds, UK: Centre for Studies in Science and Mathematics Education, University of Leeds.

LearningScience. *www.learningscience.org*

Lederman, N. G. 2007. Nature of science: Past, present and future. In *Handbook of research on science education,* ed. S. Abell and N. Lederman, 320–325, Mahwah, NJ: Lawrence Erlbaum Associates.

Lee, O., H. Quinn, and G. Valdes. 2013. Science and language for English language learners: Language demands and opportunities in relation to *Next Generation Science Standards. Educational Researcher* 42: 223–233.

Lee, S. C. 2012. Teachers feedback to foster scientific discourse in connected science classrooms. PhD diss., The Ohio State University.

Lemke, J. L. 1990. *Talking science: Language, learning, and values.* Norwood, NJ: Ablex.

Lent, R. C. 2012. *Overcoming textbook fatigue: 21st century tools to revitalize teaching and learning.* Alexandria, VA: Association for Supervision and Curriculum Development.

Lewis, J., and U. Kattmann. 2004. Traits, genes, particles, and information: Re-visiting students' understanding of genetics. *International Journal of Science Education* 26 (2): 195–206.

Lewis, J., and C. Wood-Robinson. 2000. Genes, chromosomes, cell division and inheritance—do students see any relationship? *International Journal of Science Education* 22 (2): 177–195.

Lipton, L., and B. Wellman. 1998. *Pathways to understanding: Patterns and practices in the learning-focused classroom.* 3rd ed. Sherman, CT: MiraVia.

Lockhart, A., and J. Le Doux. 2012. A partnership for problem-based learning: Challenging students to consider open-ended problems involving gene therapy. In *Integrating engineering and science in your classroom,* ed. E. Brunsell, 67–74. Arlington, VA: NSTA Press.

Loreau, M. 2006. Diversity without representation. *Nature* 442: 245–246.

Loreau, M., S. Naeem, P. Inchausti, J. Bengtsson, J. P. Grime, A. Hector, et al. 2001. Biodiversity and ecosystem functioning: current knowledge and future challenges. *Science* (294): 804–808.

Lowery, L. F. 1990. *The biological basis of thinking and learning.* Berkeley, CA: University of California Press.

Madden, A. S., M. F. Hochella Jr., G. E. Glasson, J. R. Grady, T. L. Bank, A. M. Green, M. A. Norris, A. N. Hurst, and S. C. Eriksson. 2011. *Welcome to nanoscience: Interdisciplinary environmental explorations.* Arlington, VA: NSTA Press.

Magnusson, S. J., and A. S. Palinscar. 2003. A theoretical framework for the development of second-hand investigation texts. Paper presented at the American Education Research Association Conference, Chicago.

Magnusson, S. J., and A. S. Palinscar. 2004. Learning from text designed to model scientific thinking in inquiry-based instruction. In *Crossing borders in literacy and science instruction,* ed. E. W. Saul, 316–339. Newark, DE: International Reading Association and Arlington, VA: NSTA Press.

Miller, J. D. 2004. Public understanding of, and attitudes toward, scientific research: What we know and what we need to know. *Public Understanding of Science* 13 (3): 273–294.

Mills Shaw, K. R., K. Van Horne, H. Zhang, and J. Boughman. 2008. Essay contest reveals misconceptions of high school students in genetic content. *Genetics* 178 (3): 1157–1168.

Marcarelli, K. 2010. *Teaching science with interactive notebooks.* Thousand Oaks, CA: Corwin Press.

Mansilla, V. B., and H. Gardner. 2008. Disciplining the mind. *Educational Leadership* 65 (5): 14–19.

Marzano, R. J. 1992. *A different kind of classroom: Teaching with dimensions of learning.* Alexandria, VA: Association for Supervision and Curriculum Development.

Marzano, R., and D. Pickering. 1997. *Dimensions of learning trainer's manual.* Alexandria, VA: ASCD.

Marzano, R., D. Pickering, and J. Pollock. 2001. *Classroom instruction that works: Research-based strategies for increasing student achievement.* Alexandria, VA: ASCD.

Mayr, E. 1991. *One long argument: Charles Darwin and the genesis of modern evolutionary thought.* Cambridge: Harvard University Press.

Mayr, E. 1997. *This is biology: The science of the living world.* Cambridge: Harvard University Press.

McConnell, S. 1993. Talking drawings: A strategy for assisting learners. *Journal of Reading* 36 (4): 260–269.

McKinney, D., and M. Michalovic. 2004. Teaching the stories of scientists and their discoveries. *The Science Teacher* 71 (9): 46–51.

McMahon, M., P. Simmons, R. Sommers, D. DeBaets, and F. Crawley, eds. 2006. *Assessment in science: Practical experiences and education research.* Arlington, VA: NSTA Press.

McNair, S., and M. Stein. 2001. Drawing on their understanding: Using illustrations to invoke deeper thinking about plants. Pennsylvania State University. *http://edr1.educ. msu.edu/EnvironmentalLit/publicsite/html/paper.html.*

McNeill, K. L., and J. Krajcik. 2006. Supporting students' construction of scientific explanation through generic versus context-specific written scaffolds. Paper presented at the annual meeting of the American Educational Research Association, San Francisco, CA.

McNeill, K. L., and J. Krajcik. 2008. Inquiry and scientific explanations: Helping students use evidence and reasoning. In *Science as inquiry in the secondary settings,* ed. J. Luft, R. L. Bell, and J. Gess-Newsome, 121–133. Arlington, VA: NSTA Press.

McNeill, K. L., and J. Krajcik. 2011. Claim, evidence and reasoning: Supporting middle school students in evidence-based scientific explanations. Workshop presented at the annual national meeting of National Science Teachers Association in San Francisco, CA.

McNeill, K. L., and J. S. Krajcik. 2012. *Supporting grade 5–8 students in constructing explanations in science: The claim, evidence, and reasoning framework for talk and writing.* Boston, MA: Pearson.

McTighe, J., and G. Wiggins. 1999. *Understanding by design handbook.* Alexandria, VA: ASCD.

McTighe, J., and G. Wiggins. 2004. *Understanding by design: Professional development workbook.* Alexandria, VA: ASCD.

McTighe, J., and G. Wiggins. 2013. *Essential questions: Opening doors to student understanding.* Alexandria, VA: ASCD.

Mead, L. W., and G. Branch. 2011. Overcoming obstacles to evolution education: Why bother teaching evolution in high school? *Evolution: Education & Outreach* 4 (1): 114–116.

Meadows, L. 2009. *The missing link: An inquiry approach for teaching all students about evolution.* Portsmouth: Heinemann.

Mertens, T., and J. Walker. 1992. A paper-and-pencil strategy for teaching mitosis and meiosis, diagnosing learning problems and predicting examination performance. *The American Biology Teacher* 54 (8): 470–474.

Metropolitan Educational Research Consortium (MERC). 2011. *Encouraging self-regulated learning in the classroom: A review of the literature.* Richmond, VA: Virginia Commonwealth University.

Meyer, A., D. H. Rose, and D. Gordon. 2013. *Universal design for learning: Theory and practice.* Wakefield, MA: Cast Incorporated.

Michaels, S., A. W. Shouse, and H. A. Schweingruber. 2007. *Ready, set science: Putting research to work in K–8 science classrooms.* Washington, DC: National Academies Press.

Milne, C. 2008. In praise of questions: Elevating the role of questions for inquiry in secondary school science. In *Science as inquiry in the secondary setting,* ed. J. Luft, R. L. Bell, and J. Gess-Newsome, 99–106. Arlington, VA: NSTA Press.

Mind Tools. n.d. Systems diagrams. Mind Tools. *www.mindtools.com/pages/article/ newTMC_04.htm*

Minstrell, J. 1989. Teaching science for understanding. In *Toward the thinking curriculum: Current cognitive research,* ed. L. B. Resnick and L. E. Klopfer, 150–172. Alexandria, VA: Association for Supervision and Curriculum Devemopment.

Mohan, L., J. Chen, and C. W. Anderson. Developing a multi-year learning progression for carbon cycling in socio-ecological systems. *Journal of Research in Science Teaching,* 46 (6): 675–698.

Mortimer, E. F. 1998. Multivoicedness and univocality in classroom discourse: An example from theory of matter. *International Journal of Science Education* 20: 67–82.

Mortimer, E. F., and P. H. Scott. 2003. *Meaning making in secondary science classrooms.* London: Open University Press.

Moyer, R. H., and S. A. Everett. 2012. *Everyday engineering: Putting the E in STEM teaching and learning.* Arlington, VA: NSTA Press.

Nathanson, S. 2006. Harnessing the power of story: Using narrative reading and writing across content areas. *Reading Horizons* 47 (1): 1–26.

National Academy of Sciences (NAS). 1998. *Teaching about evolution and the nature of science.* Washington, DC: National Academies Press.

National Institutes of Health (NIH). 1998. User's guide to the scientific method. Materials presented at a workshop at the University of Colorado-Boulder.

National Research Council (NRC). 1996. *National science education standards.* Washington, DC: National Academies Press.

National Research Council (NRC). 2000. *Inquiry and the national science education standards.* Washington, DC: National Academies Press.

National Research Council (NRC). 2001. *Classroom assessment and the national science education standards.* Washington, DC: National Academies Press.

National Research Council (NRC). 2005. *How students learn: Science in the classroom.* Washington, DC: National Academies Press.

National Research Council (NRC). 2007. *Taking science to school: Learning and teaching science in grades K–8.* Washington, DC: National Academies Press.

NATIONAL SCIENCE TEACHERS ASSOCIATION

National Research Council (NRC). 2011a. *Successful K–12 STEM education: Identifying effective approaches in science, technology, engineering, and mathematics.* Washington, DC: National Academies Press.

National Research Council (NRC). 2011b. *Successful K–12 STEM education: A workshop summary.* Washington, DC: National Academies Press.

National Research Council (NRC). 2012. *A framework for K–12 science education: Practices, crosscutting concepts, and core ideas.* Washington, DC: National Academies Press.

Natural Resources Defense Council. *www.nrdc.org*

Nelson-Herber, J. 1986. Expanding and refining vocabulary in content areas. *Journal of Reading* 29 (7): 626–33.

NGSS Lead States. 2013. *Next Generation Science Standards: For states, by states.* Washington, DC: National Academies Press. *www.nextgenscience.org/next-generation-science-standards*

Novak, J. D. 1996. Concept mapping: A tool for improving science teaching and learning. In *Improving teaching and learning in science and mathematics,* ed. D. F. Treagust, R. Duit, and B. J. Fraser, 32–43. New York: Teachers College Press.

Novak, J. D. 1998. *Learning, creating, and using knowledge: Concept maps as facilitative tools in schools and corporations.* Mahwah, NJ: Lawrence Erlbaum.

NSDL-AAAS. *The living environment, Diversity of life.* Science Literacy Maps. *http://strandmaps.nsdl/?id=SMS-MAP-2105*

NSDL-AAAS. *The living environment, Interdependence of life.* Science Literacy Maps. *http://strandmaps.nsdl/?id=SMS-MAP-2122*

O'Connell, D. 2008. An inquiry-based approach to teaching photosynthesis & cellular respiration. *American Biology Teacher* 70: 350–356.

O'Connell, D. 2010. Dust thou art *not* & unto dust thou shan't return: Common mistakes in teaching biogeochemical cycles. *American Biology Teacher* 72 (9): 552–556.

Organization for Economic Co-operation and Development (OECD); Centre for Educational Research and Innovation. 2005. *Formative assessment: Improving learning in secondary classrooms.* Paris, France: OECD

Padian, K. 2013. Correcting some common misrepresentations of evolution in textbooks and the media. *Evolution: Education and Outreach* 6 (11).

Palinscar, A. S., and A. L. Brown. 1985. Reciprocal teaching: Activities to promote "reading with your mind." In *Reading, thinking, and concept development,* ed. T. L. Harris and E. J. Cooper, 147–158. New York: College Board Publications.

Passmore, C., E. Coleman, J. Horton, and H. Parker. 2013. Making sense of natural selection: Developing and using the natural selection model as an anchor for practice and content. *The Science Teacher* 80 (6): 43–49.

Pitcher, S. M., L. K. Albright, D. J. DeLaney, N. T. Walker, K. Seunarinesingh, et. al. 2007. Assessing adolescents' motivation to read. *Journal of Adolescent & Adult Literacy* 50 (5): 378–396.

Pratt, H. 2013. *The NSTA reader's guide to the* Next Generation Science Standards. Arlington, VA: NSTA Press.

Pressley, M. 2002. Comprehension strategies instruction: A turn-of-the-century status report. In *Comprehension instruction: Research–based best practices,* ed. C. C. Block and M. Pressley, 11–27. New York: Guilford.

Pressley, M. 2006. *Reading instruction that works: The case for balanced teaching.* 3rd ed. New York: Guilford.

Presley, M. L., A. J. Sickel, N. Muslu, D. Merle-Johnson, S. B. Witzig, K. Izei, and T. D. Sadler. 2013. A framework for socio-scientific issues based education. *Science Educator* 22 (1): 26–32.

Rangahau, W. M. 2002. *Curriculum, learning, and effective pedagogy: A literature review in science education.* Wellington, New Zealand: Ministry of Education.

Reiser, B. J. 2013. *What professional development strategies are needed for successful implementation of the* Next General Science Standards? Austin, TX: K–12 Center at ETS.

Reiser, B. J., L. K. Berland, and L. Kenyon. 2012. Engaging students in the scientific practices of explanation and argumentation: Understanding *A Framework for K–12 Science Education. The Science Teacher* 79 (4): 34–39.

Rice, S. D. 2013. Using interactive animations to enhance teaching, learning, and retention of respiration pathway concepts in face-to-face and online high school, undergraduate, and continuing education learning environments. *Journal of Microbiology & Biology Education* 14 (1): 113–115.

Richardson, W. 2013. Students first, not stuff. *Educational Leadership* 70 (6): 10–14.

Rico, G. 2000. *Writing the natural way.* New York: Putnam. Described in D. Hyerle, *A field guide to using visual tools.* Alexandria, VA: ASCD.

Ritchart, R., and D. Perkins. n.d. Visible thinking: Engaged students, in-depth learning, better teaching. Council for Exceptional Children. *www.cec.sped.org/AM/Template.cfm?Section=Home*

Ritchart, R., and D. Perkins. 2008. Making thinking visible. *Educational Leadership* 65 (5): 57–61.

Rolheiser, C., and J. A. Ross. n.d. Student self-evaluation: What research says and what practice shows. Center for Developmental Learning. *www.cdl.org/resource-library/articles/self_eval.php?type=subject&id=4*

Rose, A., and V. Scott. 2013. Personal communication with Nancy Kellogg.

Roseman, J. E., A. Caldwell, A. Gogos, and L. Kurth. 2006. Mapping a coherent learning progression for the molecular basis of heredity. Project 2061. *www.project2061.org/publications/articles/papers/narst2006.htm*

Rotbain, Y., G. Marbach-Ad, and R. Stavy. 2005. Understanding molecular genetics through a drawing-based activity. *Journal of Biological Education* 39 (4): 174–178.

Roth, K. J., H. E. Garnier, C. Chen, M. Lemmens, J. Schwille, and N. I. Z. Wickler. 2011. Videobased lesson analysis: Effective science PD for teacher and student learning. *Journal of Research in Science Teaching* 48 (2): 117–148.

Roth, K. J., S. L. Druker, H. E. Garnier, M. Lemmens, C. Chen, T. Kawanaka, D. Rasmussen, S. Trubacova, Y . Okamoto, P. Gonzales, J. Stigler, and R. Gallimore. 2006. *Highlights from the TIMSS 1999 video study of eighth-grade science teaching* (NCES 2006-017). U.S. Department of Education, National Center for Education Statistics. Washington, DC: U.S. Government Printing Office.

Roth, K. J., C. Chen, M. Lemmens, H. E. Garnier, N. I. Z. Wickler, L. J. Atkins, A. C. Barton, J. E. Roseman, A. W. Shouse, and C. Zembal-Saul. 2009. Coherence and science content storylines in science teaching: Evidence of neglect? Evidence of effect? NARST 2009 Symposium Papers.

NATIONAL SCIENCE TEACHERS ASSOCIATION

Rowell, P. 1997. Learning in school science: The promises and practices of writing. *Studies in Science Education* 30 (1): 19–56.

Russ, R. S., and M. G. Sherin. 2013. Using interviews to explore student ideas in science. *Science Scope* 36 (5): 19–23.

Sadler, D. R. 1989. Formative assessment: Revisiting the territory. *Assessment in Education: Principles, Policy, and Practice* 5 (1): 77–84.

Sampson, V., and S. Schleigh. 2013. *Scientific argumentation in biology: 30 classroom activities.* Arlington, VA: NSTA Press.

Sandoval, W. A. 2003. Conceptual and epistemic aspects of students' scientific explanations. *The Journal of the Learning Sciences* 12 (1): 5–51.

Scherer, M. 2011. Transforming education with technology: A conversation with Karen Cator. *Educational Leadership* 68 (5): 17–21.

Scholastic. 2010. *Scholastic 2010 kids & family reading report: Turning the page in the digital age. www.scholastic.com/readingreport*

Schwendimann, B. A. 2008. Scaffolding interactive dynamic model to promote coherent connections in high school biology. Paper presented at the annual meeting of the American Educational Research Association, New York.

Scott, E. C. 2013. This I believe: We need to understand evolution, adaptation, and phenotype. *www.frontiersin.org*

Scott, P., H. Asoko, and R. Driver. 1992. Teaching for conceptual change: A review of strategies. In *Research in physics learning: Theoretical issues and empirical studies,* ed. R. Duit, F. Goldberg, and H. Niederer, 310–329. Kiel, Germany: IPN.

Scott, V. G., and M. K. Weishaar. 2008. Talking drawings as a university classroom assessment technique. *The Journal of Effective Teaching* 8 (1): 42–51.

Sejnost, R. L., and S. M. Thiese. 2010. *Building content literacy: Strategies for the adolescent learner.* Thousand Oaks, CA: Corwin Press.

Sinan, O., H. Aydin, and K. Gezer. 2007. Prospective science teachers' conceptual understanding about proteins and protein synthesis. *Journal of Applied Sciences* 7 (21): 3154–3166.

Singer, S., M. Hilton, and H. Schweingruber. 2007. *America's lab report: Investigations in high school science.* Washington, DC: National Academies Press.

Slotta, J. D., and M. T. H. Chi. 2006. The impact of ontology training on conceptual change: Helping students understand the challenging topics in science. *Cognition and Instruction* 24 (2): 261–289.

Smith, J. 2012. Natural selection. PhET Interactive Simulations. *http://phet.colorado.edu/en/simulation/natural-selection*

Smith, J. P., A. A. diSessa, and J. Roschelle. 1993. Misconception reconceived: A constructivist analysis of knowledge in transition. *The Journal of the Learning Science* 3: 115–163.

Smith, S. M. 2003. A cross-age study of students' conceptual understanding of interdependency in seed dispersal, pollination, and food chains using a constructivist theoretical framework. *http://repository.lib.ncsu.edu/ir/bitstream/1840.16/5188/1/etd.pdf*

Southwest Center for Education and the Natural Environment (SCENE). 2004. *The inquiry process.* SCENE. *http://scene.asu.edu/habitat/inquiry.html*

Southeast Comprehensive Center (SCC). 2012. *Using formative assessment to improve student achievement in the core content areas.* Austin, TX: SEDL.

Special issue on diversity and equity in science education. 2013. *Theory into Practice* 52 (1).

Spektor-Levy, O., B. Eylon, and E. Scherz. 2009. Teaching scientific communication skills in science studies: Does it make a difference? *International Journal of Science Education* 7: 875–903.

Stage, E. K., H. Asturias, T. Cheuk, P. A. Daro, and S. B. Hampton. 2013. Science education: Opportunities and challenges in next generation standards. *Science* (340): 276–277.

Stewart, J., J. L. Cartier and C. M. Passmore. 2005. Developing understanding through model-based inquiry. In *How students learn*, ed. M. S. Donovan and J. D. Bransford, 515–565. Washington, DC: National Academies Press.

Stewart, J., and M. Dale. 1989. High school students' understanding of chromosome/gene behavior during meiosis. *Science Education* 73 (4): 501–521.

Stein, M., and S. McNair. 2002. Science drawings as a tool for analyzing conceptual understanding. Paper presented at the annual meeting of the Association for the Education of Teachers of Science, Charlotte, NC.

Stow, W. 1997. Concept mapping as a tool for self-assessment? *Primary Science Review* 49: 12–15.

Strike, K., and G. Posner. 1985. A conceptual change view of learning and understanding. In *Cognitive structure and conceptual change*, ed. L. West and A. Pines, 6. Orlando, FL: Academic Press.

Sutherland, L. M., K. L. McNeill, and J. S. Krajcik. 2006. Supporting middle school students in developing scientific explanations. In *Linking science and literacy in the K–8 classroom*, ed. R. Douglas and K. Worth, with W. Binder, 163–181. Arlington, VA: NSTA Press.

Swain, J., and P. Lawrence. 1994. Learning about disability: Changing attitudes or challenging understanding? In *On equal terms: Working with disabled people*, ed. S. French, 87–102. Oxford: Butterworth Heinemann.

Tanner, K., and D. Allen. 2005. Approaches to biology teaching and learning: Understanding the wrong answers—Teaching toward conceptual change. *Cell Biology Education* 4 (2): 112–117.

TERC. 2013. *Science by design: Construct a boat, catapult, glove and greenhouse*. Arlington, VA: NSTA Press.

Texley, J., and A. Wild. 2004. *NSTA pathways to the science standards, high school edition*. 2nd ed. Arlington, VA: NSTA Press.

The Concord Consortium. 2013. Perspective: Improving STEM education with *Next Generation Science Standards*. *@Concord* 17 (1): 2–3.

Thier, M. 2002. *The new science literacy: Using language skills to help students learn science*. Portsmouth, NH: Heinemann.

Trautmann, N. M., J. Fee, T. M. Tomasek, and N. R. Bergey. 2013. *Citizen science: 15 lessons that bring biology to life, 6–12*. Arlington, VA: NSTA Press.

Trautmann, N. M., J. L. Shirk, J. Fee, and M. E. Krasny. 2012. Who poses the question? Using citizen science to help K–12 teachers meet the mandate for inquiry. In *Citizen science: Public participation in environmental research*, ed. J. L. Dickinson and R. Bonney, 179–190. Ithaca, NY: Comstock Publishing Associates.

Trunfio, P., B. Berenfeld, P. Kreikemeier, J. Moran, and S. Moodley. 2003. Molecular modeling and visualization tools in science education. Paper presented at the annual conference of the National Association for Research in Science Teaching, Philadelphia, PA.

NATIONAL SCIENCE TEACHERS ASSOCIATION

Tweed, A. 2009. *Designing effective science instruction: What works in science classrooms.* Arlington, VA: NSTA Press.

University of California Museum of Paleontology. 2013. *Understanding Evolution. http://evolution.berkeley.edu/evolibrary/misconceptions_faq.php*

University of California Museum of Paleontology (UCMP). *Thomas Malthus (1766–1834). www.ucmp.berkeley.edu/history/malthus.html*

University of Wisconsin. Bottle biology instructional materials development program. *www.bottlebiology.org*

Urquhart, V., and M. McIver. 2005. *Teaching writing in the content areas.* Alexandria, VA: Association for Supervision and Curriculum Devemopment.

Urquhart, V., and D. Frazee. 2012. *Teaching reading in the content areas: If not me, then who?* 3rd ed. Alexandria, VA: Association for Supervision and Curriculum Devemopment.

U.S. Department of Agriculture, U.S. Agricultural Research Service. *www.nps.ars.usda.gov*

U.S. Environmental Protection Agency. *www.epa.gov*

Varelas, M., and C. C. Pappas. 2006. Intertextuality in read-alouds of integrated science-literacy units in primary classrooms: Opportunities for the development of thought and language. *Cognition and Instruction* 24 (2): 211–259.

Vasquez, J., M. W. Comer, and F. Troutman. 2010. *Developing visual literacy in science, K–8.* Arlington, VA: NSTA Press.

Vasquez, J. A., C. Sneider, and M. Comer. 2013. *STEM lesson essentials: Integrating science, technology, engineering, and mathematics, grades 3–8.* Portsmouth, NH: Heinemann.

Venville, G., S. J. Gribble, and J. Donovan. 2005. An exploration of young children's understandings of genetics concepts from ontological and epistemological perspectives. *Science Education* 89 (4): 614–633.

Vitale, M. R., N. R. Romance, and M. F. Dolan. 2006. A knowledge-based framework for the classroom assessment of student understanding. In *Assessment in science: Practical experiences and education research,* ed. M. McMahon, P. Simmons, R. Sommers, D. DeBaets, and F. Crawley, 1–13. Arlington, VA: NSTA Press.

Walsh, B. *Time* 2013. The plight of the honeybee: Mass deaths in bee colonies mean disaster for farmers, and your favorite foods. August 19: 24–31.

Walsh, J. A., and B. D. Sattes. 2005. *Quality questioning: Research-based practice to engage every learner.* Thousand Oaks, CA: Corwin Press.

Wang, M. C., G. D. Haertel, and H. J. Walberg. 1993/1994. What helps students learn? *Educational Leadership* 51 (4): 74–79.

WatchKnowLearn. *WatchKnowLearn.org*

Weiss, I. R., J. D. Pasley, P. S. Smith, E. R. Banilower, and D. J. Heck. 2003. Looking inside the classroom: A study of K–12 mathematics and science education in the United States. Horizon Research International, Inside the Classroom. *www.horizon-research.com/insidetheclassroom/reports/looking*

Westcott, D. J., and D. L. Cunningham. 2005. Recognizing student misconceptions about science and evolution. *Mountain Rise Electronic Journal* 2 (2) (Spring/Summer).

Wiggins, G. P. 1998. *Educative assessment: Designing assessments to inform and improve student performance.* San Francisco: Jossey-Bass.

Wiggins, G., and J. McTighe. 1998. *Understanding by design.* Alexandria, VA: ASCD.

Williams, M., A. H. DeBarger, B. L. Montgomery, X. Zhou and E. Tate. 2012. Exploring middle school students' conceptions of the relationship between genetic inheritance and cell division. *Science Education* 96 (1): 78–103.

Wilson, E. O., ed. 1988. *Biodiversity.* Washington DC: National Academies Press.

Wilson, E. O. 2002. *The future of life.* New York: Vintage Books.

Windschitl, M. 2008. What is inquiry? A framework for thinking about authentic scientific practice in the classroom. In *Science as inquiry in the secondary setting,* ed. J. Luft, R.L. Bell, and J. Gess-Newsome, 1–20. Arlington, VA: NSTA Press.

Windschitl, M., and J. J. Thompson. 2013. The modeling toolkit: Making student thinking visible with public representations. *The Science Teacher* 80 (6): 63–69.

Windschitl, M., J. Thompson, and M. Braaten. 2008. How novice science teachers appropriate epistemic discourses around model-based inquiry for use in classrooms. *Cognition and Instruction* 26 (3): 310–378.

Windschitl, M., J. Thompson, M. Braaten, and D. Stroupe. 2012. Proposing a core set of instructional practices and tools for teachers of science. *Science Education* 96 (5): 878–903.

Wolfe, P. 2001. *Brain matters: Translating research into classroom practice.* Alexandria, VA: Association for Supervision and Curriculum Devemopment.

Woodruff, E., and K. Meyer. 1997. Explanations from intra- and inter-group discourse: Students building knowledge in the science classroom. *Research in Science Education* 27 (1): 25–39.

Wormeli, R. 2009. *Metaphors and analogies: Power tools for teaching any subject.* Portland, ME: Stenhouse Publishers.

Wright, A. W., and K. Bilica. 2007. Instructional tools to probe biology students' prior understanding. *American Biology Teacher* 69 (1): 1–5.

Young, P. n.d. Visual thinking tools. San Diego State University's Encyclopedia of Educational Technology. *http://coe.sdsu.edu/eet/articles/VisThinkTools/start.htm*

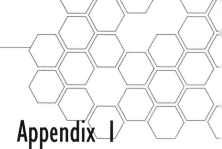

Glossary of Acronyms

AAAS	American Association for the Advancement of Science
BSCS 5E	Biological Sciences Curriculum Study 5E (Engage, Explore, Explain, Elaborate, Evaluate)
CCCs	Crosscutting Concepts
CCD	Colony Collapse Disorder
CCSS	Common Core State Standards
CCSS-ELA	Common Core State Standards—English Language Arts
CCSS-M	Common Core State Standards—Mathematics
C-U-E	Content-Understanding-Environment
DCIs	Disciplinary Core Ideas
ELA	English Language Arts
FAR Guide	FAR = Focus, Action, and Reflection
KWLs	What we **K**now, what we **W**ant to find out and what we **L**earned and still need to Learn graphic organizer
NARST	National Association for Research in Science Teaching
NGSS	Next Generation Science Standards
NOS	Nature of Science
NRC	National Research Council
NSTA	National Science Teachers Association
PEs	Performance Expectations
PhET	Physics Education Technology (now simulations across disciplines)
PLAN	Predict; Locate; Add; Note
SEPs	Science and Engineering Practices
SWH	Science Writing Heuristic

Note: The abbreviations used in the instructional tools for the practices are described in each of the tools.

- Asking questions and defining problems (QP)
- Developing and using models (M)
- Planning and carrying out investigations (I)
- Analyzing and interpreting data (D)
- Using mathematics and computational thinking (MCT)
- Constructing explanations and designing solutions (ES)
- Engaging in argument from evidence (AE)
- Obtaining, evaluating, and communicating information (OEC)

Hints and Resources for the Design Process

Stage I
Hint: It is important that you understand the unit content. Consider these resources.
Resources
• *A Framework for Science Education* (NRC 2012) should be considered a companion document to the *Next Generation Science Standards* and carefully reviewed to ensure your full understanding of the DCIs, practices, and crosscutting concepts.
• *Curriculum Topic* Study 2nd Edition: *Linking Common STEM Standards and Research on Learning* (Keeley, Tugel and Gabler, forthcoming.) provides a systematic strategy that begins with topics and connects standards and research with curriculum, instruction, and assessment.
• *Science Matters: Achieving Science Literacy* (Hazen and Trefil 1991) provides accessible explanations of core science concepts.
• National Science Teachers Association (NSTA) Learning Center resources (*http://learningcenter.nsta.org/?lid=lnavhp*) include various texts and online resources that target your particular learning needs.
Hint: Working with standards—not the textbook—is the starting place for instruction. Review the *Next Generation Science Standards* as well as your district and state standards.
Resources
• Next Generation Standards site: *www.nextgenscience.org/next-generation-science-standards*
• NGSS@NSTA site: *www.nsta.org/about/standardsupdate/standards.aspx*
Stage II
Hint: It is important that you create a learning sequence that builds on students' prior knowledge, is chunked into learnable pieces, and establishes a clear storyline leading to the essential understandings. Consider these resources.
Resources
• Begin with the life science disciplinary core idea progressions in the *Next Generation Science Standards* (*www.nextgenscience.org/sites/ngss/files/Appendix%20E%20-%20Progressions%20within%20NGSS%20-%20052213_0.pdf*).
• Refer to *A Framework for Science Education* (NRC 2012) to further study the crosscutting concepts and practices.
• Study the strand maps in the two volumes of the Atlas for Science Literacy (*http://strandmaps.nsdl.org*).

Hint: Some concepts and vocabulary in the standards are mandatory if students are to develop essential understandings. Others are not. Everything taught in an introductory biology course is not appropriate for an introductory high school course. These resources should help you separate the two and prune unnecessary concepts and vocabulary.

Resources

- Study the assessment boundaries and clarification statements for the *NGSS* performance expectations that align with your selected DCIs.
- Review *Designs for Science Literacy* (AAAS 2001b), which includes a process for and examples of how to "prune" the content in the standards.
- Chapter 2 of *Designing Effective Science Instruction: What Works in Science Classrooms* (Tweed 2009) includes a nice description, with helpful examples, of how to distinguish between mandatory and non-mandatory concepts and vocabulary.

Stage III

Hint: Develop criteria using the various facets of understanding used in "backwards design." The work by Grant Wiggins and Jay McTighe includes several valuable resources that make it clear that essential understandings must come first in the design process.

Resources

- *Understanding by Design* (Wiggins and McTighe 1998)
- *The Understanding by Design Handbook* (McTighe and Wiggins 1999)
- *The Understanding by Design Professional Development Workbook* (McTighe and Wiggins 2004)

Hint: The criteria you develop can be used to create formative and summative assessments as well as rubrics. In addition to the work by Wiggins and McTighe listed above, some excellent science-specific resources follow.

Resources

- Begin with chapter 10 of *A Framework for K–12 Science Education* (NRC 2012), which includes guidelines for implementation—curriculum, instruction, teacher development, and assessment.
- *Assessment and Inquiry-Based Science Education: Issues in Policy and Practice* (Harlen 2013)
- *Assessing Student Understanding in Science,* 2nd ed. (Enger and Yager 2009)
- *Assessment for Learning: Putting It Into Practice* (Black et al. 2003)
- *Assessment in Science: Practical Experiences and Education Research* (McMahon et al. 2006)
- *Everyday Assessment in the Science Classroom* (Atkin and Coffey 2003)

Stage V
Hint: Metacognitive abilities are best improved when taught in conjunction with content learning (cognitive focus). Carefully select strategies that align well with your students' needs, the time of year, and the particular content you are teaching.

Resources

- The three metacognitive approaches used in this book are drawn from Marzano's (1992) work *A Different Kind of Classroom: Teaching with Dimensions of Learning.* This is a wonderful resource to extend your learning about ways to improve your students' metacognitive abilities.

- The Visible Thinking website (*www.visiblethinkingpz.org*) provides excellent tools and ideas that promote metacognition.

Other good critical- and creative-thinking websites include the following:

- *www.engin.umich.edu/~problemsolving/strategy/crit-n-creat.htm*

- *http://members.optusnet.com.au/~charles57/Creative/Techniques/index.html*

- www.mindtools.com/pages/main/newMN_CT.htm

Stage VI
Hint: Research helps us determine common misconceptions that our students have. Since these are naïve or preconceptions that are developed prior to instruction, adults often have the same misconceptions. Your identification of these misconceptions will help you better understand the content and design lessons that target these misconceptions in your students.

Resources

- The National Science Digital Library (NSDL) has science literacy maps and resources online. These present the strand maps developed by Project 2061 (AAAS 2001b, 2007). If you have these resources in book form, you can use them. If not, visit the NSDL site. Go to *http://strandmaps.nsdl.org* and select a topic. For the work we did in Chapter 3, you would select "The Living Environment" and then "DNA and Inherited Characteristics." In the upper left-hand corner, you will find a link that says "View Student Misconceptions." You can also select "The Human Organism" and then "Disease" and find related misconceptions.

- An excellent resource is *Making Sense of Secondary Science: Research into Children's Ideas* (Driver et al. 1994). This book provides a very nice summary of various science topics and the related research. Note that although the word *secondary* is used in the title, this resource outlines misconceptions commonly held by students in grades 3–5 and above.

- It's worth trying to use Google Scholar (*scholar.google.com*) for misconceptions. For example, you can search for "DNA and misconceptions," "proteins and misconceptions," or "genes proteins and misconceptions."

- University research databases provide access to the most recent research.

Hint: Resources exist that will help you directly target specific misconceptions and provide good instructional ideas as you begin to plan your lessons.

Resources

- *Making Sense of Secondary Science: Research Into Children's Ideas* (Driver et al. 1994) is the classic work on this topic.
- The *Uncovering Student Ideas in Science* series by Page Keeley and her colleagues includes numerous probes that help you determine and explore your students' ideas about many science concepts. This series is available through NSTA Press (*www.nsta. org/publications/press/uncovering.aspx*).

Stage VII

Hint: Select a strategy to determine your students' preconceptions about *all* the learning targets. Use the strategy with students well ahead of instruction and then use their responses to guide unit development. Later, for each learning target, you can use strategies to elicit preconceptions related specifically to that target.

Resources

- The Instructional Tools are your primary resources. Within each tool you will find specific resources for the strategies. Study those resources to learn more about use of the strategy and to find materials that can be readily used in your classroom. For example, text and online resources for concept cartoons are listed in the Resources section as well as on page 136 in Instructional Tool 3.8.

Stages VIII and IX

Hint: There are numerous strategies for eliciting and confronting preconceptions and for sense making. Remember that your selection of strategies is influenced by the particular content you are teaching, where you are in the school year, your metacognitive focus, and the targeted practices and crosscutting concepts. Try to use a large variety of strategies over the course of the year so that you address various learning styles and avoid repetition in lessons (repetition is sure to diminish student engagement).

Resources

- *Tips for the Science Teacher: Research-based Strategies to Help Students Learn* (Hartman and Glasgow 2002) translates research into practical and easy-to-use classroom applications.
- *Classroom Instruction That Works: Research-based Strategies for Increasing Student Achievement, 2nd Ed.* (Dean, Pitler, Ross, and Stone 2012) outlines instruction strategies to use to raise student achievement.
- *http://education.jhu.edu/newhorizons/strategies/index.html* is a website that provides resources about numerous strategies, including differentiation strategies.
- *http://sydney.edu.au/science/uniserve_science/school/support/strategy.html* is an online resource for various science-teaching strategies.
- *www.muskingum.edu/~cal/database/general* provides general learning strategies as well as strategies specific to content areas, including science.
- *Differentiated Instructional Strategies for Science, Grades K–8* (Gregory and Hammerman 2008) provides a blueprint (including differentiation strategies) for improving science instruction while accommodating individual learning styles. This is a helpful resource, even if written to target the K–8 population.

Hint: There are some resources that focus specifically on graphic organizers.

Resources:

- *A Field Guide to Using Visual Tools* (Hyerle 2000) is an excellent introduction to the use of visual tools for learning.
- The Graphic Organizer Website (*www.graphic.org*) includes a rich set of resources on many different types of graphic organizers.
- The SCORE website (*www.sdcoe.k12.ca.us/SCORE/actbank/torganiz.htm*) includes both a teacher activity bank and a student activity bank about graphic organizers.
- The Education Oasis website provides PDFs of 58 different graphic organizers (*www.educationoasis.com/curriculum/graphic_organizers.htm*).
- *Developing Visual Literacy in Science, K–8* (Vasquez, Comer, and Troutman 2010), in addition to discussing and displaying graphic organizers, uses photographs, illustrations, diagrams, and student-created visual thinking tools to demonstrate how students can be taught to be better "readers" of graphic images. This also is useful, even though it targets the K–8 population.

Stage X

Hints: Formative assessments are assessments *for student learning.* They are not to be used to determine if students have the right or wrong answer but rather to determine their understanding of a concept. Formative assessments can be planned and organized or they can be quick assessments (e.g., you ask a student a question or you glance at a student's written work). You use formative assessments to modify your instruction to ensure that every student learns. Many of the strategies in the Instructional Tools can also serve as formative assessments.

Resources

- *Science Formative Assessment: 75 Practical Strategies for Linking Assessment, Instruction, and Learning* (Keeley 2008) includes specific techniques that help science teachers determine students' understanding of important ideas and design learning opportunities to meet students' needs.
- The growing NSTA Press *Uncovering Student Ideas* series provide a variety of (a) formative assessment probes you can use to determine your students' preconceptions of fundamental science concepts and (b) suggestions to address students' ideas and promote learning. See information on all books at *www.nsta.org/publications/press/uncovering.aspx*.

Planning Template for Proteins and Genes Unit

Unit Topic—Proteins and Genes

Phase 1. Identifying Essential Content		
Conceptual Target Development	Disciplinary Core Ideas Addressed	• Systems of specialized cells within organisms help them perform the essential functions of life • All cells contain genetic information in the form of DNA molecules. Genes are regions in the DNA that contain the instructions that code for the formation of proteins, which carry out most of the work of cells. • Multicellular organisms have a hierarchical structural organization, in which any one system is made of numerous parts and is itself a component of the next level.
	Crosscutting Concepts Addressed	• Models (e.g., physical, mathematical, computers models) can be used to simulate systems and interactions—including energy, matter, and information flows—within and between systems at different scales. • Investigating or designing new systems or structures requires a detailed examination of the properties of materials, the structures of different components, and connections of components to reveal the function and/or solve a problem.
	Science and Engineering Practices Addressed	• Develop and use a model based on evidence to illustrate the relationship between systems or between components of a system. • Construct and revise an explanation based on valid and reliable evidence obtained from a variety of sources (including students' own investigations, models, theories, simulations, peer review) and the assumption that theories and laws that describe the natural world operate today as they did in the past and will continue to do so in the future.

Essential Understandings	Organisms are systems made of dynamic and complex subsystems of interacting molecules in cells, tissues, and organs (levels of organization). Proteins are molecules that carry out the major biological processes of cells and impact the functioning of the tissues and organs they build. If they are not made properly the entire organisms can be affected. Changes (mutations) in the gene/DNA can impact not only the genetic code but also the protein, cells, tissues and organs, since various components of the system interact and depend on each other. Though some mutations can be helpful, they might also stop or limit the protein's ability to function and potentially lead to physiological disorder in the entire organism.
Criteria to Determine Understanding	• Develop and explain a model that illustrates the production of a protein and its action across levels of organization (cell, tissue, organ, organism), resulting in a particular phenotype. • Construct an explanation for how DNA coding determines the structure of a protein. • Construct and refine an explanation of how amino acid sequence determines the shape of a protein and thus its function. • Refine the model developed in Learning Target #1 to demonstrate the impact of a mutation on a phenotype/trait, including a discussion of the protein's role in the process. • Accurately and carefully develop explanations and models, providing evidence for each.
Performance Expectations Addressed	• Construct an explanation based on evidence for how the structure of DNA determines the structure of proteins, which carry out the essential functions of life through systems of specialized cells. • Develop and use a model to illustrate the hierarchical organization of interacting systems that provide specific functions within multicellular organisms.

Phase 2. Planning for Responsive Action	
Identifying Student Preconceptions	Use an anticipation guide (see Figure 3.4, p. 73).

Learning Sequence Targets	
Learning Target #1	Proteins carry out the major work of cells and are responsible for both the structure and function of organisms. These proteins are made based on the code found in the organism's DNA (genotype) and result in the organism' traits (phenotype). How well the cellular system functions and interacts among cells and at the various levels of organization impacts the entire organism.
Research-Identified Misconceptions Addressed	• Students demonstrate confusion over levels of organization, particularly with cells and molecules. They tend to think of molecules as related to the physical sciences and cells to life science. Some students even think that proteins are made of cells and that molecules of protein are bigger than cells (Driver et al. 1994). • The majority of upper division biology students and future science teachers recognize the physical constitution of an organism as its phenotype, yet do not understand the role of genes and proteins in producing the phenotype (Elrod n.d.). Because some students do not connect genes to proteins to phenotypes (Lewis and Wood-Robinson 2000), they assume that genes directly express traits in organisms (Lewis and Kattmann 2004). • Though students usually equate genes with traits, they do not understand that genes code for specific proteins and that the production of these proteins results in the traits (Friedrichsen and Stone 2004).

Initial Instructional Plan	*Eliciting Preconceptions:* Facilitate a brainstorming session using the probe, "What are proteins and why are they important?" This can be as a whole-class or in small groups (but share as a whole class after small-group work). You can use one of the brainstorming webs or use the understanding routine, Think/Puzzle/Explore, found at the Visible Thinking website. During the class discussion, elicit ideas about various proteins and introduce the concept that missing or ill-functioning genes can significantly impact protein function and, in many cases, cause disease. The goal is to ensure that students understand the robust and important functions of proteins. End the discussion by generating student questions. *Confronting Preconceptions:* Show the YouTube video "Protein Functions in the Body" (see Recommended Resources at the end of this chapter). After a brief class discussion, have students revisit and revise their brainstorming webs. Facilitate another class discussion about the question, "What would happen if there was a missing or ill-functioning protein?" show the YouTube video "Sickle Cell" (see Recommended Resources at the end of this chapter) and then have students again modify their brainstorming webs. *Sense Making:* Provide student groups a list of various proteins and the associated disorders that can arise if the protein is missing or malfunctioning. Have each group choose a disorder. Their task is to research the protein and disorder to determine the impact of the missing or malfunctioning protein on the various levels of organization (cells, tissues, organs, organisms). A resource that students can use as a starting point is a summary of disorders at the *Your Genes, Your Health website* (www.ygyh.org). Review with students the use of two-column notes and system diagrams before they begin research. Facilitate their use during small-group work, probing for student understanding of both concepts and use of the tools. Require students to develop individual explanations in their learning logs, share and critique their explanations (as a group), and prepare a presentation to share with the entire class. Support students during individual learning log summary development and during presentation preparation. After student presentations, facilitate a whole-class discussion, summarizing key ideas, and generating questions. Finally, ask groups to develop a single systems map that generalizes what they learned about proteins and genes from the various presentations.

Formative Assessment Plan *(Demonstrating Understanding)*	• Teacher review of brainstorming webs • Teacher review of learning logs, including two-column notes, and systems diagram • Analysis of systems diagram for inclusion of key ideas • Questions and probes during small-group work and class discussion • Final presentation critique using pre-established rubrics *Note:* Specific suggestions for questions, probes and feedback are provided in Chapter 4 in the section on formative assessment (pp. 142–149).
Learning Target #2	Genetic information (genes) coded in DNA provides the information necessary to assemble proteins. The sequence of subunits (nucleotides) in DNA determines the sequence of amino acid in proteins.
Research-Identified Misconceptions Addressed	• Some students think a gene is a trait or that the DNA makes proteins (Elrod n.d.). • Less than half of upper-division biology students and future science teachers understand the nature of the genetic code (Elrod nd). Only 22% of undergraduate students with some biological science course work defined the gene in terms of nucleotide sequences involved in protein synthesis (Chattopadhyay and Mahajan 2006). • Students often think that genes code for more than proteins. They also often think that the genes code for information at multiple levels of organization (e.g., the gene "tells" a tissue or organ to malfunction), which bypasses the need for students to provide a mechanistic explanation of molecular genetics phenomena (Duncan and Reiser 2005). • 30–52% of upper-division biology students and future science teachers do not recognize RNA as the product of transcription. 50–75% of introductory biology and genetics students in college and future science teachers do not identify proteins as the product of translation (Elrod nd; Fisher 1985).

Initial Instructional Plan	*Eliciting Preconceptions:* Begin with the idea that all the work students just completed was about genetic disorders. Ask them how genes are related to the proteins just studied. Provide students with a list of the essential vocabulary terms and ask them to use all the terms in the list to write their best explanations of the relationship. Share in small groups and then discuss selected explanations as a whole class. Next share an animation that provides an overview of protein synthesis (see Resources, below). Have small groups discuss and map connections to previous understandings, interpretations of the overview, and any questions they have. Facilitate discussions in small groups and then as whole class, focusing on the ideas in the learning target. (*Note:* If students have never written scientific explanations, provide a model – for example Claims, Evidence and Reasoning – and opportunities to develop explanations prior to beginning or in conjunction with this activity.)

Confronting Preconceptions: Have students read about protein synthesis from their text or other print resource and modify their maps and explanations based on what they learn. Next, have them complete a shockwave activity available at the DNA Workshop site, choosing the protein synthesis option (see Resources, below) or one of the activities at the Concord Consortium's Molecular Logic Project (see Resources, below). Note that students are not responsible for all vocabulary used in either resource, unless this is expected in your school district. The idea is to provide visual images of the connection between genes and proteins.

Sense Making: Have students revisit their cluster maps and systems diagrams from Learning Target #1, as well as their explanations and questions, revising based on new experiences. Assign students sections of text (print/electronic sources) on protein synthesis to read expanding their two-column notes from Learning Target #1. Once again, have them modify maps, diagrams, and explanations. Facilitate the work of each group with probes and questions. Then use student questions as guides for class discussion. |
| Formative Assessment Plan (*Demonstrating Understanding*) | • Use questions and probes during small-group work and class discussion.

• Analyze student explanations using a rubric.

• Analyze student cluster maps (peer critique of maps can also be included).

• Analyze student system diagrams.

• Use student questions to indicate ongoing areas of misunderstanding and elicit ideas using questions as a guide for class discussion. |

Learning Target #3	The sequence of amino acids determines not only the kind, but also the shape of the protein, and thus its function.
Research-Identified Misconceptions Addressed	Because students are not aware that proteins play a role that is central to living things (most/all genetic phenomena are mediated by proteins) and robust (many functions), it hampers their ability to provide mechanistic explanations of genetic phenomena (Duncan and Reiser 2005).
Initial Instructional Plan	*Eliciting Preconceptions:* Go to *http://molo.concord.org/database/activities/76.html* and show "How a Protein Gets Its Shape: The Role of Charge." Run the "original chain" as a demonstration, also showing how the simulation works. Share your mental model of the process, thinking out loud. Have students run the additional demonstrations (individual or with a partner). Then have them discuss, in small groups or as partnered problem solving, why they think protein shape is important. Share ideas with the whole class and facilitate a discussion.
	Confronting Preconceptions: Go to *http://molo.concord.org/database/activities/225.html* and have students complete "Protein Folding: Stepping Stones Full Interactive," screens #1–#4. As they complete the activities, they generate a report that is submitted at the end of the activity. However, this will not be complete until after Learning Target #4.
	Sense Making: Model for students how to develop an analogy for protein synthesis. Make sure to identify the concept and the analog, discuss how the two are similar and how they are different, and reflect on the effectiveness of the analogy. What about it was effective and what was confusing? "Building a house" is one analog for protein synthesis, recommended by Harrison and Coll (2008). They outline an application of the FAR Guide in their book (see Resources, below). Once you model this, have students identify and develop an analogy.
Formative Assessment Plan *(Demonstrating Understanding)*	• Use questions and probes during small-group work and class discussion.
	• Even though the student reports for the protein-folding interaction are not complete until the completion of Learning Target #4, you can review student work online during the interactions and use this information to gauge level of understanding.
	• Do critiques of student analogies by self, peers, and teacher, using the FAR Guide if available.

Hard-to-Teach Biology Concepts, Revised 2nd Edition

Learning Target #4	Mutations, changes in the DNA, impact protein production. Errors in the DNA (mutation) can result in missing proteins or ones that function inadequately. Since actions in the cell impact the other levels of organization in the organism, changes in the DNA and proteins can result in a change in phenotype/trait.
Research-Identified Misconceptions Addressed	• Though 80% of undergraduates knew that a disease could be linked to a gene, only 35% correctly represented a flow diagram between the genes and disease. Even if they could explain the concept of the central dogma, they could not extrapolate their understanding to a real-life situation (Chattopadhyay and Majahan 2006). • Most students are unable to explain a situation where a change to the DNA sequence does not change the protein sequence (Eklund et al. 2007).
Initial Instructional Plan	*Eliciting Preconceptions:* Have students continue the protein-folding activity started in Learning Target #3. It now introduces mutation, focusing on sickle-cell anemia. Students are asked to predict the impact of a change to the DNA sequence before completing the simulation. Have them make this prediction (claim) and share their explanations as a class. Recall that scientific explanations require students to make claims, provide evidence for their claims, and link the claims to the evidence. *Confronting Preconceptions:* Have students complete screens #5–#8. They can now complete and generate their report. They can also complete the activity, Modeling Mutations, further described at *www.concord.org* *Sense Making:* Then have students refer to their textbooks or other print resources to research other types of mutations and their potential impact. They should connect this information back to their original research on their selected disorder. They should also revise their systems diagrams from Learning Target #1 and their explanations and analogies developed during Learning Target #3 to include what they learned in this part of the lesson.
Formative Assessment Plan (Demonstrating Understanding)	• Use questions and probes during small-group work and class discussion. • Ask for student reports generated by protein-folding interaction • Do critiques of scientific explanations by students and teacher or use the "Explanation Analysis" strategy (Keeley 2008) • Do critiques of student analogies by self, peers, and teacher, using the FAR Guide if available • Do final review of systems diagrams begun in Learning Target #1 • Re-administer the anticipation guide used to identify student preconceptions

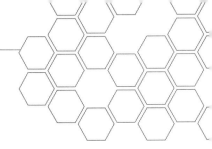

Index

*Page numbers in **boldface** type refer to tables, figures, or Instructional Tools.*

A

A Different Kind of Classroom: Teaching with Dimensions of Learning, 313
A Field Guide to Using Visual Tools, **118, 133,** 314
A Framework for K–12 Science Education: Practices, Crosscutting Concepts, and Core Ideas, ix, 4, 14, 20, 40, 48, 174, 201, 230, 260, 311
    on argumentation, 224
    Common Core State Standards and, 150
    conceptual shifts in, 25
    diverse learners and, 158
    integrating three dimensions of, xii, 13, 16, 21, 24, 25, 150
    learning progressions in, 28
A Private Universe: Minds of Our Own, xiv
Accountable talk, **110, 112**
Acronyms, 309
Adaptation. *See* The Role of Adaptation in Biological Evolution unit
AdLit.Org website, **103, 106.108**
Alsos Digital LIbrary for Nuclear Issues, **107**
Alternative conceptions of students, 18. *See also* Misconceptions of students; Preconceptions of students
American Association for the Advancement of Science (AAAS), 52
Analogies, **68, 70, 121–122**
Animations, **53, 70, 122, 125, 139,** 165, **182,** 183, 187, 188, 322
Anticipation guide, 72–73, **105**
    for Matter and Energy in Organisms and Ecosystems unit, 181
    for Proteins and Genes unit, 73, **73,** 81, **82,** 319, 324
Argumentation, xi, 14, 17, 35, **99, 111,** 152–153, 199, **201,** 224, 260
Assessing Science Learning: Perspectives from Research and Practice, 166
Assessing Student Understanding in Science, 2nd ed., 312
Assessment, x, 14
    formative, 14, 16, 142–148 (*See also* Probes)
        benefits of, 142
        curriculum-embedded, 143, 145
        definition of, 142–143
        determining responsive actions based on, 78–81
        for Ecosystems: Interactions, Energy, and Dynamics unit, 226
        hints and resources for, 315
        in Instructional Planning Framework, 3, **7,** 10, **12**
        for Matter and Energy in Organisms and Ecosystems unit, 184, **184**
        on-the-fly, 143, 146
        planned-for, **36–37,** 143, 146
        for Proteins and Genes unit, 54–55, 145–149,

        321, 322, 323, 324
        putting into practice, 143
        strategies and rationales for, **79**
        for Variations of Traits unit, 253
    high-stakes tests, x, xi
    hints and resources for, 166, 312
    peer- and self-assessment, **12,** 18, 35, 61–62, 67, 78, **79,** 144–145
    preassessment for The Role of Adaptation in Biological Evolution unit, 270–271, **271**
    summative, 14, 16, **23,** 143, 256, 270, 272, 279, **312**
        for Proteins and Genes unit, 54, 55
Assessment and Inquiry-Based Science Education: Issues in Policy and Practice, 312
Assessment for Learning: Putting It Into Practice, 166, 312
Assessment in Science: Practical Experiences and Education Research, 312
Atlas for Science Literacy, 311
Authentic science, **96, 110,** 157, 192, 193

B

Backwards design process, 24, 54, 312
Benchmarks for Science Literacy, 211
Big ideas, 13, 14, 38, 54–55, 172, 174, 181, 186, 237
Biodiversity in ecosystems, x, 34, 190–227. *See also* Ecosystems: Interactions, Energy, and Dynamics unit
    definition of, 194–195
    misconceptions about, 195–196, **196,** 211, **214–215**
Biological literacy, x, 13, 268. *See also* Science literacy
Black, P., 92, 144, 167
Blakey, E., **61**
Brainstorming webs, **68,** 72, **80, 127**
    for Ecosystems: Interactions, Energy, and Dynamics unit, 212
    formative assessments based on, 146, 147, 148, **184,** 321
    for Matter and Energy in Organisms and Ecosystems unit, **184**
    for Proteins and Genes unit, 81, **84,** 86, 163, 320, 321
    thinking-process maps and, **131, 132**
    for Variation of Traits unit, 248
Branch, G., 257
Bransford, J., 4
Braun, G., 248
Brody, M. J., 211
BSCS 5E Instructional Model, 10, **11–12, 23**
Bybee, Rodger, 20, 44, 154

C

C-U-E lesson framework, 35
Calhoun, E., 142

Campbell, T., **120**
Careers in science, 25, 27, 107, 155, 156, 231
CCs. *See* Crosscutting concepts
Cellular respiration, 4, **29**, 30, 172–188, **173**. *See also*
    Matter and Energy in Organisms and Ecosystems unit
Challenges in teaching biology, x–xi
Chemical Heritage Foundation, **107**
Churchill, Winston, ix
Circle maps, 72, **127, 128, 132, 133**
Citizen Science: 15 Lessons That Bring Biology to Life,
    K–12, 228
Citizen science projects, 190–191, **191**, 220, **222**, 228
Classroom Instruction That Works: Research-Based
    Strategies for Increasing Student Achievement, 2nd ed.,
    314
Climate Change From Pole to Pole, 227
Clustering, 72, 81, **127, 128**, 212, 216
Colony Collapse Disorder (CCD) in honeybees, 190–226,
    222. *See also* Ecosystems: Interactions, Energy, and
    Dynamics unit
Comer, M., 154
Comic Creator, **136**
ComicLife, **136**
Common Core State Standards (CCSS), English Language
    Arts and *Mathematics,* xvi, 20
        NGSS and, 25, 27, 35, **116**, 142, 150–153
            implications for Proteins and Genes unit, 153–154
            research on, 150–153
Communication strategies, **67, 111, 115–118**
        kinesthetic, **137, 138**
        technology and, **118**, 156, 157, 158, 161
The Comprehension ToolKit: Language and Lessons for
    Active Literacy, **106**
Computations Science Education Reference Desk, **124**
Concept cartoons, **58, 68, 75, 76, 79, 110, 111, 135–136**
        for Variations of Traits unit, 247, 251, **251**, 252
Concept Cartoons in Science Education, **136**
Concept maps, **58, 131, 132**, 163
        for Matter and Energy in Organisms and Ecosystems
            unit, 180, 183
        for Variation of Traits unit, **247**, 248
Conceptual change, xii–xiii, xv
        addressed in *A Framework for K–12 Science*
            Education, 4
        definition of, 5
        Instructional Planning Framework for addressing,
            3–18
        process of, 5–6
Conceptual Change Model, 10, **11–12**
Conceptual shifts in *NGSS,* 25–27
Concord Consortium resources, 90, 91, 254
        Molecular Logic Project, 91, 322
        Molecular Workbench, 90, **126**
Content Clips, **128, 133**
Content storyline, 14, 24, 27, **39**, 40, 44
        description of, 52

for Ecosystems: Interactions, Energy, and Dynamics
    unit, 190, 202, **205–206**
for implementation of Instructional Planning
    Framework, 40
for Matter and Energy in Organisms and Ecosystems
    unit, 177–178
for photosynthesis, 29–31, **30**
for Proteins and Genes unit, 52, **53**
for The Role of Adaptation in Biological Evolution
    unit, 263, **264–266**
for Variations of Traits unit, 238–240
Course maps, 38, 46
Creative thinking and learning, 18, 56, **62**, 127, 209, **210**, 313
Critical thinking and learning, 13, 18, 56, **57–58**, 63, 313
Crosscutting concepts (CCs), ix, x, 24
        for Ecosystems: Interactions, Energy, and Dynamics
            unit, 190, 197, **198–200**, 202, **205–206**, 220–221
        and implementation of Instructional Planning
            Framework, 40
        incorporation of, 26–27
        in Instructional Planning Framework, 3, **7**, 8
        integrating with other dimensions of *NGSS,* xii, 13,
            16, 21, 24, 25, 32, 150
        for Matter and Energy in Organisms and Ecosystems
            unit, 174, **175, 179**
        performance expectations for, 19, **22**
        for Proteins and Genes unit, **47**, 48, 52, **53, 317**
        for The Role of Adaptation in Biological Evolution
            unit, **258**, 259, **265–266**
        for Variations of Traits unit, 235, 240
Curriculum maps, 21, 26, 38
Curriculum revision, ix, 20
Curriculum Topic Study, 2nd edition: *Linking Common*
    STEM Standards and Research on Learning, 311

D

Darwin's theory of evolution, 257. *See also* The Role of
    Adaptation in Biological Evolution unit
DCIs. *See* Disciplinary core ideas
Dean, C. B., **122**
Debrief the thinking process, **61**
Demonstrating understanding, **67–68**
Designing Effective Science Instruction: What Works in
    Science Classrooms, 35, 312
Designs for Science Literacy, 312
Developing Assessments for the Next Generation Science
    Standards, 230
Developing Visual Literacy in Science, K–8, 315
Diagrams, 6, **70**, **123–125**, **131**
        concept, 72, **105**
        cycle, 130
        for Ecosystems: Interactions, Energy, and Dynamics
            unit, **213, 219**
        Frayer, **219**
        for Matter and Energy in Organisms and Ecosystems
            unit, **179**, 181, **182**, 183, **184**, 185, 186, 188

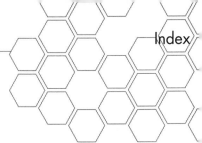

for Proteins and Genes unit, 320, 321, 322, 324
systems, **75,** 77, **79, 80, 84,** 87–88, **125, 131, 132, 133,** 143, 146, 149, 320, 321, 324
for Variation of Traits unit, 253
Venn, 150
Venn-Euler, **129, 130**
Differentiated Instructional Strategies for Science, Grades K–8, 314
Digital libraries, **107,** 313
Disciplinary core ideas (DCIs), ix, x, xi, 4, 24
curriculum mapping for, 38
for Ecosystems: Interactions, Energy, and Dynamics unit, 190, 196–197, **198–200**
deconstruction of, 201–202, **203–204**
and implementation of Instructional Planning Framework, 40
incorporation of, 26–27
in Instructional Planning Framework, 3, **7,** 8
integrating with other dimensions of *NGSS,* xii, 13, 16, 21, 24, 25, 32, 150
learning progressions for, 28–32, **29**
for Matter and Energy in Organisms and Ecosystems unit, 174–177, **175**
deconstruction of, 177–178
performance expectations for, 19, **22**
for Proteins and Genes unit, 46–48, **47,** 317
deconstruction of, 49–51, **50, 51**
pruning of, **50,** 51–52
for The Role of Adaptation in Biological Evolution unit, 257–258, **258–259**
deconstruction of, 260–263, **261–262**
for Variation of Traits unit, 231, 232–233
deconstruction of, 235–238
Discourse, xi
small- and large-group, **75, 110–112,** 216, 218, **219,** 280
Diverse science learners, 158–165
NGSS and, 158–160
Proteins and Genes unit for, 163–165
Universal Design for Learning and, xvi, 160–162, **164**
DNA. *See* Proteins and Genes unit; Variations of Traits unit
DNA Workshop website, 91
Dobzhansky, T., 256
Donovan, S., 4
Drawing Out Thinking (Instructional Tool 3.8), **68, 134–136,** 181, 251, **314**
Drawings, **68,** 69, **70,** 72, 74, **80, 96, 127, 131, 134–135.** *See also* Posters
annotated, **68,** 72, 74, **75, 80, 94, 134,** 181
before-during-after, **124**
for Ecosystems: Interactions, Energy, and Dynamics unit, **213,** 216, **219**
for Matter and Energy in Organisms and Ecosystems unit, 181
for The Role of Adaptation in Biological Evolution unit, 272

"talking," **135**
technical, **97**
Driver, R., 211
Drosophila Virtual Lab, 166
Duncan, Ravit, 230
Dynamic models, **68, 76,** 89, **125–126, 275**

E
E + S: Integrating Engineering + Science in Your Classroom, 167
ebird, 228
Ecosystems: Interactions, Energy, and Dynamics—An Issue-Based Approach unit, 189–228
alignment with *NGSS,* **33,** 190, 192
application of Phase 1 to, 196–209
stage I: identify DCIs, SEPs, and CCs, 196–200, **198–201**
stage II: deconstruct DCIs, create storyline, and align SEPs and CCs, 201–202, **203–206**
stage III: determine PEs and identify criteria to determine student understanding, 202, **207–208**
stage IV: determine NOS connections, 202, 209
stage V: identify metacognitive goals and strategies, 209, **210**
application of Phase 2 to, 210–226
stage VI: research student misconceptions, 210–212, **213–215**
stage VII: identify students' preconceptions, 212, 216
stage VIII: elicit and confront students' preconceptions, 216–218, **217, 219**
stage IX: determine sense-making strategies, 218–226
stage X: determine responsive actions based on formative assessment, 226
content storyline for, 190, 202, **205–206**
overview of: honeybees and Colony Collapse Disorder, 190–192
personal experience and student tips for implementation of, 192–194
real-world issues and resources for, **222,** 226–228
relevance of topic for, 194–196, **196**
teacher reflection on, 226
vs. traditional teaching approach, 32–34
Education Oasis website, 315
Engineering education, xiii, 4, 154–157. *See also* STEM education
in Ecosystems: Interactions, Energy, and Dynamics unit, 199–200
in *NGSS,* 25, 27
in Proteins and Genes unit, 157–158
resources for, 167
for students with diverse needs, 158–159
Engineering practices. *See* Science and engineering practices

Environmental Literacy Council, 211

Essential Questions: Opening Doors to Student Understanding, **115**

Essential understandings, 8, **11,** 13, 24, 28, 29, **30**

Evaluate Alternatives Instructional Model, 224–225

Everyday Assessment in the Science Classroom, 312

Everyday Engineering: Putting the E in STEM Teaching and Learning, 167

Everyday Science Mysteries series, **100**

Evolution and Nature of Science Institutes (ENSI/SENSI), 284

Evolutionary biology, x, 24, 89, 255–285. *See also* The Role of Adaptation in Biological Evolution unit

Explanation Analysis, 90, 324

F

Feedback to students, 4, 9, 10, 143–145

    form and content of, 144

    recommendations for, 144–145

FieldNotes LT, **139,** 166

Fixsen, Dean, 26

Focus, Action, and Reflection (FAR) Guide, 89, 91, 92, **122,** 323, 324

Forest Watch, 228

Formative assessment, 14, 16, 142–148

    benefits of, 142

    curriculum-embedded, 143, 145

    definition of, 142–143

    determining responsive actions based on, 78–81

    for Ecosystems: Interactions, Energy, and Dynamics unit, 226

    hints and resources for, 315

    in Instructional Planning Framework, 3, **7,** 10, **12**

    for Matter and Energy in Organisms and Ecosystems unit, 184, **184**

    on-the-fly, 143, 146

    planned-for, **36–37,** 143, 146

    for Proteins and Genes unit, 54–55, 145–149, 321, 322, 323, 324

    putting into practice, 143

    strategies and rationales for, **79**

    for Variations of Traits unit, 253

Frayer Model of vocabulary development, **103,** 163, 218, **219,** 276, **277**

FreeMind, **128**

G

Gallery walk, 216, **219,** 248

Genetic engineering, x, 157, **215**

Genetics. *See* Proteins and Genes unit; Variations of Traits unit

Genetics Science Learning Center, 254

Global Rivers Environmental Education Network (GREEN), 228

Glogster, 165

Google Scholar, 313

Google SketchUp, **124**

Graphic Organizer website, 315

Graphic organizers, **58, 60, 62,** 69, **70, 79, 105**

    categorical, 77, **130**

    comparison or relational, **130**

    descriptive, **130**

    for Ecosystems: Interactions, Energy, and Dynamics unit, 218, 223

    for Matter and Energy in Organisms and Ecosystems unit, 180

    problem-solution, **130**

    process or cause-and-effect, **130**

    for Proteins and Genes unit, 163

    resources for, **131,** 315

    for The Role of Adaptation in Biological Evolution unit, **275,** 282, 283

    sequential, **130**

    Speaking to Learn and, **111, 118**

    task-specific, **129–131**

    thinking-process maps and, **131, 132**

Graphs, 6, **70, 96, 119–120, 123–125**

    for Ecosystems: Interactions, Energy, and Dynamics unit, 194, 202, **205, 207, 208,** 221, 223

    for The Role of Adaptation in Biological Evolution unit, 282

H

Hard-to-teach biology concepts, ix–xvi, 4–6, **6**

Hardy-Weinberg equation, 282

Harlen, W., 143, 144

Harrison, C., 92, 167

Hazen, R. M., 194

Heredity. *See* Proteins and Genes unit; Variations of Traits unit

Heritage, Margaret, 14, 28, 143, 145

High-stakes tests, x, xi

Hints and resources for the design process, 311–315

Honeybees and Colony Collapse Disorder, 190–226, **191, 222.** *See also* Ecosystems: Interactions, Energy, and Dynamics unit

How People Learn: Brain, Mind, Experience, and School, xiv, 31

How Students Learn: Science in the Classroom, xiv, 15

Hubbell, E. R., **122**

Hyerle, David, **133**

I

iLabCentral, 166

iMind Map, **128**

Information Services and Technology (IST), 156–157

Informational text strategies, **67,** 72, 74, 77, **80,** 87, **104–106**

    for The Role of Adaptation in Biological Evolution unit, **275,** 280, 281

Inquiry, xii, xiii, 10, 16, 27, 44

    citizen science projects and, 228

concept cartoons and, **136**
for Ecosystems: Interactions, Energy, an Dynamics unit, 192, 193
guided, **100, 102**
kinesthetic strategies and, **137, 138**
linguistic representations of knowledge and, 69
    communication, **115–116**
    discussion, **111,** 159
    questioning, **113, 114, 115**
    reading, **102, 104**
for Matter and Energy in Organisms and Ecosystems unit, **182, 183,** 187
metacognitive approaches and, **57, 61**
nature of science and, 55, 56, 269
for The Role of Adaptation in Biological Evolution unit, 269
scientific explanations and, **75, 98, 99**
for Variations of Traits unit, 247
Inspiration software, **59, 103, 115, 128, 133**
Instructional approach(es), ix, xv, xvi, 8, 13
    vs. instructional strategies, 66
Instructional Planning Framework, ix, xi, 3–18, **7,** 287
    compared with 5E Instructional Model and Conceptual Change Model, 10, **11–12**
    connecting to *NGSS,* 17, 24–25
    formative assessment and, 144
    lesson design process aligned with *NGSS* and, 35, **36–37**
    Phase 1: identifying essential content, 7–8, 13–15, 18, 40–41
        for Ecosystems: Interactions, Energy, and Dynamics unit, 196–209
        for Matter and Energy in Organisms and Ecosystems unit, 172–180
        for Proteins and Genes unit, 45–63
        for The Role of Adaptation in Biological Evolution unit, 257–275
        for Variations of Traits unit, 232–242
    Phase 2: planning for responsive action, 8–10, 15–17, 41–42
        for Ecosystems: Interactions, Energy, and Dynamics unit, 210–226
        for Matter and Energy in Organisms and Ecosystems, 180–184
        for Proteins and Genes unit, 63–81
        for The Role of Adaptation in Biological Evolution unit, 276–284
        for Variations of Traits unit, 243–253
    reflections on use of, 287–289
    research basis of, 10–17
Instructional Planning Framework implementation, 43–139, 287
    design process for, 38–42, **39**
        hints and resources for, 311–315
    for Ecosystems: Interactions, Energy, and Dynamics unit, 189–228

Instructional Tools for, 10, 16, 44, 57–139
for Matter and Energy in Organisms and Ecosystems unit, 171–188
overview of, 44–45
for Proteins and Genes unit, 44–92
for The Role of Adaptation in Biological Evolution unit, 255–285
for Variations of Traits unit, 229–254
Instructional strategies, xi, xiv, 5, 16
    vs. instructional approach, 66
Instructional Strategy Selection Tool (Instructional Tool 3.2), 15, 66, **67–68,** 69, 72, 74, 76, 78, 90, 212, 221
Instructional Tools, 10, 16, 44, 57–139, 287
    connections to *NGSS,* 17 (*See also specific Instructional Tools*)
    Drawing Out Thinking (3.8), **68, 134–136,** 181, 251, **314**
    Instructional Strategy Selection Tool (3.2), 15, 66, **67–68,** 69, 72, 74, 76, 78, 90, 212, 221
    Sense-Making Approaches: Linguistic Representations—Reading to Learn (3.4), **67,** 69, 77, **102–109**
    Sense-Making Approaches: Linguistic Representations—Speaking to Learn (3.5), **67,** 69, **110–118**
    Sense-Making Approaches: Linguistic Representations—Writing to Learn (3.3), **67,** 69, 74, **93–101,** 221
    Sense-Making Approaches: Nonlinguistic Representations—Kinesthetic Strategies (3.9), **68,** 69, **70, 76, 137–139**
    Sense-Making Approaches: Nonlinguistic Representations—Six Kinds of Models (3.6), **68,** 69, **70,** 89, **119–126,** 253
    Sense-Making Approaches: Nonlinguistic Representations—Visual Tools (3.7), **58, 68, 70,** 77, **127–133,** 212, 218, 248
    Three Approaches That Support Metacognition (3.1), **57–62**
Invasive Species, 227
Investigating and Questioning Our World through Science and Technology (IQWST), 230
Investigations, x, xv, 14, 16

J
Jones, John Stephen, 230
Journaling, 18, **61, 62, 93, 94, 95, 109**
Journey North, 228
Joyce, B., 142

K
Kattmann, U., **66**
Keeley, Page, 59, 90, **94, 95, 100**
Kinesthetic activities, 69, **70, 137–139,** 280, 282
Kinesthetic Strategies (Instructional Tool 3.9), **68,** 69, **70, 76, 137–139**

Hard-to-Teach Biology Concepts, Revised 2nd Edition

Koba, Susan, 256
Konicek-Moran, R., **100**
Krajcik, J. S., 90
KWL charts, **58**, 72, **105**
 for The Role of Adaptation in Biological Evolution
  unit, 272, **275**, 279
 for Variation of Traits unit, 248

L

LabShare, **126**, 166
Lamarckian thinking, **265, 271**, 276, **277**, 280, 281
Language literacy, xi, 69, 72, 150–152. *See also* Vocabulary
 development
 Common Core State Standards for English language
  arts, 25, 27, 150–151
 for Proteins and Genes unit, 153–154
 Reading to Learn (Instructional Tool 3.4), **67**, 69, 77,
  102–109
 Speaking to Learn (Instructional Tool 3.5), **67**, 69,
  110–118
 Writing to Learn (Instructional Tool 3.3), **67**, 69, 74,
  93–101, 221
Learning
 challenges for, xiv–xv, 5
 cognitive, 13, **75, 124, 134**
 collaborative, xv, 162
 conceptual, xxxi, xxxii, **120**
 conceptual change for, 5–7
 creative thinking and, 18, 56, **62**, 127, 209, **210**, 313
 critical thinking and, 13, 18, 56, **57–58**, 63, 313
 environment for, xiv, 41
 experiential, 27, 192, 30
 fact-based, xv, 4, 16
 kinesthetic, **138–139**
 metacognitive, **75, 120**, 143 (*See also* Metacognitive
  goals and strategies)
 motivation for, 5
 self-regulated thinking and, 18, 41, 56, **58–61**, 180,
  242, 270, **275**
*Learning, Creating, and Using Knowledge: Concept Maps
 as Facilitative Tools in Schools and Corporations,* **133**
Learning goals, 10, 13–15, 17, 24, 26
 for *NGSS*, 20–21, **23**, 32
Learning logs, **67**, 74, 77, **84, 93–94**, 109
 for Proteins and Genes unit, 146, 149, 320, 321
Learning progressions, 14, 17, 24, 26, 28–32
Learning sequence(s), 8, 10, 14, 15, **23**, 26, 28
 hints and resources for creating, 311–312
 for implementation of Instructional Planning
  Framework, 40
 NGSS curriculum mapping for, 21, 26, 38
 for photosynthesis, 29–31, **30**
 planned vs. implemented, **145**, 145–146
 for Proteins and Genes unit, **51**
 for The Role of Adaptation in Biological Evolution
  unit, **264**

Learning targets, 3, **8**, 13, 14, 30–31
 for Ecosystems: Interactions, Energy, and Dynamics
  unit, 202, **203–206**
  criteria to determine student understanding of,
   202, **207**
 for implementation of Instructional Planning
  Framework, 40
 for Matter and Energy in Organisms and Ecosystems
  unit, 177, **179**
  criteria to determine student understanding of,
   179
  formative assessments and responsive actions
   for, **184**
  strategies and activities for, **182**
 for Proteins and Genes unit, 49–51, **50, 51, 53**
  criteria to determine student understanding of,
   54–55, 318
  Strategy Selection Template for, 71, **71, 80**
 for The Role of Adaptation in Biological Evolution
  unit, **264–266**
  criteria to determine student understanding of,
   267, **267**
  Strategy Selection Template for, **274–275**
  study guide for, 272, **273**
 in Unit Planning Template, **36–37**
 for Variations of Traits unit, 235–238
  criteria to determine student understanding of,
   240–241
Lesson planning, 8, 17, 21, 35, 78, 160, 163, 267. *See also*
 Unit planning
Lesson Plans, Inc., 138, 139, **138, 139**
Lewis, J., **66**
Library of Congress, **107**
Linguistic representations of knowledge, 69
 Reading to Learn (Instructional Tool 3.4), **67**, 69, 77,
  102–109
 Speaking to Learn (Instructional Tool 3.5), **67**, 69,
  110–118
 Writing to Learn (Instructional Tool 3.3), **67**, 69, 74,
  93–101, 221

M

*Making Sense of Secondary Science: Research Into
 Children's Ideas,* 313, 314
Malthus, Thomas, 223
Maps, 69, **70**
Marzano, R., 313
Math: Stop Faking It!, **120**
Mathematical models, 6, **68, 70, 119–120**, 153, **201**
 for Ecosystems: Interactions, Energy, and Dynamics
  unit, 221, 223, 224
Mathematics, xiii, 154–155. *See also* STEM education
 Common Core State Standards for, 25, 27, 150–151
 in Ecosystems: Interactions, Energy, and Dynamics
  unit, 197, **201**, 202, 221, 223, 224
 NGSS and standards for, 150–153

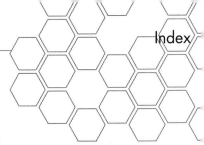

in Proteins and Genes unit, 153–154
in The Role of Adaptation in Biological Evolution unit, 282–283, 284
Matter and Energy in Organisms and Ecosystems unit, 171–188
  application of Phase 1 to, 172–180
    stage I: identify DCIs, SEPs, and CCs, 174–177, **175**
    stage II: deconstruct DCIs, create storyline, and align SEPs and CCs, 177–178
    stage III: determine PEs and identify criteria to determine student understanding, 178, **179**
    stage IV: determine NOS connections, 178–180
    stage V: identify metacognitive goals and strategies, 180
  application of Phase 2 to, 180–184
    stage VI: research student misconceptions, 180–181
    stage VII: identify students' preconceptions, 181, **182**
    stage VIII: elicit and confront students' preconceptions, 181–183, **183**
    stage IX: determine sense-making strategies, 183–184
    stage X: determine responsive actions based on formative assessment, 184, **184**
  content storyline for, 177–178
  resources for, 188
  teacher instructions for, 185–188
    final assessment, 188
    lesson 1, 185–186
    lesson 2, 186
    lesson 3, 186–187
    lesson 4, 187–188
McIver, M., **94**
McNeill, K. L., 90
McTighe, Jay, 23, 312
Mead, L. W., 257
Meaning Making in Secondary Science Classrooms, **112**
Mendeley app, 166
MendelWeb, **107, 108**
Metacognitive goals and strategies, xvi, 9, **11,** 13, 14, 18, 41
  for Ecosystems: Interactions, Energy, and Dynamics unit, 209, **210**
  hints and resources for, 313
  for Matter and Energy in Organisms and Ecosystems unit, 180
  for Proteins and Genes unit, 55–56, 63
  for The Role of Adaptation in Biological Evolution unit, 270–272
  Three Approaches That Support Metacognition (Instructional Tool 3.1), **57–62**
  for Variations of Traits unit, 242
Metaphors, **68, 70, 122–123,** 253
Metaphors and Analogies: Power Tools for Teaching Any Subject, **122, 123**

Michaels, Sarah, 254
Mind mapping, 72, **127, 128**
Misconceptions of students, xiii, xv, **6.** *See also* Preconceptions of students
  definition of, 18
  identification of, 9
    hints and resources for, 313–314
  related to Ecosystems: Interactions, Energy, and Dynamics unit, 210–212, **213–215**
    about biodiversity, 195–196, **196,** 211, **214–215**
  related to Matter and Energy in Organisms and Ecosystems unit, 179–180
  related to Proteins and Genes unit, 63–66, **64–66,** 319, 321, 323
  related to The Role of Adaptation in Biological Evolution unit, 276, **277–278**
  related to Variations of Traits unit, 243–245
Model-It, **125, 133**
The Modeling Tookkit, **124**
Models
  dynamic, **68, 76,** 89, **125–126, 275**
  mathematical, 6, **68, 70, 119–120,** 153, **201**
    for Ecosystems: Interactions, Energy, and Dynamics unit, 221, 223, 224
  physical, **68, 70, 120–121, 137**
  Six Kinds of Models (Instructional Tool 3.6), **68,** 69, **70,** 89, **119–126,** 253
  verbal
    analogies, **68, 70, 121–122**
    metaphors, **68, 70, 122–123,** 253
  visual, **68, 70, 123–125, 134,** 253
Models-Based Science Teaching, **121, 122, 123, 124**
Molecular Logic Project, 91, 322
Molecular Workbench, 90, **126**
Monarch Larva Monitoring Project, 228
Moore, John A., ix

N

Naïve conceptions of students, 18. *See also* Misconceptions of students; Preconceptions of students
Nanoscale Science: Activities for Grades 6–12, 91
Nanotechnology, 155–156
Narrative text strategies, **67, 106–108**
National Association of Biology Teachers (NABT), 256, 284
National Center for Science Education (NCSE), 284
National Evolution Synthesis Center (NESCent), 284
National Research Council (NRC), 154–155, 230, 231
National Science Digital Library (NSDL), **107–108, 128, 133,** 195, **277,** 311, 313
National Science Education Standards (NSES), 32, **116,** 194, 211
National Science Foundation, 166
National Science Teachers Association (NSTA) resources, 21, 63, 91, **100, 106,** 166–167, 226, 284, 311, 314, 315
Natural selection. *See* The Role of Adaptation in Biological Evolution unit

Nature of science (NOS) connections, xiii, 9, 13, **22**, 25, 27, 32, 35
    for Ecosystems: Interactions, Energy, and Dynamics unit, 202, 209
    and implementation of Instructional Planning Framework, 40–41
    for Matter and Energy in Organisms and Ecosystems unit, 178–180
    for Proteins and Genes unit, 55
    for The Role of Adaptation in Biological Evolution unit, 269–270
    teaching about, 26
    for Variations of Traits unit, 241–242
Neilson, D., **120**
Next Generation Science Standards (NGSS), ix–xii, xv–xvi, 4, 8, 19–42, 174, 230
    Appendix A of, 25
    Appendix D of, 158
    Appendix E of, 28, **29**
    Appendix F of, 40
    Appendix G of, 40
    Appendix H of, 40–41, 178
    Appendix K of, 38
    Appendix L of, 153
    Appendix M of, 152
    Common Core State Standards and, 25, 27, 150–153
        implications for Proteins and Genes unit, 153–154
    conceptual shifts in, 25–27
    connecting Instructional Planning Framework with, 17, 24–25
    connecting Instructional Tools with, 17
    course scope and sequence implications of, 38
    diverse learners and, 158–165, **164**
    foundation boxes in, **22**, 23, 24, 32
    integrating three dimensions of, xii, 13, 16, 21, 24, 25, 150
    interpretation of, 32–35, **33**
    learning goals for, 20–21, **23**, 32
    learning progressions in, 14, 24, 26, 28–32
    lesson design process aligned with Instructional Planning Framework and, 35, **36–37**
    The NSTA Reader's Guide to design process and, 21–25, **23**
    overview of, 20–21
    performance expectations for, 8, 19, **22**, 23, 26, 32, **33**, 34–35, 40
    purposes of, 20–21
    state adoption of, x, xiii, xx, 20
    website for, 311
The NMC Horizon Report: 2013 K–12 Edition, 167
Nonlinguistic representations of knowledge, **68**, 69, **70**
    Drawing Out Thinking (Instructional Tool 3.8), **68**, **134–136**, 181, 251, **314**
    Kinesthetic Strategies (Instructional Tool 3.9), **68**, 69, **70, 76, 137–139**

Six Kinds of Models (Instructional Tool 3.6), **68**, 69, **70**, 89, **119–126**, 253
Visual Tools (Instructional Tool 3.7), **58, 68, 70**, 77, **127–133**, 212, 218, 248
NOS. *See* Nature of science connections
NoteStar, **61**
The NSTA Reader's Guide to the Next Generation Science Standards, 20, 21–25, **23**
NYU-Poly Virtual Lab, **126**, 166

O
O'Connor, Cathy, 254
Oh, P. S., **120**
Open-Ended Questioning: A Handbook for Educators, **94, 112, 115**
Overcoming Textbook Fatigue: 21st Century Tools to Revitalize Teaching and Learning, **103**

P
Paired problem solving, **59, 114**
Passmore, C., **124**
Peer- and self-assessment, **12**, 18, 35, 61–62, 67, 78, **79**, 144–145
Pellegrino, J., 4
Performance expectations (PEs), 8, 19, **22**, 23, 26, 32, **33**, 34–35
    for Ecosystems: Interactions, Energy, and Dynamics unit, 202, **208**
    and implementation of Instructional Planning Framework, 40
    for Matter and Energy in Organisms and Ecosystems unit, 174–176, **175, 179**
    for Proteins and Genes unit, 53–54, 318
    for The Role of Adaptation in Biological Evolution unit, 267–268, **267–268**
    for Variations of Traits unit, 232–234, **233**, 237, 240–241
Photosynthesis, xiv, **29, 61, 107, 130**
    cellular respiration and, 172–176, **175, 182**, 185, 188
    concept cartoon for, **136**
    learning sequence for, 29–31, **30**
Physical models, **68, 70, 120–121, 137**
Physical movements, **68, 79, 137, 138–139, 275**
Pictures, **70, 123–125**. *See also* Drawings
Pitler, H., **122**
Posters, 88, **137**, 153
    gallery walk of, 216, **219**, 248
    interactive, 165
Pratt, Harold, 20
Preconceptions of students, xiv, xv–xvi, 4, 5–6, **6**. *See also* Misconceptions of students
    definition of, 17, 18
    desired learning goals and, 14
    eliciting and confronting of, 3, **7**, 9, **11**, 16, 41
        for Ecosystems: Interactions, Energy, and Dynamics unit, 216–218, **217, 219**

hints and resources for, 314–315
for Matter and Energy in Organisms and
Ecosystems unit, 181–183, **183**
for Proteins and Genes unit, 74, 320, 322, 323,
324
for The Role of Adaptation in Biological Evolution
unit, 280–281
strategies for, **67–68,** 74, **75–76**
for Variations of Traits unit, 247–252
identification of, 3, **7,** 9, **11,** 15, 41
for Ecosystems: Interactions, Energy, and
Dynamics unit, 212, 216
hints and resources for, 314
for Matter and Energy in Organisms and
Ecosystems unit, 181, **182**
for Proteins and Genes unit, 72–74, **73,** 319, 322,
323, 324
for The Role of Adaptation in Biological Evolution
unit, 279
strategies for, **67–68,** 72–73
for Variations of Traits unit, 243–245
sense-making strategies for addressing, 3, **7,** 9,
41–42
Probes, 16, **67, 75, 95, 111, 116.** *See also* Formative
assessment
concept maps and, **132**
for Ecosystems: Interactions, Energy, and Dynamics
unit, 216–218, **217,** 226
formative assessment and, 143, 146, 226
for Proteins and Genes unit, **80,** 81, **84, 94, 95,** 320,
321, 322, 323, 324
for The Role of Adaptation in Biological Evolution
unit, **275, 277, 278,** 280, 281
simulations as, **125**
in *Uncovering Student Ideas in Science* series, **59,**
91, **94, 95, 100,** 166, 226, 314, 315
Probeware, **137, 138,** 156
Project-based science, 247
Project BudBurst, 228
ProjectGLOBE, 228
Proteins and Genes unit, xvi, xx, 13, 44–92
application of Phase 1 to, **45,** 45–63, 317–318
stage I: identify DCIs, SEPs, and CCs, 46–49, **47**
stage II: deconstruct DCIs, create storyline, and
align SEPs and CCs, 49–54, **50, 51, 53**
stage III: determine PEs and identify criteria to
determine student understanding, 54–55
stage IV: determine NOS connections, 55
stage V: identify metacognitive goals and
strategies, 55–63, **56–62**
application of Phase 2 to, 63–81, 319–324
identifying strategies for Phase 2 planning, 66–
71, **67–68, 70–71**
stage VI: research student misconceptions,
63–66, **64–66**
stage VII: identify students' preconceptions,

67–68, 72–74, **73**
stage VIII: elicit and confront students'
preconceptions, 74, **75–76,** 81
stage IX: determine sense-making strategies,
76–78
stage X: determine responsive actions based
on formative assessment evidence, 78–81,
79–80
content storyline for, 52, **53**
for diverse learners, 163–165
formative assessments for, 54–55, 145–149
lessons for learning targets #1–4, 46, 81–90
learning target #1, 81–88, **82–84**
learning targets #2–4, 88–90
sample cluster map, **85**
planning template for, 317–324
reflection on design process for, 90
relevance of topic for, 45
resources for, 90–92
STEM education and, 157–158
Strategy Selection Template for, 71, **71, 80**
Punnett squares, 243, 253

Q

*Quality Questioning: Research-Based Practice to Engage
Every Learner,* **112, 115**
Questioning by students, 35, **67, 84, 104, 113–115,** 322
Questioning tree, **114, 115**
Questions, Claims, and Evidence, **100**

R

*Reading, Writing, & Inquiry in the Science Classroom
Grades 6–12,* **106, 108**
Reading Educator website, **103**
Reading to Learn (Instructional Tool 3.4), **67,** 69, 77,
102–109
*Realizing the Promise of 21st-Century Education: An
Owner's Manual,* 167
Reciprocal teaching, **59–60, 105, 114,** 153
Reflection strategies, **67, 108–109**
Reflections on use of Instructional Planning Framework,
287–289
Reiser, Brian J., 230
The Role of Adaptation in Biological Evolution unit, 255–285
application of Phase 1 to, 257–278
stage I: identify DCIs, SEPs, and CCs, 257–260,
258–259
stage II: deconstruct DCIs, create storyline, and
align SEPs and CCs, 260–263, **261–266**
stage III: determine PEs and identify criteria to
determine student understanding, 267–268,
267–268
stage IV: determine NOS connections, 269–270
stage V: identify metacognitive goals and
strategies, 270–272, **271, 273–275**
application of Phase 2 to, 279–284

stage VI: research student misconceptions, 276, **277–278**

stage VII: identify students' preconceptions, 279

stage VIII: elicit and confront students' preconceptions, 280–281

stage IX: determine sense-making strategies, 281–284

content storyline for The, 263, **264–266**

relevance of topic for, 256–257

resources for, 284–285

Strategy Selection Template for, **274–275**

Rolheiser, C., **62**

Ross, J. A., **62**

Roth, Kathy, 14, 31, 52, 192

Rubrics, 40, **60, 84,** 90, 146, 147, 148–149, 312, 321, 322

S

Sadler, D. R., 143

Sampson, V., 152, 224, 283–284

Scaffolding learning, 16

formative assessment and, 162

graphic organizers for, 77, **129**

models for, 89, **123,** 148

for Proteins and Genes unit, **64,** 66

reading strategies for, **105, 108**

for The Role of Adaptation in Biological Evolution unit, 267, 281, 284

for self-regulation, **60**

for writing scientific explanations, **98, 99**

Schleigh, S., 152, 224, 283–284

Schmitzer, H., 248

Science and engineering practices (SEPs), ix, x, 4, 16, **22,** 24

for Ecosystems: Interactions, Energy, and Dynamics unit, 190, 192, **193,** 197–199, **198–199, 201,** 202, **205–206,** 220–221

and implementation of Instructional Planning Framework, 40

incorporation of, 25–27

in Instructional Planning Framework, 3, **7,** 8

integrating with other dimensions of *NGSS,* xii, 13, 16, 21, 24, 25, 32, 150

leaner experiences from socio-scientific issues and, 192, **193**

as learning goals, 32

for Matter and Energy in Organisms and Ecosystems unit, 174, **175, 179**

nature of science connections with, 41

performance expectations for, 19, **22**

for Proteins and Genes unit, **47,** 48–49, 317, 52, **53**

for The Role of Adaptation in Biological Evolution unit, **258–259,** 260, **265–266**

for Variations of Traits unit, 234–235, 240

Science by Design: Construct a Boat, Catapult, Glove, and Greenhouse, 167

Science education reform, xi–xiii

Science Formative Assessment: 75 Practical Strategies for Linking Assessment, Instruction, and Learning, **62,** 91, **135, 136,** 166, 315

Science Inside the Black Box: Assessment for Learning in the Science Classroom, 92, 167

Science literacy, 21, 27, 52, 69, 98, 102, 192, 230–231. *See also* Biological literacy

Science Literacy Map, 195

Science Matters: Achieving Science Literacy, 311

Science notebooks, **67,** 77, **96–97, 124,** 183, 221, 226

interactive, **118**

Science Notebooks: Writing About Inquiry, 97

Science Writing Heuristic (SWH), **67, 75,** 77, **100–101**

Scientific explanations, xiv, 5, 8, **75,** 77, 86, 155

for Ecosystems: Interactions, Energy, and Dynamics unit, 220, 221, 223

for Proteins and Genes unit, 51, 322, 324

for The Role of Adaptation in Biological Evolution unit, 269, **275**

writing, **98–100**

SCORE website, 315

Self-assessment, **12,** 18, 35, 61–62, 67, 78, **79,** 144–145

Self-evaluation strategies, **59, 61–62,** 242

Self-regulated thinking and learning, 18, 41, 56, **58–61,** 180, 242, 270, **275**

Sense-Making Approaches: Linguistic Representations—Reading to Learn (Instructional Tool 3.4), **67,** 69, 77, **102–109**

Sense-Making Approaches: Linguistic Representations—Speaking to Learn (Instructional Tool 3.5), **67,** 69, **110–118**

Sense-Making Approaches: Linguistic Representations—Writing to Learn (Instructional Tool 3.3), **67,** 69, 74, **93–101,** 221

Sense-Making Approaches: Nonlinguistic Representations—Kinesthetic Strategies (Instructional Tool 3.9), **68,** 69, **70, 76, 137–139**

Sense-Making Approaches: Nonlinguistic Representations—Six Kinds of Models (Instructional Tool 3.6), **68,** 69, **70,** 89, **119–126,** 253

Sense-Making Approaches: Nonlinguistic Representations—Visual Tools (Instructional Tool 3.7), **58, 68, 70,** 77, **127–133,** 212, 218, 248

Sense-making strategies, 3, **7,** 9, **12,** 41–42, **67–68,** 76–78

for Ecosystems: Interactions, Energy, and Dynamics unit, 218–226

for Matter and Energy in Organisms and Ecosystems unit, 183–184

for Proteins and Genes unit, 76–78, 320, 322, 323, 324

for The Role of Adaptation in Biological Evolution unit, 281–284

for Variations of Traits unit, **247,** 247–252

SEPs. *See* Science and engineering practices

Six Kinds of Models (Instructional Tool 3.6), **68,** 69, **70,** 89, **119–126,** 253

Small- and large-group discourse, **75, 110–112,** 216, 218, **219,** 280
Smith, S. M., 212
Sneider, C., 154
Socio-scientific issues (SSI), 192, **193**. *See also* Citizen science projects
Southwest Center for Education and the Natural Environment (SCENE) website, **114, 115**
Speaking to Learn (Instructional Tool 3.5), **67,** 69, **110–118**
Spence, S., **61**
State science standards, x, 4, 8, 14, 20, 28, 32, 38, 311
STEM: Student Research Handbook, 167
Stem cell research, x, 45
STEM education, xiii, xvi, 4, 27, 142, 154–157
    definitions of, 154
    implications for Proteins and Genes unit, 157–158
    resources for, 166–167
STEM Lesson Essentials: Integrating Science, Technology, Engineering, and Mathematics, 167
Stone, B., **122**
Storyline. *See* Content storyline
Strand maps, 311, 313
Strategies That Work: Teaching Comprehension for Understanding and Engagement, **106**
Strategy Selection Template, 78
    for Proteins and Genes unit, 71, **71, 80**
    for The Role of Adaptation in Biological Evolution, **274–275**
Student proficiencies for science, 8
Successful K–12 STEM Integration, 154
Summary frames, **99**
Summative assessment, 14, 16, **23,** 143, 256, 270, 272, 279, **312**
    for Proteins and Genes unit, 54, 55
Sutherland, L. M., 90
Systems diagrams, **75,** 77, **79, 80, 84,** 87–88, **125, 131, 132, 133,** 143, 146, 149, 320, 321, 324

T
Taking Science to school, 8, 230
Talk Science Primer, 254
Teachers Domain, 91
The Teacher's Guide to Leading Student-Centered Discussions, **112**
TeacherVision website, **122**
Teaching, Learning & Assessing Science 5–12, **94**
Teaching Reading in Science: A Supplement to the Second Edition of Teaching Reading in the Content Areas Teacher's Manual, **103, 106, 109**
Teaching Reading in the Content Areas, **103, 106, 109**
Teaching Science with Interactive Notebooks, **97**
Teaching-with-Analogies Model, **122**
Teaching With the Brain in Mind, 5
Teaching Writing in the Content Areas, **94, 97**
Technology applications, 166. *See also specific Instructional Tools*

Technology education, xx, 154–157
    implications for Proteins and Genes unit, 157–158
    resources for, 166–167
Think, Pair, Share activities, **105, 275,** 281
Thinking. *See also* Metacognitive goals and strategies
    creative, 18, 56, **62,** 127, 209, **210,** 313
    critical, 13, 18, 56, **57–58,** 63, 313
    Lamarckian, **265, 271,** 276, **277,** 280, 281
    self-regulated, 18, 41, 56, **58–61,** 180, 242, 270, **275**
Thinking-process maps, **68, 75,** 77, **79, 80, 131–133**
ThinkingMaps software, **133**
Thompson, J. J., **124**
Three Approaches That Support Metacognition (Instructional Tool 3.1), **57–62**
Tierney, D., 248
Tips for the Science Teacher: Research-Based Strategies to Help Students Learn, 314
"Tools for Ambitious Science Teaching" website, **112, 118,** 254
Transmissive writing, **98**
Trautmann, N. M., 228
Tree maps, 77, 78, **79, 80, 132, 133**
Trefil, J. S., 194
Tweed, Anne, 256
2010 Kids and Family Reading Report, 152
Two-column notes, 77, **80, 84,** 87, **105,** 146, 149, 320, 321, 322

U
Uncovering Student Ideas in Science series, **59,** 91, **94, 95, 100,** 166, 226, 314, 315
Understanding by Design, 23, 54, 312
Understanding by Design Handbook, 312
Understanding by Design Professional Development Workbook, 312
Understanding Models in Earth and Space Science, **121, 123**
Unit planning, 14–15, 20, 21, 28, 35
    for Ecosystems: Interactions, Energy, and Dynamics unit, 189–228
    for Matter and Energy in Organisms and Ecosystems unit, 171–188
    for Proteins and Genes unit, xvi, xx, 13, 44–92
    for The Role of Adaptation in Biological Evolution unit, 255–285
    for Variations of Traits unit, 229–254
    web-based, 162
Unit Planning Template, 35, **36–37**
Universal Design for Learning (UDL), xvi, 160–163
Universal Design Teacher Planning Template, **164**
Urquhart, V., **94**
Using Analogies in Middle and Secondary Science Classrooms: The FAR Guide, 91, **122**

V
Variations of Traits unit, 229–254

application of Phase 1 to, 232–242
    stage I: identify DCIs, SEPs, and CCs, 232–235
    stage II: deconstruct DCIs, create storyline, and align with SEPs and CCs, 235–240
    stage III: determine PEs and identify criteria to determine student understanding, 240–241
    stage IV: determine NOS connections, 241–242
    stage V: identify metacognitive goals and strategies, 242
application of Phase 2 to, 243–253
    stage VI: research student misconceptions, 243–245, **244**
    stage VII: identify students' preconceptions, 245–246, **246**
    stages VIII and IX: elicit and address preconceptions: sense-making strategies, **247**, 247–252, **251**
    stage X: determine responsive actions based on formative assessment, 253
content storyline for, 238–240
overview of, 231
relevance of topic for, 231–232, 253–254
resources for, 254
Vasquez, J. A., 154
Verbal models
    analogies, **68, 70, 121–122**
    metaphors, **68, 70, 122–123**, 253
Visible Thinking Creativity Routines, 62
Visible Thinking Fairness Routines, 58
Visible Thinking Truth Routines, 57
Visible Thinking website, 57, 58, 60, 62, 81, 84, 313, 320

Visual models, **68, 70, 123–125, 134**, 253
Visual Tools (Instructional Tool 3.7), **58, 68, 70**, 77, **127–133**, 212, 218, 248
Visuwords website, **103**
VocabAhead website, **103**
Vocabulary development, x, xv, xxi, 27, 51–52, 54–55, 150, 160, 162
    for Ecosystems: Interactions, Energy, and Dynamics unit, 202, 218, 221, 226
    Frayer Model of, **103**, 163, 218, **219**, 276, **277**
    hints and resources for, 312
    for Matter and Energy in Organisms and Ecosystems unit, 183, 186
    for Proteins and Genes unit, 52, 163, 322
    for The Role of Adaptation in Biological Evolution unit, 263, 270, 276, **277**, 278, 280, 281
    strategies for, **67, 102–103, 105, 115**
    for Variations of Traits unit, 248
VoiceThread, **60, 62, 112**

W

Welcome to Nanoscience: Interdisciplinary Environmental Explorations, 167
Whiteboards, 272, 281, 283
    interactive, 81, 87, **128, 133**
Wiggins, Grant, 23, 312
Wilson, E. O., 190, 194
Windschitl, M., **124**
Writing to Learn (Instructional Tool 3.3), **67**, 69, 74, **93–101**, 221

5380013